T0214561

Communications in Computer and Information Science 921

Commenced Publication in 2007
Founding and Former Series Editors:
Phoebe Chen, Alfredo Cuzzocrea, Xiaoyong Du, Orhun Kara, Ting Liu,
Dominik Ślęzak, and Xiaokang Yang

More information about this series at http://www.springer.com/series/7899

Brian Donnellan · Cornel Klein
Markus Helfert · Oleg Gusikhin
António Pascoal (Eds.)

Smart Cities, Green Technologies, and Intelligent Transport Systems

6th International Conference, SMARTGREENS 2017
and Third International Conference, VEHITS 2017
Porto, Portugal, April 22–24, 2017
Revised Selected Papers

 Springer

Editors
Brian Donnellan
Innovation Value Institute
National University of Ireland, Maynooth
Maynooth, Ireland

Cornel Klein
Siemens Corporate Technology
Munich, Germany

Markus Helfert
School of Computing
Dublin City University
Dublin, Ireland

Oleg Gusikhin
Ford Research and Advanced Engineering
Dearborn, MI, USA

António Pascoal
University of Lisbon
Lisbon, Portugal

ISSN 1865-0929   ISSN 1865-0937 (electronic)
Communications in Computer and Information Science
ISBN 978-3-030-02906-7   ISBN 978-3-030-02907-4 (eBook)
https://doi.org/10.1007/978-3-030-02907-4

Library of Congress Control Number: 2018958516

This Springer imprint is published by the registered company Springer Nature Switzerland AG
The registered company address is: Gewerbestrasse 11, 6330 Cham, Switzerland

Preface

This book includes extended and revised versions of a set of selected papers from SMARTGREENS 2017 (6th International Conference on Smart Cities and Green ICT Systems) and VEHITS 2017 (3rd International Conference on Vehicle Technology and Intelligent Transport Systems), held in Porto, Portugal, during April 22–24.

SMARTGREENS 2017 received 70 paper submissions from 33 countries, of which 11% are included in this book. VEHITS 2017 received 77 paper submissions from 25 countries, of which 12% are included in this book.

The papers were selected by the event chairs of both events and their selection is based on a number of criteria that include the classifications and comments provided by the Program Committee members, the session chairs' assessment, and also the program chairs' global view of all papers included in the technical program. The authors of selected papers were then invited to submit a revised and extended version of their papers having at least 30% innovative material.

The purpose of the 6th International Conference on Smart Cities and Green ICT Systems (SMARTGREENS) was to bring together researchers, designers, developers, and practitioners interested in the advances and applications in the field of smart cities, green information and communication technologies, sustainability, energy-aware systems and technologies.

The purpose of the Third International Conference on Vehicle Technology and Intelligent Transport Systems (VEHITS) was to bring together engineers, researchers, and practitioners interested in the advances and applications in the field of vehicle technology and intelligent transport systems. This conference focuses on innovative applications, tools, and platforms in all technology areas such as signal processing, wireless communications, informatics, and electronics related to different kinds of vehicles, including cars, off-road vehicles, trains, ships, underwater vehicles, or flying machines, and the intelligent transportation systems that connect and manage large numbers of vehicles, not only in the context of smart cities but in many other application domains.

The papers selected to be included in this book contribute to the understanding of relevant trends of current research on smart cities, green ICT systems, vehicle technology and intelligent transport systems including: smart grids, monitoring data, Internet of Things, electric vehicles, intelligent transportation systems, transportation planning, and traffic operation.

With the advances of new and innovative technologies, the field of smart and connected cities is expected to grow even further. Topics such as data privacy, Internet of Things, and architecture or business models for smart cities are becoming increasingly important for both researchers and practitioners. At the same time sustainability and energy are two crucial aspects to consider for the advances and applications in the field of vehicle technology and intelligent transport systems as well as smart cities. In the next few years we can expect a range of innovative technologies and research results for these topics within the area of smart cities and intelligent transportation systems such as energy and vehicle analytics and autonomous and connected vehicles.

We would like to thank all the authors for their contributions and also the reviewers who have helped ensure the quality of this publication.

April 2017

Brian Donnellan
Cornel Klein
Markus Helfert
Oleg Gusikhin
António Pascoal

Organization

SMARTGREENS Conference Chair

Markus Helfert   Dublin City University, Ireland

VEHITS Conference Chair

Oleg Gusikhin   Ford Motor Company, USA

SMARTGREENS Program Co-chairs

Cornel Klein   Siemens AG, Germany
Brian Donnellan   Maynooth University, Ireland

VEHITS Program Co-chairs

Markus Helfert   Dublin City University, Ireland
António Pascoal   Instituto Superior Técnico, Portugal

SMARTGREENS Program Committee

Javier M. Aguiar	Universidad de Valladolid, Spain
Nour Ali	University of Brighton, UK
Carlos Henggeler Antunes	University of Coimbra/INESC Coimbra, Portugal
Siegfried Benkner	University of Vienna, Austria
Nik Bessis	Edge Hill University, UK
Riccardo Bettati	Texas A&M University, USA
Dumitru Burdescu	University of Craiova, Romania
Bong Jun Choi	The State University of New York Korea, South Korea
Ken Christensen	University of South Florida, USA
Calin Ciufudean	Stefan cel Mare University, Romania
Georges Da Costa	IRIT, Paul Sabatier University, France
Amélie Coulbaut-Lazzarini	Université Versailles Saint Quentin en Yvelines, France
Thomas Dandres	CIRAIG, Canada
Margot Deruyck	Ghent University, Belgium
Yong Ding	Karlsruhe Institute of Technology, Germany
Venizelos Efthymiou	University of Cyprus, Cyprus
Tullio Facchinetti	University of Pavia, Italy
Cléver Ricardo Guareis de Farias	University of São Paulo, Brazil
Eugene A. Feinberg	Stony Brook University - New York, USA
Bela Genge	University of Targu Mures, Romania

Afshin Tafazzoli   Vestas, Spain
Paolo Tenti   University of Padua, Italy
Dimitrios Tsoumakos   Ionian University, Greece
Athina Vakali   Aristotle University, Greece
Silvano Vergura   Polytechnic University of Bari, Italy
Shengquan Wang   University of Michigan-Dearborn, USA
Igor Wojnicki   AGH University of Science and Technology, Poland
Yinlong Xu   University of Science and Technology of China, China
Ramin Yahyapour   University Göttingen, Germany
Chau Yuen   Singapore University of Technology and Design,
                              Singapore
Yayun Zhou   Siemens AG, Germany
Sotirios Ziavras   New Jersey Institute of Technology, USA

SMARTGREENS Additional Reviewers

Tobias Küster   DAI-Labor, TU Berlin, Germany
Mariusz Zal   Poznan University of Technology, Poland

VEHITS Program Committee

Carlos Abreu   Instituto Politécnico de Viana do Castelo, Portugal
Konstantinos Ampountolas   University of Glasgow, UK
Ramachandran Balakrishna   Caliper Corporation, USA
Francesco Basile   Università degli Studi di Salerno, Italy
Mohamed Benbouzid   University of Brest, France
Sandford Bessler   Austrian Institute of Technology, Austria
Ghulam H. Bham   University of Alaska Fairbanks, USA
Neila Bhouri   IFSTTAR, France
Christine Buisson   Université de Lyon, France
Catalin Buiu   Universitatea Politehnica din Bucuresti, Romania
Gihwan Cho   Chonbuk University, South Korea
Seibum Choi   Korea Advanced Institute of Science and Technology,
                              South Korea
Andy Chow   City University of Hong Kong, SAR China
Baldomero Coll-Perales   Universidad Miguel Hernandez de Elche, Spain
Ned Djilali   University of Victoria, Canada
Mariagrazia Dotoli   Politecnico di Bari, Italy
Bertrand Ducourthial   Université de Technologie de Compiegne, France
Mehmet Onder Efe   Hacettepe University, Turkey
Nicola Epicoco   Politecnico di Bari, Italy
Peppino Fazio   University of Calabria, Italy
Lino Figueiredo   Instituto Superior de Engenharia do Porto, Portugal
Yi Guo   The University of Texas at Dallas, USA
Oleg Gusikhin   Ford Motor Company, USA
Markus Helfert   Dublin City University, Ireland

Yoichi Hori	University of Tokyo, Japan
William Horrey	AAA Foundation for Traffic Safety, USA
Zhongsheng Hou	Beijing Jiaotong University, China
Zechun Hu	Tsinghua University, China
Hocine Imine	IFSTTAR, France
Chul-Goo Kang	Konkuk University, South Korea
Hakil Kim	Inha University, South Korea
Xiangjie Kong	Dalian University of Technology, China
Anastasios Kouvelas	EPFL, Switzerland
Zdzislaw Kowalczuk	Gdansk University of Technology, Poland
Milan Krbálek	Czech Technical University, Czech Republic
Wei Liu	University of Glasgow, UK
José Santa Lozano	University of Murcia, Spain
Reinhard Mahnke	University of Rostock, Germany
Johann Marquez-Barja	University of Antwerp, Belgium
Zeljko Medenica	Altran, Germany
Youcef Mezouar	IFMA/Institut Pascal, France
Lyudmila Mihaylova	University of Sheffield, UK
Pedro Moura	Institute of Systems and Robotics University of Coimbra, Portugal
Mirco Nanni	Italian National Research Council, Italy
Katsuhiro Nishinari	University of Tokyo, Japan
Alfredo Núñez	Delft University of Technology, The Netherlands
Dario Pacciarelli	Roma Tre University, Italy
Sara Paiva	Instituto Politécnico de Viana do Castelo, Portugal
Markos Papageorgiou	Technical University of Crete, Greece
Ioannis Papamichail	Technical University of Crete, Greece
Brian Park	University of Virginia, USA
Paulo Pereirinha	INESC Coimbra/Polytechnic of Coimbra/APVE, Portugal
Fernando Pereñiguez	University Centre of Defence, Spanish Air Force Academy, Spain
Xiaobo Qu	Chalmers University of Technology, Sweden
Claudio Roncoli	Aalto University, Finland
Oyunchimeg Shagdar	Institute VEDECOM, France
Sanjay Sharma	Plymouth University, UK
Shih-Lung Shaw	University of Tennessee, USA
Silvia Siri	University of Genoa, Italy
Uwe Stilla	Technische Universität München, Germany
Tatsuya Suzuki	Nagoya University, Japan
Wai Yuen Szeto	The University of Hong Kong, SAR China
Richard Tay	RMIT University, Australia
C. James Taylor	Lancaster University, UK
Tomer Toledo	Technion - Israel Institute of Technology, Israel
Esko Turunen	Tampere University of Technology, Finland

István Varga	Budapest University of Technology and Economics, Hungary
Francesco Viti	University of Luxembourg, Luxembourg
Eleni Vlahogianni	National Technical University of Athens, Greece
Peter Vortisch	Karlsruhe Institute of Technology, Germany
Anil Vuppala	International Institute of Information Technology Hyderabad, India
Peng Zhang	Shangai University, China

VEHITS Additional Reviewers

Marija Bezbradica	Lero, DCU, Ireland
Thomas Burrell	Lancaster University, UK
Alexander Hanel	Technische Universität München, Germany
Antonio Meireles	ISEP, Portugal
Majid Rostami	Shiraz University, Iran
Mohammadreza Saeedmanesh	EPFL, Switzerland

Invited Speakers

Cristina Olaverri Monreal	University of Applied Sciences Technikum Wien, Austria
Emil Vassev	Lero - The Irish Software Research Centre, UL, Limerick, Ireland
Gerard Smit	University of Twente, The Netherlands
João Barros	Veniam, Portugal
Sandra Gannon	IBM Internet of Things, Ireland

Contents

Energy-Aware Systems and Technologies

Towards an Integrated Development and Sustainability Evaluation of Energy Scenarios Assisted by Automated Information Exchange

Jan Sören Schwarz[1] ⓘ, Tobias Witt[2] ⓘ, Astrid Nieße[3] ⓘ, Jutta Geldermann[2] ⓘ,
Sebastian Lehnhoff[1](✉) ⓘ, and Michael Sonnenschein[3]

[1] Department of Computing Science, University of Oldenburg, Oldenburg, Germany
{jan.soeren.schwarz,sebastian.lehnhoff}@uni-oldenburg.de
[2] Chair of Production and Logistics, University of Goettingen, Goettingen, Germany
{tobias.witt,geldermann}@wiwi.uni-goettingen.de
[3] OFFIS - Institute for Information Technology, Oldenburg, Germany
{niesse,sonnenschein}@offis.de

Abstract. Today, decision making in politics and businesses should aim for sustainable development, and one field of action is the transformation of energy systems. To reshape energy systems towards renewable energy resources, it has to be decided today on how to accomplish the transition. Energy scenarios are widely used to guide decision making in this context. While considerable effort has been put into developing energy scenarios, researchers have pointed out three requirements for energy scenarios that are not fulfilled satisfactorily yet: The development and evaluation of energy scenarios should (1) incorporate the concept of sustainability, (2) provide decision support in a transparent way, and (3) be replicable for other researchers. To meet these requirements, we combine different methodological approaches: story-and-simulation (SAS) scenarios, multi-criteria decision making (MCDM), information modeling, and co-simulation. We show how the combination of these methods can lead to an integrated approach for development and sustainability evaluation of energy scenarios assisted by automated information exchange. We concretize this approach with a sustainability evaluation process (SEP) model and an information model. We highlight, which artifacts are developed during the SEP and how the information model can help to automate the information exchange in this process. The objectives are to facilitate a sustainable development of the energy sector and to make the development and decision support processes of energy scenarios more transparent.

Keywords: Co-simulation · Energy scenarios · Information model
Multi-criteria decision making (MCDM) · Ontology
Scenario planning · Story-and-simulation (SAS)
Sustainability evaluation

© Springer Nature Switzerland AG 2019
B. Donnellan et al. (Eds.): SMARTGREENS 2017/VEHITS 2017, CCIS 921, pp. 3–26, 2019.
https://doi.org/10.1007/978-3-030-02907-4_1

1 Introduction

The intended phase-out from nuclear and fossil power and the transition to renewable energy resources in Germany pose new challenges. With the EU and the federal government of Germany having set targets for reducing energy demand and greenhouse gas (GHG) emissions until 2030 and 2050 [5,12], decision makers in politics need to initiate and project the transition process today to achieve these targets. The goal of this transition process is to reshape the energy infrastructure and related planning and operation processes while also considering sustainable development. Thus, an approach to evaluate and compare future scenarios and corresponding transition paths regarding their sustainability characteristics is needed to provide guidance to the politically intended transition process.

Long-term energy scenarios, which are a tool to investigate the *"possible future development (or a future state) of the energy system"* [15, p. 9], have long been used to guide decision making in this context. Researchers have already put considerable effort into developing these scenarios. For example, the German database "Forschungsradar Energiewende"[1] lists more than 920 publications on energy transition research in Germany from 2011 to 2017. This database also includes publications on EU and global levels, which are also relevant for the German debate, e.g., the World Energy Outlook [17]. Based on these energy scenarios, different transition paths can be distinguished for the development of the future energy system. Therefore, energy scenarios help to develop long-term strategies by describing how the set targets of the energy transition can be achieved. However, the development and evaluation of energy scenarios should meet some basic requirements:

- A concept of sustainability should be defined and operationalized with relevant dimensions in the evaluation process.
- The decision support process should be transparent and easy to understand.
- The development and decision support processes of energy scenarios should be replicable.

In the article at hand, we will show an integrated approach for development and sustainability evaluation of energy scenarios assisted by automated information exchange addressing these requirements. Our approach is based on the sustainability evaluation process (SEP) developed in the research project NEDS [4]. The SEP combines qualitative future scenarios with quantitative simulation and multi-criteria decision making (MCDM)[2]. To facilitate the required replicability, we introduce an information model, which supports the automation of information exchange in the SEP.

The remainder of this article is structured as follows: In Sect. 2, we elaborate on the requirements for energy scenarios. In Sect. 3, we describe related methodologies that we will combine to meet these requirements: We describe

---

[1] http://www.forschungsradar.de/studiendatenbank.html (accessed August 17, 2017).
[2] Also called multi-criteria decision aiding/analysis (MCDA).

story-and-simulation (SAS) scenarios, review approaches for integrating scenarios with methods from multi-criteria decision making, and describe the applicability of information modeling in the context of energy scenarios. In Sect. 4, we combine the reviewed methodologies in our conceptual solution to set up the SEP as an integrated process with continuous tool support. After that, we give details on this solution regarding the SEP in Sect. 5, and regarding the information model supporting the SEP in Sect. 6. In Sect. 7, we show how the three above-mentioned requirements are met by our solution.

2 Requirements for Development and Evaluation of Energy Scenarios

The development and evaluation of energy scenarios should satisfy three basic requirements: They should include a definition and operationalization for sustainability, provide decision support in a transparent way, and be replicable for other researchers. In this section, we elaborate on these requirements in more detail and point out that many energy scenarios do not satisfactorily fulfill them.

Usually, a triad of objectives including supply security, economic viability, and environmental compatibility is used to guide decision making in the energy sector. For example, in Germany these objectives are stated in the "Energiewirtschaftsgesetz" (Energy Sector Act) [10]. Meanwhile, the concept of sustainable development is also used to guide decisions in multiple fields of politics. According to the triple-bottom-line interpretation of sustainability, economic prosperity, social justice, and environmental quality need to be achieved simultaneously [11]. If both approaches are integrated, technical, economic, environmental, and social criteria are all relevant to operationalize sustainability in the energy sector. However, the majority of recent research on energy transitions focuses on selected aspects and consequently fails to consider all relevant aspects simultaneously - see [16,18,21] for reviews of considered aspects in Germany and the UK. Therefore, these studies might not be suitable to guide political decision making towards sustainable development of the energy sector, due to the multi-criteria nature of sustainability.

Many studies aim at providing decision support for political decisions to promote sustainable development of the energy system. However, there are obstacles impairing transparent decision support: Firstly, most energy scenarios do not explicitly evaluate energy scenarios' contributions to sustainable development, e.g., by conceptualizing the sustainability of energy scenarios as a multi-criteria decision problem. For example, in their review on 24 energy scenarios in Germany, Kronenberg et al. [21] point out that an integrated sustainability assessment would increase the usefulness for decision support, but is missing in most of the reviewed energy scenarios. Secondly, most energy scenarios do not specify decision alternatives and delineate them from external uncertainties. Thirdly, shareholders' preferences are not included in the evaluation of energy scenarios. Overall, this means that the process of generating recommendations from energy scenarios, i.e., the underlying decision support process, is not transparent [15]. To this end, transparency of the methods and models used in the scenario evaluation process is needed. To overcome these obstacles, Grunwald et al. [15] proposed to introduce standards for energy scenarios in terms of scientific validity, transparency and openness of the results.

Replicability is crucial to achieve scientific validity of energy scenario studies. It allows other researchers to repeat calculations and simulations of scenarios with the same parameters to replicate the results. Therefore, all information about the scenarios should be documented and published including input data, models, and assumptions [15]. This enables researchers to take an energy scenario and vary input data, models, and assumptions. For example, newer data or data from another country could be used. Models could also be interchanged or new models could be added to expand the focus of the scenario. This way, replicability also allows to reuse scenarios and to compare different scenario studies. Furthermore, in combination with a transparent decision making process, it allows other researchers to replicate scenarios while also considering further criteria. During scenario development, simulation, and evaluation, various data is exchanged between different actors, models, and software. Due to the complexity of these processes (in particular the integration of simulation models from various domains), the information exchange is highly prone to error. This calls for tool-support and automation with all data flows and dependencies being well defined and documented. This would significantly improve the replicability, as requested in [15].

3   Related Work

Since the publication of the Brundtland report [7] in 1987, many sustainability assessment tools have been introduced with different focus. As can be seen in the review from Singh et al. [30], the main work has focused on developing indicators, composite indicators, and appropriate scaling and weighting systems.

In July 2017, the United Nations Development Group published an online compendium on different sustainability development goals assessment methods, tools, and indicator systems as one of the actions within the 2030 agenda[3]. With the collected tools, all relevant aspects of sustainability-related decision making and planning are supposed to be covered. Nevertheless, we argue that the presented tools do not fulfill the requirements for the evaluation of energy scenarios. For example, the Integrated Sustainable Development Model simulation tool (iSDG[4]) developed by the Millenium Institute complies with the first and second requirements as defined in Sect. 2 only in parts, as there is no support for the definition and weighting of shareholders' preferences. Furthermore, the iSDG needs input values that can be derived only from an integrated simulation of the underlying system (in our case: the energy system comprising a vast amount

---

[3] https://undg.org/2030-agenda/sdg-acceleration-toolkit (accessed August 17, 2017).
[4] http://www.isdgs.org (accessed August 17, 2017).

of components). An appropriate automation of the data transfer from simulation models to an evaluation system – as requested in the third requirement in Sect. 2 – is not supported within any of the presented tools. Additionally, the integration of a scenario planning tool is missing.

To address this issue, we suggest integrating methodologies for scenario planning and sustainability evaluation based on system simulations. We will describe this conceptual solution further in Sect. 4. In the remainder of this section, we will introduce appropriate methodologies and discuss possibilities to integrate them:

Firstly, we describe the concept of story-and-simulation scenarios, which are based on qualitative future scenarios and quantitative simulation. Secondly, we discuss the integration of multi-criteria decision making (MCDM) and scenarios. Thirdly, we point out how information models are used to support simulation and evaluation processes.

3.1 Story-and-Simulation (SAS) Scenarios

One methodology to set up energy scenarios is the SAS approach [1]. While this approach stems from environmental modeling, it is increasingly used in the energy context [37]. SAS scenarios combine qualitative stories and quantitative data for simulation studies. On the one hand, the storyline of SAS scenarios allows the involvement of shareholders in the scenario development process. On the other hand, the technical simulation can provide numerical estimates of the future development. The variation of quantitative attributes within the SAS scenarios leads to adapted simulation model parametrization. To this end, it should be clear, why certain quantitative parameters of a simulation are varied and others are not, based on a storyline.

For the sake of transparency, a well-defined process for creating these stories is necessary, which can be found, for example, in scenario planning[5]. This is an expert-based management tool, which originates from strategic planning on the company level in the 1970s [34]. Based on this prior work, Gausemeier et al. [14] proposed the following process for the development of future scenarios (we will call this **scenario planning process**)[6]:

Firstly, participating domain experts discuss factors influencing a scenario and define the so-called "decision and scenario fields": The factors are separated into internal and external factors: If the development of factors can be influenced (**decision alternatives**), these are called internal factors and form the decision field. If the development of factors cannot be influenced (**external uncertainties**), these are called external factors and form the scenario field. As an example, the German Federal State of Baden-Württemberg defined special regulations that apply when a new heating system is installed in a house. These force the owner to either base the new system at least partly on renewables or to

---

[5] Also called scenario management.

[6] In the remaining sections, definitions are highlighted in bold typesetting at their first occurrence.

take substantial steps to increase energy efficiency (e.g., roof insulation). But if subsidies are allocated by the Federal Government, no decisions can be taken in that field by one of the federal states. Thus, these subsidies are excluded from the decision alternatives. Nevertheless, there are enormous effects of these external uncertainties. Therefore, they have to be taken into account during evaluation of the decision alternatives, but do not themselves define decision alternatives. With this classification, three different types of future scenarios can be distinguished: *External scenarios* comprise only external factors, and can therefore be used to describe the development of the environment of decision alternatives. *Internal scenarios* comprise only internal factors, and can therefore be used to derive possible decision alternatives. *Systems scenarios* include both external and internal factors, and can therefore be used to describe the development of complex systems.

Having described the decision and scenario fields, the experts systematically identify the key factors. Afterward, they identify the key factors' projections, which are possible developments up to a certain point in time to span a broad range of possible future developments, and describe them. These different projections are checked by the experts for consistency and the results are recorded in a consistency matrix. Based on this matrix, scenario software uses cluster analysis to build projection bundles, which represent consistent combinations of projections and thus possible future scenarios. In the last step, the domain experts write storylines, i.e., textual descriptions, for all future scenarios.

The second part of the SAS scenarios is simulation, which is defined by Robinson [26] as *"Experimentation with a simplified imitation (on a computer) of an operations system as it progresses through time, for the purpose of better understanding and/or improving that system"* [26, p. 4]. In SAS scenarios, simulation is used to substantiate the previously introduced future scenarios with numerical estimates of the future state for better understanding. While future scenarios give a textual description of possible developments in the future, simulation scenarios describe the configuration of a concrete simulation, which includes simulation models and how they are connected and parametrized.

Typically, one single simulation software, e.g., Matlab, is used. This makes it hard to integrate simulation models from different domains, because most domains use specific software and languages [27]. For a holistic simulation of the energy system, material flows (e.g., coal or biogas for power plants), information flows (e.g., in smart grid control strategies), and electric power flows have to be considered. Additionally, the behavior of consumers should be included to achieve realistic results. Therefore, various simulation models have to be coupled to represent this complex, dynamic, sociotechnical system. An approach for solving this issue is co-simulation, which is defined as *"an approach for the joint simulation of models developed with different tools (tool coupling) where each tool treats one part of a modular coupled problem"* [2]. An example for a co-simulation framework with focus on the energy domain is mosaik[7] [23, 25]. It allows to couple different simulators, provides an application programming inter-

---

[7] http://mosaik.offis.de (accessed August 17, 2017).

face (API) for different programming languages, and handles the scheduling and information exchange between the simulation models during runtime.

3.2 Integration of Scenario Planning and MCDM

According to Belton and Stewart [3, p. 2], MCDM is *"an umbrella term to describe a collection of formal approaches, which seek to take explicit account of multiple criteria in helping individuals or groups explore decisions that matter"*. We shall highlight three characteristics in this definition, which show the applicability of MCDM for energy systems planning: Firstly, as stated in Sect. 2, technical, social, environmental, and economic aspects are relevant for energy systems planning (*"multiple criteria"*). Secondly, the decision for a future energy system affects many shareholders with conflicting objectives. In this context, MCDM can help to structure and inform debates (*"individuals or groups"*). Thirdly, the energy system is fundamental for sustaining modern societies (*"decisions that matter"*).

Researchers have already applied MCDM extensively in sustainable energy planning, e.g., Wang et al. [36] and Oberschmidt et al. [24] provide overviews. Surprisingly, the integration with scenario planning has not been in the focus of research. Stewart et al. [32, p. 682] identify four concepts for scenarios, which are more or less well-suited for integrating MCDM and scenarios:

1. external situations affecting consequences of policy actions (this concept equals the aforementioned "external scenarios", see Sect. 3.1),
2. exploration of future conditions or environments (this concept equals the aforementioned "future scenario", see Sect. 3.1),
3. advocacy of particular courses of action,
4. representative sample of future states.

Stewart et al. [32] also provide general guidelines for integrating MCDM and scenario planning. Most notably, they point out that *"it is essential that the scenarios reflect external driving forces (events, states) which are separated from the policies or actions under consideration"* [32, p. 683]. This implies the external scenario concept. However, they do not provide a procedure for integrating MCDM and future scenarios, and therefore it remains unclear, how external scenarios can be generated in energy scenarios.

For example, Kowalski et al. [20] integrate scenario planning and MCDM and apply it to energy systems planning in Austria. However, they do not differentiate between decision alternatives and external uncertainties, but rather use the scenarios from scenario planning directly as decision alternatives. This is identical to the "internal scenarios" described by Gausemeier et al. [14]. This means that Kowalski et al. [20] do not consider external uncertainties (and therefore also do not use the external scenario concept). Particularly for long-term decisions, we argue that this does not adequately reflect planning uncertainties for such a complex system.

3.3   Information Modeling

In the aforementioned development, simulation, and evaluation of energy scenarios, various model parameters, dependencies, and data flows have to be defined. Because of the energy system's complexity, this modeling has to include multiple domains and is crucial for the replicability. As multi-domain modeling is a complex task, automating the exchange of data would highly improve the usability. In other fields of application, this is usually done by defining an information model, which is described in [22, p. 1] as *"a representation of concepts, relationships, constraints, rules, and operations to specify data semantics for a chosen domain of discourse."*

Caused by the involvement of different domain experts in the scenario development, the challenge is not only to model information but also to collect domain knowledge, as a single term can be used in different domains with different meanings. Therefore, the terms and relationships between them should be defined. A common concept for this representation of domain knowledge is the use of ontologies, which *"have been developed to provide a machine-processable semantics of information sources that can be communicated between different agents (software and humans)"* [13, p. 3].

Information models are widespread in industry to allow interoperability, e.g., the Common Information Model (CIM) in the energy domain [33]. It contains a data model (domain ontology), various interface specifications, technology-specific instantiations of the ontologies (communication and serialization), and allows automated communication between components of smart grids.

The common processes and languages for information modeling and ontology development require the user to be experienced in their usage. To avoid this barrier for experts from different domains, who most likely do not have this expertise, some approaches use concept maps for the instantiation of ontologies [8,29].

The knowledge collected in an ontology should also be usable for the data management and evaluation of an energy scenario. In this context, ontology-based data management (OBDM) is a relatively new approach [9]. It is based on a three-level architecture containing an ontology, sources, and a mapping between them. This way, OBDM faces the challenge of data heterogeneity by replacing a global scheme in data management with the ontology describing the domains. Kontchakov et al. [19] describe an ontology-based data access with databases, which uses an ontology to query databases. In this approach, queries can use the vocabulary of the ontology and are rewritten according to the vocabulary of the relational database (RDB).

4   Conceptual Solution

In Sect. 2, we have defined three main requirements for the development and evaluation process of energy scenarios: A definition and operationalization of sustainability, a transparent and easily understandable decision support process, and replicability to ensure comparability.

Against this background, the main purpose of the process is to deliver a sustainability evaluation of decision alternatives (see Sect. 3.1). On the left-hand side of Fig. 1, we can see possible configurations of the future energy system, which are described within qualitative descriptions of future scenarios. These future scenarios are systems scenarios and therefore also include external uncertainties. On the right-hand side, the desired output of the process is displayed: A sustainability-based order of decision alternatives. We assume that the configuration of the future energy system can be influenced and thus, different configurations of the future energy system form the decision alternatives. We consequently interpret factors outside the future energy system as external uncertainties.

The main task to be solved in this context is to define a process that leads to a sustainability-based order of decision alternatives by (1) defining the different inputs needed for this process and (2) defining the process steps to generate the outputs from these inputs. Part of this task is to define, which shareholders are included in such a sustainability evaluation, a crucial aspect, when societal aspects are supposed to be part of the evaluation result as well.

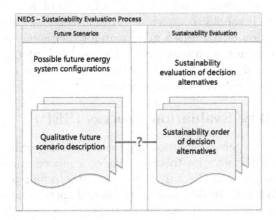

Fig. 1. Methodological challenge: from qualitative future scenarios to a sustainability-based order of decision alternatives [4, adapted].

To close the gap between qualitative future scenarios and a sustainability-based order of decision alternatives (see Fig. 1), we propose to use SAS scenarios in combination with the scenario planning process for the story and integrate the following two approaches:

Firstly, to increase the transparency of the decision support process and to cope with the multi-criteria nature of the decision problem, we propose to integrate SAS scenarios and MCDM. This allows to simultaneously consider the relevant (and usually conflicting) criteria that make a sustainable energy system.

Furthermore, decision support can be given in a transparent way by structuring the decision problem and taking into account shareholders' preferences. To quantify the relevant sustainability criteria from different domains, we suggest using multi-domain co-simulation for the simulation part of SAS scenarios (see Sect. 3.1). For the integration of scenarios and MCDM, Stewart et al. [32] provide an overview of approaches from the 1990s and early 2000s aiming at an integrated sustainability assessment. However, today only few energy scenarios build upon these approaches and differentiate between decision alternatives and external uncertainties. One obstacle might be the lack of a process model, which integrates SAS scenarios and MCDM. While there already exist some guidelines for constructing SAS scenarios [1,37], these guidelines do not integrate MCDM. In the research project NEDS, a novel process model has been introduced for the integration of MCDM and SAS scenarios for sustainability evaluation: the sustainability evaluation process (SEP) [4]. We will describe a revised version of the SEP and its resulting artifacts in more detail in Sect. 5.

Secondly, the complexity of energy scenarios calls for tool-support and automation of the process. Therefore, we propose an information model to structure the data flows and dependencies in the scenario development and evaluation processes, organize the communication between experts from different domains, and collect their knowledge. To allow the participation of domain experts without having them to learn new complex description languages and techniques in detail, the information model uses a mind map for the representation of knowledge. For integration in the process, we use the modeled information to instantiate an ontology and make the information available in a machine readable format, which is the basis for data management.

5   Sustainability Evaluation Process (SEP)

Having presented our general solution approach for integrating SAS scenarios and MCDM, we now concretize this in terms of a process model for the SEP [4]. An overview is given in Fig. 2. It is subdivided into four parts that will be explained in the following subsections. Two different entry points are given, each on the left-hand side of Fig. 2. In the various steps[8] of the SEP, different artifacts are created. Table 1 provides an overview of the different types of artifacts with the respective types of data, short descriptions, required inputs, and related outputs. The first digit of the numeration indicates the part of the SEP, in which the artifact is listed. For the types of data, we differentiate between unstructured data, such as text documents, and structured data. Structured data can be stored in the information model, a scenario database, or used by an MCDM tool. The information model and the scenario database will be described in Sect. 6.

5.1   Preparatory Steps

The first part of the SEP serves to prepare for the sustainability evaluation of potential future energy systems, and is therefore separated into two sub-parts:

---

[8] For better readability we use the term "step" to describe subprocesses of the SEP.

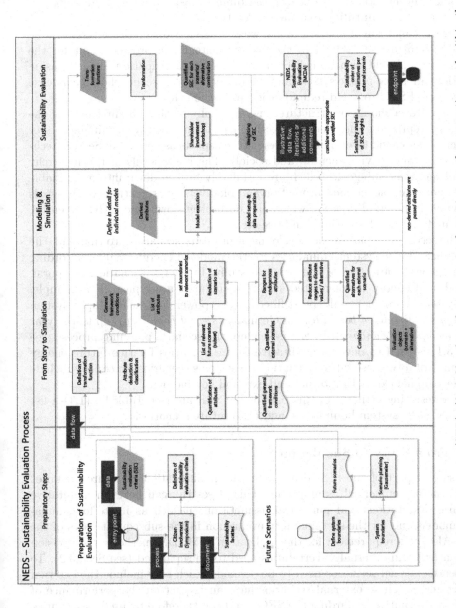

Fig. 2. Sustainability evaluation process (SEP) *(dark blue legend icons give information on the semantics used within the diagram).* (Color figure online)

The first sub-part is the preparation of the sustainability evaluation. As a requirement for any MCDM method, the researchers need to define the criteria, in terms of which the performance of alternative energy systems can be measured [3]. Since the criteria should reflect the different aspects of sustainability, we name them **sustainability evaluation criteria** (SECs).

For the collection of relevant SECs, we propose a two-step procedure: Firstly, related literature on MCDM in the energy context can provide input for the SECs. SECs should also be collected with public participation, e.g., by using questionnaires in a public symposium. Secondly, the researchers should condense the collected SEC to avoid redundancies and define them.

The criteria should be structured hierarchically so that the first level of the hierarchy represents the overall objective, e.g., identifying a sustainable power system. The second level should represent the aspects as defined in Sect. 2 (technical, economic, environmental, and social). The lower levels of the hierarchy should represent sub-goals, which are ultimately broken down into quantifiable SECs. For example, climate protection and biodiversity protection are sub-goals in the environmental domain, which can be broken down into GHG potential, particulate matter formation, land use etc.

The second sub-part is the development of future scenarios. To that end, the researchers define system boundaries for the modeled energy system, according to the general objective of the scenario study. In energy scenarios, temporal, spatial, and energy sector-related system boundaries are typical. For example, the system boundary could be defined as the power and heat supply system (*energy-sector related boundary*) in Germany (*spatial boundary*), up to the year 2050 (*temporal boundary*). In the next step, the scenario planning process (see Sect. 3.1) is used to develop qualitative future scenarios for this future energy system. Notably, these future scenarios are systems scenarios, since they include the development of both internal and external key factors.

The resulting artifacts of this part of the SEP are (see Table 1): (1.1) a list of SECs, (1.2) system boundaries, and (1.3) future scenarios.

5.2   From Story to Simulation

The second part of the SEP deals with the quantification of future scenarios with concrete numerical assumptions. This closes the gap between qualitative scenario descriptions and numerical assumptions needed as input for simulation models and for the sustainability evaluation in the subsequent parts of the SEP. Also, as a requirement for the application of MCDM methods, the decision alternatives and external uncertainties need to be separated (see Sect. 3.2). To that end, external scenarios need to be defined, which only include the external uncertainties. These external scenarios have an impact on the performance of decision alternatives regarding the SECs, and can therefore be used as scenarios in MCDM methods.

The first step is the *deduction* and *classification* of **attributes**, which will be used to quantify the future scenarios developed in the first part. There are two ways to *deduce* attributes: Firstly, the attributes can be deduced from key factors of the future scenarios. Secondly, attributes can also be deduced by inferring the attributes that will be needed to calculate the values of the SECs.

As systems scenarios contain both decision alternatives and external uncertainties, a *classification* of attributes is needed. On its left-hand side, Fig. 3 provides an overview on the different types of attributes.

In terms of *key factors*, the definition of decision and scenario fields in the scenario planning process allows to differentiate between decision alternatives and external uncertainties (see Sect. 3.1). We can also use this definition to allow to differentiate between decision alternatives and external uncertainties in terms of *attributes*: As stated in Sect. 4, we assume that the configuration of the future energy system can be influenced. Thus, attributes describing the configuration of the future energy system are classified as **endogenous attributes**, e.g., the shares of renewable energy technologies in the energy mix. A combination of values for all endogenous attributes consequently constitutes a decision alternative. In contrast, attributes, which characterize the environment of the future energy system, reflect external uncertainties, e.g., the development of prices for crude oil on the world market. We further distinguish external uncertainties into **scenario-specific framework conditions** and **general framework conditions**. A combination of values for all scenario-specific framework conditions constitutes an external scenario. For example, prices for crude oil might be a scenario-specific framework condition, while the demographic development might be a general framework condition, i.e., one value applies to all scenarios. Since general framework conditions are not included in the external scenarios, they also do not have an impact on the decision between alternatives. While endogenous attributes, scenario-specific framework conditions, and general framework conditions all provide input for simulation models, the values of **derived attributes** are calculated as output of simulation models. All attributes are collected in a list of attributes, which is implemented in the information model.

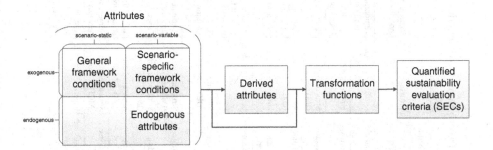

Fig. 3. Overview of attribute classification in different types and the data flows of their values (Source: [28]).

The next step is to define a **transformation function** for each SEC to determine the quantified SECs for each decision alternative (see the right-hand side of Fig. 3). This function transforms derived and non-derived attributes into

Table 1. Description of artifacts developed during the SEP (IM = information model; DB = scenario database).

#	Artifact Name	Type of Data	Description	Required Input	Input for
1.1	Sustainability Evaluation Criteria (SECs)	Structured data (IM)	Criteria, which are used to evaluate the sustainability of future energy systems with MCDM methods	Definition and concept for sustainability, shareholder involvement	2.2, 4.1
1.2	System Boundaries	Unstructured data	Temporal, spatial, and energy-sector related boundaries of the modeled energy system	General objectives of energy scenario study	1.3, 2.1
1.3	Future Scenarios	Unstructured data	Qualitative future scenarios described with a storyline, detailing the interplay between key factor projections	1.2, Scenario planning process	2.2, 2.3
2.1	General Framework Conditions	Structured data (IM)	Boundary conditions narrowing down valid decision alternatives	1.2	2.3
2.2	List of Attributes	Structured data (IM)	Collection of (derived and non-derived) attributes quantifying the development of qualitative key factors in numerical terms	1.1, 1.3, 4.1, classification of attributes	2.4, 4.1
2.3	List of Relevant Future Scenarios	Unstructured data	Reduced set of future scenarios, after compliance with general framework conditions is checked	1.3, 2.1	2.4
2.4	Quantified Scenario Assumptions	Unstructured data	References and rationale for numeric values for all attributes, including general framework conditions, scenario-specific framework conditions, ranges for endogenous attributes, and final discrete decision alternatives	2.2, 2.3, expert involvement, related energy scenarios	2.5
2.5	Evaluation Objects	Structured data (DB)	Specification of single decision alternatives for each external scenario, also including general framework conditions to provide parametrization for simulation scenarios	2.4	3.1, 4.2
3.1	Derived Attributes	Structured data (IM and DB)	Attributes, whose values are calculated in simulation scenarios, with simulation models of the energy system	2.5, simulation models	4.2
4.1	Transformation Functions	Structured data (IM)	Functions mapping attributes and derived attributes onto SECs	1.1, 2.2	2.2, 4.2
4.2	Quantified SECs	Structured data (DB)	Values for the SECs describing the performance of alternatives regarding sustainability within external scenarios	1.1, 2.5, 3.1, 4.1	4.4
4.3	Weighting of SECs	Structured data (MCDM tool)	Subjective weights of SECs representing shareholders' preferences between criteria	Choice of weighting method, shareholder involvement	4.4
4.4	Sustainability Order of Alternatives	Unstructured data	Ranking of decision alternatives and recommendations according to MCDM aggregation method, given the decision table and shareholders' preferences	4.2, 4.3, choice of MCDM method, shareholder involvement	—

concrete values for the SECs. To define the transformation functions, it is important that the SECs fit the simulation models or rather that this fit is established: In the best case, researchers can choose, expand, or design simulation models in such a way that all SECs can be calculated. If this is not possible, external studies might provide input for the transformation functions. For example, studies on life-cycle assessments could provide input for the environmental criteria. However, in this case, the researchers need to check whether the system boundaries of the life-cycle assessment fit to the system boundaries of the future scenarios (see Sect. 5.1).

Reducing the scenario set is the next step. To that end, scenarios are evaluated against the general framework conditions, e.g., targets for reducing GHG emissions, which set boundaries for identifying valid decision alternatives. This is necessary, because the future scenarios are designed to reflect a broad range of future projections. In this step, future scenarios may be discarded, if it is obvious that they will not comply with all general framework conditions. For example, if a future scenario exists, in which the shares of renewable energy technologies stay on today's levels and energy demand increases, it is quite obvious that GHG reduction targets cannot be met. However, such a scenario may be used as a reference scenario, depending on the objective of the evaluation.

The next step is to quantify the different types of attributes in such a way that they reflect a future scenario. Accordingly, attribute specifications can be gained from related quantitative energy scenarios, if the assumptions fit the selected future scenarios. For general framework conditions, a single value has to be defined. For scenario-specific framework conditions, multiple values have to be defined (a single value for each external scenario). For endogenous attributes, ranges of possible values have to be defined for each external scenario, so that multiple discrete alternatives can be determined afterward. This discretization is important, because it limits the minimum number of simulations, which need to be run to provide input for the sustainability evaluation of decision alternatives. The discrete attribute values should be chosen in such a way that they reflect discrete decision alternatives. According to the SAS procedure proposed by Alcamo [1], fuzzy set theory can be used for this step.

In the last step of this part, the quantified general framework conditions, quantified external scenarios, and quantified decision alternatives are combined to form the **evaluation objects**. These will be used as input parameters for the simulation models to simulate a particular configuration of the energy system (decision alternative) within a defined environment (general framework conditions and external scenario).

The resulting artifacts of this part of the SEP are (see Table 1): (2.1) general framework conditions, (2.2) a list of attributes, (4.1) a set of transformation functions, (2.3) a list of relevant future scenarios, (2.4) quantified scenario assumptions, and (2.5) evaluation objects with quantified decision alternatives and quantified external scenarios illustrating the associated uncertainties.

5.3 Modeling and Simulation

In the previous part of the SEP, attributes have been deduced, classified, and quantified with the help of related literature. This part deals with simulations of the future energy system, which provide the values of derived attributes. These attributes are produced by simulation models, which are parametrized with the evaluation objects. As mentioned in Sect. 3.1, co-simulation facilitates the integration of simulation models from different domains. Since modeling is done differently in every domain and this is not the focus of this paper, only a rough specification of this part is shown here.

Firstly, the different simulation models have to be set up and the input data has to be prepared, e.g., scaled to the scope of the simulation scenario. As indicated by a feedback loop in the process, models may have dependencies between each other that have to be considered. After simulation is set up, it can be started and the values of derived attributes are provided.

The resulting artifacts of this part of the SEP are (see Table 1): (3.1) values of derived attributes.

5.4 Sustainability Evaluation

The last part of the SEP is the sustainability evaluation, which is assisted by MCDM methods. The first step in this part is to provide a decision table for a MCDM method in the form of quantified SECs. This step is assisted by the transformation functions defined earlier (see Sect. 5.2). The quantified SECs represent the performance of a decision alternative within a given external scenario in terms of these criteria for a certain year. For example, a typical SEC for energy systems is "CO_2-emissions". To calculate CO_2-emissions for a specific decision alternative, the installed capacities of different power plants and a specified demand curve for a year, e.g., 2050, are given as input for a simulation of this alternative. Then, schedules of these power plants in 2050 can be determined in the simulation by matching supply with demand. From these schedules, the CO_2-emissions in 2050 can be calculated for this decision alternative (installed capacities of power plants) with a transformation function, on condition that the CO_2-emissions for specific power plants are also known and normalized to a reference unit, e.g., [t CO_2-eq/MWh]. In this example, the demand curve could be a scenario-specific framework condition and thus altered to reflect different future scenarios.

With the researchers having collected and structured the SECs, shareholders need to assign weights to reflect their preferences [3]. Wang et al. [36] provide an overview on standard procedures for assigning weights in MCDM. While directly asking single particular shareholders is the straightforward approach, involving different shareholder groups is also possible [31].

Lastly, the actual evaluation of the decision alternatives is performed by aggregating the quantified SECs of the different alternatives with the weights associated with them. To that end, different aggregation methods exist in

MCDM. A suitable method in this context is the Preference Ranking Organization Method for Enrichment Evaluation (PROMETHEE) [6], since it is easily understandable and therefore also transparent. The result of this method is a (partial or total) ranking of the decision alternatives. This order can then be used to generate recommendations. Furthermore, sensitivity analysis of the SEC weights can be used to check the robustness of the results.

The resulting artifacts of this part of the SEP are (see Table 1): (4.2) quantified SECs for each combination of alternatives and scenarios, (4.3) a weighting of the SECs, and (4.4) a ranking of alternatives and recommendations for further action.

6 Information Model

Having presented the SEP of energy scenarios, we shall point out how an information model can help to automate the information exchange during this process. Our approach for the information model implemented in a mind map is explained in the following sections. For our implementation, we chose the mind mapping tool XMind. It can be extended with plug-ins and is available as open source software. In the following subsections, we describe the information model's general structure and concretize it with an example. To allow the automated integration of the modeled information, we also show its machine readable representation as an ontology and explain the automated generation of a scenario database schema.

6.1   Structure of the Information Model

The information model links future scenarios and simulation scenarios to the sustainability evaluation. The connections of the quantified attributes from both of these scenarios to the evaluation are implemented with transformation functions, which provide the dependencies and mathematical descriptions by mapping derived and non-derived attributes onto the SECs.

The structure of the information model is depicted in Fig. 4. The boxes with rounded edges are the main nodes for the different objects and the rectangular boxes represent the instances of the objects.

The left-hand side represents the domains of interest within the future scenarios and simulation scenarios and consists of different levels ordered from left to right. On the first level, the domains are listed. Each domain is subdivided into domain objects (second level), which represent objects of the real world. The domain objects consist of derived and non-derived attributes (third level), which describe them. Attributes have certain units and are instantiated with values for each external scenario and decision alternative. To represent the connection between attributes and the scenario planning process, the key factors influencing an attribute are annotated with their number. Also, the different categories of attributes are annotated by their first letter to document their definition in the SEP: general framework condition (G), scenario-specific framework condition (S), and endogenous attribute (E).

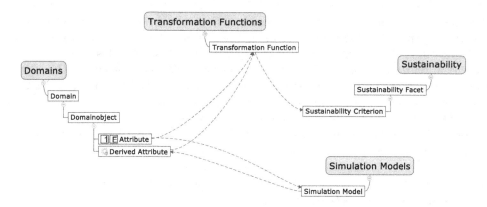

Fig. 4. Structure of the information model.

The next part at the bottom represents the simulation models. Their data flows are modeled by arrows. The inputs of a simulation model are connected with incoming arrows, which show the data flows from the attributes to a simulation model. The outputs of a simulation model are connected with reversed arrows from the simulation model to the derived attributes. These derived attributes are annotated with cogwheel icons to allow to distinguish them from other attributes.

The right-hand side represents the evaluation part of the SEP. On the first level, the major objective is defined – in our case, sustainability. The objective consists of different facets on the second level (e.g., technical, economic, environmental, and social as introduced in Sect. 2). Every facet is subdivided into various SECs on the third level as described in Sect. 5.1.

The last part at the top represents the transformation functions, which may have derived and non-derived attributes from the left-hand side of the information model as input and map these onto the SECs.

6.2   Example

An exemplary, simplified excerpt of the information model from the project NEDS is shown in Fig. 5 to illustrate the information model on practical level. On the left-hand side, the domains ICT, policy, market, user, and energy are depicted.

In this excerpt, the SEC "GHG emission" is chosen, which is part of ecological sustainability. To give an example of the interdisciplinary simulation models, four simplified models are shown: a smart home model from electrical engineering, a macro-economic market model, a bottom-up power system market model, and an operational control model from energy informatics. These models from different domains are coupled for a holistic simulation of a future power system.

The domain "user" contains the domain object "lifestyle" with the attribute "utilization frequency of electrical equipment" representing the user behavior.

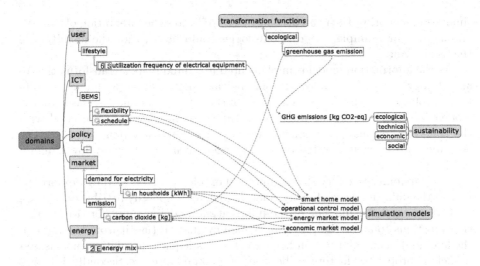

Fig. 5. Example of the information model.

This attribute is scenario-specific and based on a key factor from the scenario planning process. It represents possible changes in the user behavior in the future and is used as input for the smart home model, which contains various models of consumer devices encapsulated in a smart home. Based on the user behavior the smart home model offers "flexibility", which is used by the operational control to schedule the Building Energy Management Systems' (BEMS) electricity consumption or production. The schedule yields the "demand for electricity in households", which is input for the economic market model and the energy market model. Both market models also have the "energy mix" as input, which consists of the amount of energy generated by the different primary energy sources. As output they provide the "carbon dioxide emissions" of the power plants and the overall economy. The "carbon dioxide emissions" is input of the transformation function for the SEC "GHG emissions".

This example illustrates how modeling scenarios with the information model works. However, modeling a complete scenario is highly complex and therefore needs assistance and automation. Thus, the next section describes the ontological representation of the information model.

6.3 Ontological Representation

To allow the automated integration of the modeled information in the SEP, the information has to be made available in a machine readable format. As described in Sect. 3.3, ontologies are a frequently used concept for a machine readable representation of knowledge. Thus, we aim to allow the representation of the information model's content in an ontology. By doing so, the ontology provides a structure for reasoning with the data to infer implicit knowledge and querying the data with languages such as SPARQL to support the development of simulation scenarios. Using an ontology also facilitates the integration of existing ontologies. For example, external ontologies could be used for the definition of the used terms.

As the information model is implemented in a mind map, a transformation to an ontology is needed. To allow the ontological representation, we implemented a base ontology with the general structure of the information model. Based on the modeled information in the mind map, the base ontology is instantiated with the concrete information by an extension of the mind mapping software. The described process is illustrated in the upper left corner of the data flow diagram in Fig. 6.

The capability to query the instantiated ontology can be used to support the user in the following ways. Firstly, querying allows to identify attributes that are missing on the left-hand side of the information model as input for transformation functions to allow the previous defined evaluation (use it from the right to the left side). This information helps to choose adequate simulation models and to include proper key factors in the scenario planning process. Secondly, it allows to identify evaluation criteria that can be added on the right-hand side of the information model to make sure that all relevant results from future scenarios and simulation are used (use it from the left to the right side). This way, the instantiated ontology supports the definition of simulation scenarios as depicted in Fig. 6.

Additionally, the modeled information allows to examine the dependencies between simulation models. More specifically, the data flows between models can be analyzed to find mutual or cyclic dependencies to get information about the order in which the models have to be executed. Combined with an ontological definition of the simulation models, the integration of simulation models in the information model would be possible. This would enable the information model to suggest suitable simulation models for a simulation, based on the modeled information and available simulation models. This integration would be a step towards automation of simulation.

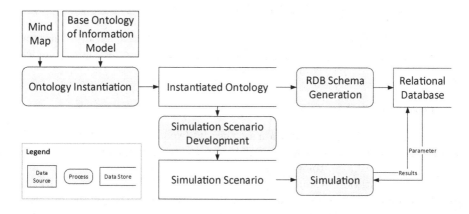

Fig. 6. Data flow diagram: ontology instantiation and generation of RDB schema.

6.4   Generating a Scenario Database Schema

As presented in the previous sections, the information model describes the data flows between different models and processes. Especially in projects with many partners and models, the data flows become highly complex and the users benefit from assistance and a centralized data store. In our approach the data store is implemented as a RDB called scenario database. The schema for this scenario database is based on the instantiated ontology introduced in the previous section, as illustrated in Fig. 6.

We developed an algorithm automating the generation of a RDB schema from the ontological representation of the information model. A similar approach is described by Vysniauskas et al. [35]. They mapped the concepts of ontologies defined in the Web Ontology Language (OWL) to the concepts of RDB. For example, an OWL class is represented by a table in RDB schema. Additionally, they developed an algorithm for transforming an OWL ontology into a RDB schema. Our approach is similar to this, but also takes into account the annotations. This way, it can generate a RDB schema, which allows to store the attribute values defined in the future scenarios and literature research and also the results from simulation. To facilitate the cooperative work of different experts, different views can be defined on the schema. This way, every expert sees only the relevant data. As shown in Fig. 6, the RDB is supposed to be used for parametrization of the simulation and also serves as data store for the results of simulation.

7   Conclusion

In this article, we have shown how the SEP developed in the project NEDS [4] leads towards an integrated development and sustainability evaluation of energy scenarios. The SEP comprises the development of qualitative future scenario descriptions, the quantification and simulation of these scenarios, and evaluation using MCDM. We explained a revised version of the process, illustrated the resulting artifacts, and showed how these are connected. Additionally, we have introduced an information model, which leads towards an automation of information exchange in the SEP. In this information model, future and simulation scenarios are linked to the sustainability evaluation via attributes, transformation functions, and SECs. We shall highlight how the three requirements for the development and evaluation of energy scenarios – integrate a sustainability definition, add transparency to the decision support process, and allow for replicability – are met by this integrated approach:

Firstly, integrating MCDM into the approach allows to consider relevant sustainability criteria in the evaluation of energy scenarios. Multi-domain co-simulation can assist this, because it allows to couple simulation models from different domains, modeling tools, and programming languages to provide the quantified values for the evaluation criteria. Thereby, not only technical or economic, but also environmental and social criteria can be considered simultaneously in energy scenarios. The information model facilitates the integration of different dimensions of sustainability by modeling the data flows between different models, software, and actors in the SEP. Overall, it allows to handle the complexity of multi-domain co-simulation and supports the communication between experts from different domains.

Secondly, integrating MCDM into the approach makes the decision support processes more transparent, because it facilitates problem structuring: The structured nature of MCDM approaches challenges shareholders to think about their own preferences and make them explicit. Additionally, the proposed SEP fosters the communication of uncertainties by differentiating between decision alternatives and external uncertainties. Thereby, the impacts of decisions in highly uncertain environments can be analyzed. This way, decision alternatives representing configurations of the future energy system can be identified and evaluated.

Thirdly, the information model addresses both the transparency and replicability of the development and evaluation of energy scenarios by modeling the data flows and dependencies between different models, software, and actors in the SEP. The ontological representation of information leads towards an automation of scenario definition, simulation preparation, and evaluation of the results in the context of energy scenarios. This is crucial to handle the complexity of energy scenarios.

In summary, the defined requirements are addressed by our approach. The presented integrated development and sustainability evaluation of energy scenarios will be subject to future refinements.

Acknowledgements. The research project 'NEDS – Nachhaltige Energieversorgung Niedersachsen' acknowledges the support of the Lower Saxony Ministry of Science and Culture through the 'Niedersächsisches Vorab' grant program (grant ZN3043).

References

1. Alcamo, J.: The SAS approach: combining qualitative and quantitative knowledge in environmental scenarios. In: Alcamo, J. (ed.) Environmental Futures: The Practice of Environmental Scenario Analysis, Developments in Integrated Environmental Assessment, vol. 2, pp. 123–150. Elsevier, Amsterdam and Boston (2008). https://doi.org/10.1016/S1574-101X(08)00406-7
2. Bastian, J., Clauß, C., Wolf, S., Schneider, P.: Master for co-simulation using FMI. In: Proceedings of the 8th International Modelica Conference, pp. 115–120 (2011). https://doi.org/10.3384/ecp11063115
3. Belton, V., Stewart, T.J.: Multiple Criteria Decision Analysis: An Integrated Approach, 2nd edn. Kluwer Academic Publishers, Boston (2003)
4. Blank, M., et al.: Process for simulation-based sustainability evaluation of smart grid future scenarios, version 1.0 (2016). http://neds-niedersachsen.de/uploads/tx_tkpublikationen/Whitepaper-SEP-V1.pdf. Accessed 17 Aug 2017
5. BMWi: Federal Ministry of Economics and Technology. Energiekonzept für eine umweltschonende, zuverlässige und bezahlbare Energieversorgung (2010). https://www.bmwi.de/Redaktion/DE/Downloads/E/energiekonzept-2010.pdf. Accessed 17 Aug 2017. (in German)

6. Brans, J.P., Vincke, P.: Note–a preference ranking organisation method: (the PROMETHEE method for multiple criteria decision-making). Manag. Sci. **31**(6), 647–656 (1985). https://doi.org/10.1287/mnsc.31.6.647

7. Brundtland, G.H.: Our Common Future. Oxford Paperbacks. Oxford University Press, Oxford (1987)

8. Castro, A.G., et al.: The use of concept maps during knowledge elicitation in ontology development processes-the nutrigenomics use case. BMC Bioinform. **7**, 267 (2006). https://doi.org/10.1186/1471-2105-7-267

9. Daraio, C., Lenzerini, M., Leporelli, C., Naggar, P., Bonaccorsi, A., Bartolucci, A.: The advantages of an Ontology-Based Data Management approach: openness, interoperability and data quality. Scientometrics **108**(1), 441–455 (2016). https://doi.org/10.1007/s11192-016-1913-6

10. Deutscher Bundestag: German Bundestag. Gesetz über die Elektrizitäts- und Gasverordnung (Energiewirtschaftsgesetz): EnWG (2005). http://www.gesetze-im-internet.de/enwg_2005/BJNR197010005.html. Accessed 17 Aug 2017. (in German)

11. Elkington, J.: Cannibals with Forks: The Triple Bottom Line of 21st Century Business. Capstone, Oxford (2002). Reprint edn

12. European Commission: Communication from the European Commission: A policy framework for climate and energy in the period from 2020 to 2030 (2014). http://eur-lex.europa.eu/legal-content/EN/ALL/?uri=CELEX:52014DC0015. Accessed 17 Aug 2017

13. Fensel, D.: Ontologies: A Silver Bullet for Knowledge Management and Electronic-Commerce. Springer, Berlin (2004). https://doi.org/10.1007/978-3-662-09083-1

14. Gausemeier, J., Fink, A., Schlake, O.: Scenario management: an approach to develop future potentials. Technol. Forecast. Soc. Change **59**, 111–130 (1998). https://doi.org/10.1016/S0040-1625(97)00166-2

15. Grunwald, A., Dieckhoff, C., Fischedick, M., Höffler, F., Mayer, C., Weimer-Jehle, W.: Consulting with energy scenarios: requirements for scientific policy advice. In: acatech/Leopoldina/Akademienunion (eds.) Series on Science-Based Policy Advice (2016)

16. Hughes, N., Strachan, N.: Methodological review of UK and international low carbon scenarios. Energy Policy **38**(10), 6056–6065 (2010). https://doi.org/10.1016/j.enpol.2010.05.061

17. International Energy Agency (IEA): World Energy Outlook 2016: Executive Summary (2016)

18. Keles, D., Möst, D., Fichtner, W.: The development of the German energy market until 2030—A critical survey of selected scenarios. Energy Policy **39**(2), 812–825 (2011). https://doi.org/10.1016/j.enpol.2010.10.055

19. Kontchakov, R., Rodríguez-Muro, M., Zakharyaschev, M.: Ontology-based data access with databases: a short course. In: Rudolph, S., Gottlob, G., Horrocks, I., van Harmelen, F. (eds.) Reasoning Web 2013. LNCS, vol. 8067, pp. 194–229. Springer, Heidelberg (2013). https://doi.org/10.1007/978-3-642-39784-4_5

20. Kowalski, K., Stagl, S., Madlener, R., Omann, I.: Sustainable energy futures: methodological challenges in combining scenarios and participatory multi-criteria analysis. Eur. J. Oper. Res. **197**(3), 1063–1074 (2009). https://doi.org/10.1016/j.ejor.2007.12.049

21. Kronenberg, T., et al.: Energieszenarien für Deutschland: Stand der Literatur und methodische Auswertung. In: Bruhns, H. (ed.) Energiewende - Aspekte, Optionen, Herausforderungen, pp. 132–168. Deutsche Physikalische Gesellschaft - Arbeitskreis Energie, Berlin (2012). (in German)

22. Lee, Y.T.: Information modeling: from design to implementation. In: Proceedings of the Second World Manufacturing Congress, pp. 315–321 (1999)
23. Lehnhoff, S., et al.: Exchangeability of power flow simulators in smart grid co-simulations with mosaik. In: 2015 Workshop on Modeling and Simulation of Cyber-Physical Energy Systems, MSCPES, pp. 1–6, April 2015. https://doi.org/10.1109/MSCPES.2015.7115410
24. Oberschmidt, J., Geldermann, J., Ludwig, J., Schmehl, M.: Modified PROMETHEE approach for assessing energy technologies. Int. J. Energy Sect. Manag. 4(2), 183–212 (2010). https://doi.org/10.1108/17506221011058696
25. Rehtanz, C., Guillaud, X.: Real-time and co-simulations for the development of power system monitoring, control and protection. In: 2016 Power Systems Computation Conference, PSCC, pp. 1–20, June 2016. https://doi.org/10.1109/PSCC.2016.7541030
26. Robinson, S.: Simulation: The Practice of Model Development and Use. Wiley, Hoboken (2004)
27. Schlögl, F., Rohjans, S., Lehnhoff, S., Velasquez, J., Steinbrink, C., Palensky, P.: Towards a classification scheme for co-simulation approaches in energy systems. In: International Symposium on Smart Electric Distribution Systems and Technologies, pp. 2–7. IEEE/IES (2015). https://doi.org/10.1109/SEDST.2015.7315262
28. Schwarz, J.S., Witt, T., Nieße, A., Geldermann, J., Lehnhoff, S., Sonnenschein, M.: Towards an integrated sustainability evaluation of energy scenarios with automated information exchange. In: Proceedings of the 6th International Conference on Smart Cities and Green ICT Systems, pp. 188–199. ScitePress (2017). https://doi.org/10.5220/0006302101880199
29. Simon-Cuevas, A., Ceccaroni, L., Rosete-Suarez, A., Suarez-Rodriguez, A.: A formal modeling method applied to environmental-knowledge engineering. In: International Conference on Complex, Intelligent and Software Intensive Systems (2009). https://doi.org/10.1109/CISIS.2009.55
30. Singh, R.K., Murty, H.R., Gupta, S.K., Dikshit, A.K.: An overview of sustainability assessment methodologies. Ecol. Indic. 15(1), 281–299 (2012)
31. Steinhilber, S.: Exploring options for the harmonisation of renewable energy support policies in the EU using multi-criteria decision analysis. Dissertation, Karlsruher Institut für Technologie (2015)
32. Stewart, T.J., French, S., Rios, J.: Integrating multicriteria decision analysis and scenario planning - review and extension. Omega 41(4), 679–688 (2013). https://doi.org/10.1016/j.omega.2012.09.003
33. Uslar, M., Specht, M., Rohjans, S., Trefke, J., Vasquez González, J.M.: The Common Information Model CIM. Springer, Heidelberg (2012). https://doi.org/10.1007/978-3-642-25215-0
34. van der Heijden, K.: Scenarios: The Art of Strategic Conversation. Wiley, Chichester, England and New York (1996)
35. Vysniauskas, E., Nemuraite, L., Paradauskas, B.: Hybrid method for storing and querying ontologies in databases. Elektronika ir Elektrotechnika 115(9), 67–72 (2011). https://doi.org/10.5755/j01.eee.115.9.752
36. Wang, J.J., Jing, Y.Y., Zhang, C.F., Zhao, J.H.: Review on multi-criteria decision analysis aid in sustainable energy decision-making. Renew. Sustain. Energy Rev. 13(9), 2263–2278 (2009). https://doi.org/10.1016/j.rser.2009.06.021
37. Weimer-Jehle, W., et al.: Context scenarios and their usage for the construction of socio-technical energy scenarios. Energy 111, 956–970 (2016). https://doi.org/10.1016/j.energy.2016.05.073

Day-Ahead Scheduling of Electric Heat Pumps for Peak Shaving in Distribution Grids

Marco Pau[1(✉)], Jochen Lorenz Cremer[2], Ferdinanda Ponci[1], and Antonello Monti[1]

[1] Institute for Automation of Complex Power Systems, E.ON Energy Research Center, RWTH Aachen University, Mathieustrasse 10, 52074 Aachen, Germany
mpau@eonerc.rwth-aachen.de
[2] Department of Electrical and Electronic Engineering, Imperial College, London, UK

Abstract. In future electric distribution networks, demand flexibility offered by controllable loads will play a key role for the effective transition towards the smart grids. Electric heat pumps are flexible loads whose operation can be controlled, to some extent, to foster the efficient operation of the distribution grids. This paper presents an optimization algorithm that defines a smart day-ahead scheduling of electric heat pumps aimed at achieving power peak shaving in the distribution grid, while providing customers with the desired thermal comfort over the day. The proposed optimization relies upon a Mixed Integer Linear Programming approach and allows defining both the time schedule and the operating points of the heat pump, guaranteeing an energy efficient solution for the customers. Performed tests show the benefits achievable by means of the proposed optimal scheduling both at the distribution grid level and at the customer side, proving the goodness of the conceived solution.

Keywords: Electric heat pumps · Demand Side Management
Power peak shaving · Optimal scheduling · Mixed integer linear programming

1 Introduction

Electric systems are rapidly evolving and many important changes are reshaping the power system scenario [1]. The increasing penetration of Distributed Generation (DG) is transforming the distribution systems from simple passive grids with unidirectional flows into active complex networks with possibility of reverse power flows. The un-predictable behaviour of Renewable Energy Sources (RES) is making more difficult to achieve the balance between generation and demand, creating new challenges for the real-time management of the system. At the same time, despite the improvements in energy efficiency, the electricity consumption is constantly growing and the upcoming electrification of the mobility and the heating sector is expected to further boost this trend, leading many grids close to (or beyond) their capacity limits [2]. Due to the challenges brought by these changes, new solutions have to be devised for the management of future electric grids, in order to guarantee the efficient, reliable and secure supply of energy to the final customers [1, 3, 4].

© Springer Nature Switzerland AG 2019
B. Donnellan et al. (Eds.): SMARTGREENS 2017/VEHITS 2017, CCIS 921, pp. 27–51, 2019.
https://doi.org/10.1007/978-3-030-02907-4_2

Differently from the past, where the control of the electric system was mainly performed through a *"generation follows demand"* approach, in future Smart Grids (SG) the direct or indirect control of specific loads is expected to be used to provide an additional degree of flexibility in the management of the electric system. The set of actions aimed at modifying the original pattern of consumption of the end users goes under the name of Demand Side Management (DSM) [5]. Possible DSM measures include the interruption, activation, reduction, shift or scheduling of specific loads whose operation is not critical for the customer or whose flexible management is accepted.

Several DSM schemes based on different approaches, targeting different devices or pursuing different objectives can be found in the literature [6–13]. A common classification of the DSM programs is between *price-based* and *incentive-based* approaches. In price-based approaches, the modification of the daily consumption pattern is generally pursued through the application of different electricity prices over the day, in order to disadvantage the usage of energy during peak times (higher prices) and to foster the consumption during off-peak periods (lower tariffs). In this case, the benefits at grid level are strictly dependent on the adopted pricing strategy and on the reaction of end users to such pricing scheme. In the incentive-based approaches, a more direct control of the demand is guaranteed by the possibility to interrupt or schedule specific loads or by asking the customers to cut their consumption according to given curtailment signals. In this case, the offered flexibility can be rewarded through special incentives in the tariff or paying the customers for their response to the curtailment signals. Further classifications of the DSM schemes also exist, for example depending on the timing of the action (e.g. day-ahead scheduling or real-time curtailment to support unexpected contingencies on the grid), on the type of service provided (e.g. reduction of the consumption through energy efficiency improvements, load shifting to modify the pattern of power consumption, or offer of flexibility to relieve possible stress conditions in the grid), etc.

A large set of benefits can be obtained, at all the levels of the electricity chain, through the application of smart DSM solutions [14]. However, at the moment, several challenges still prevent a wide deployment of DSM schemes. Main obstacles are associated to the absence of a proper measurement and communication infrastructure at the distribution level of the grid, to the lack of a suitable market framework, but also to the lack of understanding and poor acceptance from the final customers [14, 15]. While these issues are likely to slow down the uptake of DSM, first initiatives have been already launched in several countries. In US, DSM programs are already available for both residential and non-residential customers since several years. According to [16], direct load control, interruptible/curtailable rates, Time of Use (ToU) rates and real-time pricing are available in most of the states and produce a potential peak load reduction of more than 50 GW. In Europe, the situation changes in each country due to the different regulatory frameworks, but DSM is commercially viable in several nations and measures aimed at fostering the application of DSM are being developed in most of the European Union state members [17].

In this paper, a DSM scheme providing the day-ahead scheduling of electric heat pumps used for space heating is considered. Electric heat pumps are expected to spread in the near future due to the incentives present in many countries to foster the electrification of the heating sector. Their operation allows some degrees of flexibility, thanks to the thermal inertia of the buildings and the relatively slow dynamics characterizing thermal phenomena. Moreover, final customers can provide additional flexibility by accepting the possible reduction of the temperature in their house within certain limits. In the proposed DSM scheme, an optimization algorithm based on a Mixed Integer Linear Programming (MILP) formulation is used to define the smart scheduling of the heat pumps for a district of houses. The objective of the optimization is to schedule the operation of the heat pumps in order to fulfill the requirements of the final customers in terms of thermal comfort, providing at the same time the minimization of the power peaks in the distribution grid during the day.

The remainder of this paper is structured as follows. Section 2 presents an overview of the proposals already available in the literature aimed at exploiting electric heat pumps for DSM purposes, emphasizing the differences and novelties of the approach proposed in this paper. In Sect. 3, the mathematical framework associated to the thermal and the heat pump model, which represents the basis for the MILP formulation, is given. Section 4 describes the details of the centralized framework and of the objective function used for the optimization. Section 5 shows the simulations performed to assess the operation of the designed algorithm and highlights the potential benefits achievable, both from a distribution grid point of view and from a customer perspective, thanks to the conceived DSM scheme. Finally, Sect. 6 summarizes the obtained results and provides the final remarks.

2   Use of Heat Pumps Flexibility for DSM

The idea to use electric heat pumps for DSM purposes is not new and has been investigated in a number of proposals in the literature. Two main reasons justify the research efforts for DSM solutions based on the use of heat pumps. First, due to regulations pushing to enhance energy efficiency, electric heat pumps are expected to largely spread in the near future [18]. As their penetration grows, the electricity demand will be significantly affected: the study in [19] shows that during the winter months the consumption of residential customers could double, while [20] indicates that a heat pump penetration of 20% could lead to an increase in the peak demand of nearly 15%. Such an increase in the power demand has to be suitably managed to avoid issues in the normal operation of the grid. Secondly, since heat pumps are mainly used for space heating, the thermal inertia of the buildings and the slow rate of the thermal variations can be exploited to conceive a smart management of the heat pump. Such flexibility can be advantageously used by customers in price-based DSM programs to minimize their energy costs and can be used by Distribution System Operators (DSO) to relieve possible issues occurring in the distribution grid.

Many research works are available in the literature, which focus on different aspects of the heat pump based DSM. A first problem is the definition of suitable thermal models to represent the thermal behaviour of buildings and heating devices. In [21], the modelling of an air source heat pump coupled with an underfloor heating system is presented; in [22, 23], a more generic framework to represent building and heating system by means of an electrical analogue is described. In some scenarios, thermal storage is also adopted to achieve additional flexibility in the definition of the DSM scheme. Some papers therefore focus on the assessment of the additional benefits offered by the use of thermal storage [24] or on the analysis of specific storage technologies [25].

Given the thermal models, many different DSM approaches can be defined with different objectives, like the minimization of the energy costs for the final customer or the maximization of the self-consumption in presence of renewable energy sources [26, 27]. In many cases, however, the DSM programs proposed in the literature do not allow a clear understanding of the benefits arising for the distribution grid operation or do not duly consider the comfort of the final users, which is an essential pre-requisite for the acceptance of the DSM scheme. In [28], different pricing schemes are analyzed together with different operating modes of the heat pumps to assess the benefits coming for the customer from the smart management of the available flexibility. Results show that the customers can obtain energy and financial savings, but these are achieved at the expense of a reduction in the indoor temperature, which could lead to thermal discomfort. The impact of the heat pumps scheduling on the grid behaviour is instead not analyzed. In [29], another price-based DSM scheme is investigated; here the grid operator provides dynamic prices to the users and can command power limitations in case of contingencies in the grid: if these limitations are not respected, the customer incurs in penalties. In such a scenario, the intelligent scheduling of the heat pumps allows minimizing the costs for the end user, but the details on their comfort are not provided.

In other proposals, the management of the heat pumps is more directly linked to the technical benefits desired in the distribution grid. In [30], a two-step process handles the heat pumps operation. First, the schedule of the heat pumps is defined, aiming at minimizing the costs given the variable price of energy over time. Then, during the real-time operation, the predefined scheduling can be overridden, in case of voltage issues in the grid, by shifting the operation of the heat pumps. In this way, possible contingencies in the grid are alleviated, but thermal discomfort could be experienced by the end users in the meanwhile. In [31], a DSM scheme based on real-time local prices is presented, where the prices are defined depending on the loading conditions of the grid. Through this approach, heat pumps operation is discouraged in those zones (or during those periods) where heavy loading conditions are present. The method allows providing a service to DSOs, but does not ensure thermal comfort for the end user. The risk to have thermal discomfort is also evaluated and confirmed in [32], where heat pumps flexibility is used to follow the generation from renewable energy sources.

To face the problem of thermal comfort, some proposals consider indoor temperature boundaries as constraints during the optimization process. In [33], temperature boundaries are included in the optimization performed by a home management system that schedules the operation of flexible loads and storage systems. In this case, the optimization goal is to minimize the costs for the final users while ensuring thermal comfort, but details on the impact at grid level of the DSM scheme are not provided. In [34], price-based schemes are analyzed together with direct control approaches in a DSM model where the objective is to balance demand and renewable generation to minimize the costs for the energy aggregator providing the DSM service. Also in this case, thermal comfort to the customers is guaranteed by constraining the indoor temperature of the houses within given boundaries, but a clear picture of the benefits achieved at grid level is not available.

To combine the benefits given by the DSM scheme for both DSO and final customer, in [35], an optimal day ahead scheduling of the heat pumps has been proposed. The goal is to minimize the power peaks in the distribution grid and, similarly to [33, 34], to ensure the thermal comfort for the final customers through the definition of indoor temperature boundaries. From a grid perspective, the minimization of the power peaks is one of the most important objectives that can be pursued through the smart allocation of flexible demand. As described in [14], through the reduction of the power peaks, utilities can reduce the energy losses, improve the voltage profile of the grid, lower the risk for contingencies and postpone network reinforcement required due to the increase of electricity demand. At system level, this also allows avoiding the use of expensive generation units to cover peak demand and reducing the need for spinning reserve, bringing an overall reduction of the system costs. Concerning the customers, the proposed DSM scheme allows guaranteeing a minimum thermal comfort (that can be chosen by the customer) and the participation to the DSM program could be rewarded through an incentive-based mechanism, thus also providing economic benefits to the users. In [36], the same DSM model is used to classify the customers based on their flexibility, and it is shown how their choices can affect some flexibility indexes: the flexibility they can offer can be thus an aspect to be considered for the definition of the incentives to the single customers. In this paper, the main concepts of the idea in [35] are presented with additional details on the models and equations used in the optimization process. Moreover, simulations are integrated with new results to provide a better overview of the benefits provided by the conceived DSM scheme.

3   Model Formulation

The optimization algorithm presented in this paper is constrained by equality and inequality constraints given in the thermal and the heat pump model. The thermal model mainly defines the equations needed to describe the evolution of the indoor temperature (within the house), whereas the heat pump model links the electrical and the thermal world, providing the equations needed to map the heat provided by the heat pump to the associated electrical power requirement. In the following, the mathematical relationships describing these models are provided together with the additional constraints that are taken into account in the optimization process.

3.1   Thermal Model

Energy Balance. The energy balance equation defines the evolution over the time of the indoor temperature in the house depending on the operating mode of the heat pump and other relevant external factors (e.g. outdoor temperature and thermal characteristics of the house). The energy balance equation used in this paper is based on the thermal model already presented in [33, 35] and it is:

$$T_{h,t}^{IN} = T_{h,t-1}^{IN} + \frac{\Delta t}{C_h} \left(Q_{h,t}^{HP} - Q_{h,t}^{L} \right) \qquad (1)$$

where: $T_{h,t}^{IN}$ is the indoor temperature of the house h at the generic instant of time t; Δt is the step of the time discretization used in the optimization; C_h is the thermal capacity of the house; and $Q_{h,t}^{HP}$ and $Q_{h,t}^{L}$ are the heat provided by the heat pump and the heat losses, respectively, for house h at time t.

In (1), the thermal capacity C_h of the house represents the thermal inertia of the building: the higher the capacity, the lower the thermal variation (for a given condition of heat flow from the heat pump and heat losses). The thermal capacity is dependent on the geometrical characteristics of the house and is given by:

$$C_h = \gamma_{air} \cdot \mu_h \qquad (2)$$

where γ_{air} is the specific heat capacity of the air (equal to 1.005 $\frac{kJ}{kg \cdot K}$ at 300 K and atmospheric pressure), while μ_h is the indoor air mass of the house, which can be calculated as:

$$\mu_h = \rho_{air} \cdot V_h \qquad (3)$$

with ρ_{air} being the air density (equal to 1.2041 $\frac{kg}{m^3}$ at standard conditions) and V_h representing the total volume of the house. Similarly to [33], V_h can be calculated considering the length, width and height of the house ($L_{1,h}$, $L_{2,h}$ and $L_{3,h}$, respectively) and the roof pitch β_h, using the following relationship:

$$V_h = L_{1,h}L_{2,h}L_{3,h} + L_{1,h}L_{2,h} \cdot \tan \beta_h \qquad (4)$$

In (1), the difference between the heat flow $Q_{h,t}^{HP}$ provided by the heat pump and the heat losses $Q_{h,t}^{L}$ is determining the increase or decrease of indoor temperature. The equations describing the heat flow $Q_{h,t}^{HP}$ will be described in the subsection on the heat pump model. As for the heat losses $Q_{h,t}^{L}$, they can be expressed as [35]:

$$Q_{h,t}^{L} = \kappa_h \left(T_{h,t-1}^{IN} - T_{h,t-1}^{OUT} \right) \qquad (5)$$

where $T_{h,t-1}^{OUT}$ is the outdoor temperature, while κ_h is the heat loss factor of the house. The heat loss factor depends on the thermal characteristics of the materials composing the house, and it can be calculated according to the following:

$$\kappa_h = \upsilon_{wa} \left[2 \left(L_{1,h} + L_{2,h} \right) L_{3,h} - n_{wi}A_{wi} \right] + n_{wi}\upsilon_{wi}A_{wi} \qquad (6)$$

where υ_{wa} and υ_{wi} are the thermal transmittances of wall and windows, respectively, A_{wi} is the average area of the windows in the house and n_{wi} is their number.

Thermal Comfort Constraints. In the conceived optimization, thermal comfort constraints have been defined to ensure that the scheduling of the heat pumps will always allow fulfilling the requirements of the customer in terms of desired indoor temperature. To this purpose, in the DSM scheme, it is assumed that each customer provides a lower and an upper boundary for the temperature he would like to have at home. Indicating with $\Gamma^{LB}_{h,t}$ and $\Gamma^{UB}_{h,t}$ the lower and the upper boundary, respectively, the comfort constraints automatically translate into the following:

$$\Gamma^{LB}_{h,t} \leq T^{IN}_{h,t} \leq \Gamma^{UB}_{h,t} \tag{7}$$

In (7), it is possible to observe that the temperature boundaries vary for each house (each customer can have different preferences) and that they can be variable in time. This variability allows accommodating different requirements during the day, like for example the need to differentiate the desired temperature profile between day and night or to guarantee minimum thermal requirements only for the periods of effective occupancy of the home. Figure 1 shows, as an example, two possible temperature profiles corresponding to the above-mentioned scenarios.

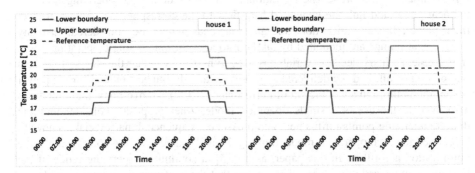

Fig. 1. Example of thermal comfort settings for different customers.

In Fig. 1, beyond the temperature limits, also a reference temperature is indicated, which is calculated as the average value between lower and upper boundary. This reference temperature can be seen as the temperature profile desired by the customer, while the lower and upper temperature boundaries are the temperature deviations that the customer offers as flexibility and that he is thus willing to accept [36].

Temperature Boundaries Constraints. In the designed optimization, two additional constraints are considered for each house indoor temperature. These refer to the temperature at the initial and the final time step of the optimization. Calling $T^{IN}_{h,0}$ the temperature at the first time step of the day, the following constraint is imposed:

$$T^{IN}_{h,0} = \Gamma^{Start}_{h} \tag{8}$$

where Γ_h^{Start} represents a starting value of temperature that, in general, will correspond to the final value of temperature obtained in the optimization for the previous day.

As for the final state of temperature, the following constraint is considered:

$$T_{h,f}^{IN} \geq \frac{\Gamma_{h,f}^{LB} + \Gamma_{h,f}^{UB}}{2} \tag{9}$$

Equation (9) forces the optimization to give a final temperature larger than the reference temperature defined at the end of the day. This is done because the optimization will tend to minimize the energy consumption of the customer. If no constraint were used, the optimization would deliver a final temperature very close to the lower boundary, thus removing the flexibility for the first time steps of the optimization performed for the following day (the heat pump would be forced to be on in order to keep the indoor temperature above the lower boundary).

3.2 Heat Pump Model

Operational Constraints. The operational constraints define the operating mode of the heat pump and link the generated heat flow to the corresponding request of electrical power. In this paper, domestic air-to-air heat pumps are considered. In many papers, heat pumps are modelled by taking into account only a single operating point. In this case, a binary decision variable is used to indicate whether the status of the heat pump is on or off, and the delivered heat flow is directly linked to the electrical power consumption through a Coefficient of Performance (COP). To overcome the limitations associated to the use of a single operation mode, other proposals consider multiple discrete operating points for the heat pump. As a consequence, binary variables have to be defined for each operation mode, thus significantly increasing the complexity of the optimization algorithm. In this paper, as in [35], a continuous operating mode of the heat pump is instead considered. This means that the heat pump can operate with any heat flow between a minimum and a maximum limit. The electrical power required to generate a specific heat flow is obtained through a mapping function, which can be, in first approximation, linearized through a given number of segments. Figure 2 shows, as an example, the power to air mass flow characteristic curve of a heat pump built by using the discrete operating points indicated in [33].

In Fig. 2, it is possible to note that the heat pump can operate for any value of air mass flow between Φ_0 and Φ_2 and that the power to air mass flow characteristic is described through two linear segments. The operating point at minimum air mass flow and the linear segments of the characteristic curve define the operating modes m (three in the example shown in Fig. 2) of the heat pump in the conceived model.

As for the point at minimum air mass flow, indicating with $m = 0$ the associated operation mode, the following constraint is imposed:

$$\Delta F_{h,0,t} = y_{h,t} \cdot \Phi_{h,0} \tag{10}$$

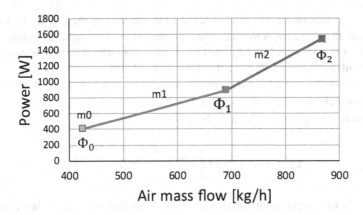

Fig. 2. Power to air mass flow characteristic of a heat pump with indication of the different operating modes m.

where $\Delta F_{h,0,t}$ is a continuous variable (defined for each house h and time step t) associated to the operating mode $m = 0$ of the heat pump, $y_{h,t}$ is a binary decision variable and $\Phi_{h,0}$ is the value of the minimum air mass flow. From (10), it is possible to observe that the operation mode $m = 0$ is discrete since the variable $\Delta F_{h,0,t}$ can only assume a value equal to 0 or $\Phi_{h,0}$ depending on the decision variable $y_{h,t}$ (which indicates if the heat pump is on or off).

The remaining operation modes m of the heat pump are, instead, all associated to a linear segment of the power to air mass flow characteristic. For each of them, the following inequality constraints are used:

$$\Delta F_{h,m,t} \leq y_{h,t} \cdot \Delta \widehat{\Phi}_{h,m} \tag{11}$$

where, similarly to (10), $\Delta F_{h,m,t}$ is a continuous variable associated to the operation mode m, whereas $\Delta \widehat{\Phi}_{h,m}$ is defined as the maximum increment of air mass flow for mode m with respect to mode m-1, and it is expressed as:

$$\Delta \widehat{\Phi}_{h,m} = \widehat{\Phi}_{h,m} - \widehat{\Phi}_{h,m-1} \tag{12}$$

with $\widehat{\Phi}_{h,m}$ being the maximum value of air mass flow for the linear segment associated to mode m.

All the variables $\Delta F_{h,m,t}$ are given as output of the optimization algorithm presented in Sect. 4 (together with the binary variables $y_{h,t}$) and they allow calculating the total air mass flow provided by the heat pump through the following:

$$\Phi_{h,t} = \sum_m \Delta F_{h,m,t} \tag{13}$$

In general, since the total air mass flow is given by the contribution of different incremental values, it is important that the variables $\Delta F_{h,m,t}$ become larger than zero only if the previous mode $\Delta F_{h,m-1,t}$ is equal to its maximum $\widehat{\Phi}_{h,m-1}$. In the following optimization, the objective is to minimize a linearization of the squared power and thus this condition is guaranteed if the power to air mass flow characteristic is composed of segments with increasing values of the derivative $\delta P/\delta \Phi$.

Given the above model for the computation of the total air mass flow at a given instant t, the heat flow $Q_{h,t}^{HP}$ required in (1) for the computation of the indoor temperature evolution is defined through the following relationship:

$$Q_{h,t}^{HP} = \gamma_{air} \cdot \Phi_{h,t} \cdot \left(\Gamma^{HP} - \frac{\Gamma_{h,t-1}^{LB} + \Gamma_{h,t-1}^{UB}}{2} \right) \tag{14}$$

where Γ^{HP} is the output temperature of the air from the heat pump. In (14), Γ^{HP} is assumed as constant, for the sake of simplicity. Furthermore, it is worth to note that, for a proper evaluation of the heat flow in (14), the actual indoor temperature of the house at the time $t - 1$ should have been used in place of the reference temperature. The approximation introduced by (14) obviously leads to some inaccuracies in the computation of the heat flow (and consequently of the indoor temperature profile), but it also allows not introducing nonlinearities in the model, giving important modeling and computational advantages at the expense of a negligible reduction in the accuracy.

Finally, the coupling of the thermal model with the electrical world is given by the power to air mass flow characteristic; the electric power required to generate a certain amount of air mass flow is computed through the following:

$$P_{h,t}^{HP} = \sum_m \beta_m \Delta F_{h,m,t} \tag{15}$$

where β_m is the power per air mass flow derivative associated to the mode m.

Time Constraints. When dealing with the scheduling of devices, time constraints can be often required to avoid multiple switch-on and switch-off events in a reduced time frame, which can lead to a fast deterioration of the performance or of the functionalities of the same device. In the defined heat pump model, this aspect is taken into account by considering the minimum number of time steps τ_{min} for which the heat pump has to work once switched on. The resulting constraint is modelled adopting the same technique used in [37] for a similar goal, but in the unit commitment problem. To this purpose, the following constraints are introduced:

$$W_{h,t} \geq y_{h,t} - y_{h,t-1} \tag{16}$$

$$W_{h,t} \leq y_{h,\tau} \ with \ \tau \in t, \ldots, \min(t + \tau min - 1, f) \tag{17}$$

where $W_{h,t}$ is a continuous variable lower bounded by zero. In (15), it is possible to observe that whenever a switch-on event occurs, the variable $W_{h,t}$ becomes larger or equal to 1. Through (17), $W_{h,t}$ will be upper bounded by 1 and, at the same time, all the binary variables between t and $t + \tau_{min}$ will be forced to be equal to 1, obtaining the constraint that the heat pump will work for at least τ_{min} time steps in a row.

4 DSM Scheme

4.1 DSM Heat Pump Scheduling

The DSM scheme presented in this paper relies upon the thermal and the heat pump model presented in Sect. 3 and is based on a centralized optimization algorithm providing the day ahead scheduling of the considered heat pumps. As discussed in Sect. 2, one of the main goals to pursue through the smart scheduling of flexible appliances is the reduction of the peak demand in the electric grid. For this reason, the optimization algorithm here presented considers a set of loads subtended by a distribution feeder or a distribution substation and aims at minimizing their peak of power demand over the day. To this purpose, an immediate solution would be to minimize the sum of the squared values of power demand over the day. However, an approach of this kind would lead to a Quadratic Programming (QP) problem that, considering the presence of binary variables (associated to the status of the heat pumps), would result in a complex and computationally demanding solution. To overcome this issue, the approach presented in [35, 36] (and replicated here) uses a linearized objective function, which, together with the linear equations and constraints presented in Sect. 3, allows formulating the optimization as a Mixed Integer Linear Programming (MILP) problem.

To understand the concept behind the proposed linearized objective function, let us refer to Fig. 3.

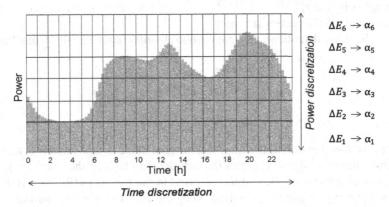

Fig. 3. Linearization concept behind the optimization objective function.

Figure 3 depicts the forecast of inflexible demand over the day, which is a necessary input for the optimization. The goal of the optimization is clearly to allocate the flexible demand on top of the inflexible power, possibly trying to fill the valleys and to avoid the peaks. To perform the optimization, a discretization of the time scale is required, and this is represented by the partition of the axis x. To define a linearized objective function, an additional discretization is also performed in the axis y, as shown in Fig. 3. This discretization creates a set of energy boxes for each time step t, which are at the core of the optimization approach. In fact, to disadvantage the allocation of flexible demand at the peak times, the energy boxes can be weighted by using larger weights α for increasing values of the associated power (with reference to Fig. 3, $\alpha_1 < \alpha_2 < \cdots < \alpha_6$). In this way, the optimization can be formulated as the minimization of the sum of the weighted *energies* and, since the weights can be defined a priori as constant values, the corresponding objective function will be linear.

From a mathematical perspective, the approach conceptually defined above can be defined as follows. Each energy box b is represented by a continuous variable ΔE_b lower bounded by zero and upper bounded by the corresponding maximum value of energy $\widehat{\varepsilon}_b = \Delta P \cdot \Delta t$ (it is worth noting that the discretization step can vary with b to introduce better resolution for certain levels of power). The following constraint thus holds, for each instant of time t:

$$0 \leq \Delta E_{b,t} \leq \widehat{\varepsilon}_b \tag{18}$$

Moreover, at each instant of time t, the sum of the single contributions of energy $\Delta E_{b,t}$ has to represent the sum of flexible and inflexible load. This can be expressed through the following constraint:

$$\sum_b \Delta E_{b,t} \geq \sum_h P_{h,t}^{HP} \Delta t + \varepsilon_t^{INFLEX} \tag{19}$$

where $P_{h,t}^{HP}$ is the flexible power of the heat pumps (resulting from the optimal scheduling), whereas $\varepsilon_t^{INFLEX} = P_t^{INFLEX} \cdot \Delta t$ is the inflexible energy associated to the non-controllable loads present at time t.

Given the above definitions, the optimization algorithm can be finally defined as [35]:

$$\underset{y_{h,t}, \Delta F_{h,m,t}}{minimize} \sum_t \sum_b \alpha_b \Delta E_b \tag{20}$$

$$s.t. \quad Eqs.(1) - (19)$$

The objective function in (20) indicates that the goal of the optimization is to minimize the overall sum of weighted energies through the definition of the variables $y_{h,t}$ and $\Delta F_{h,m,t}$, while fulfilling all the constraints given in the Eqs. (1)–(19). As a result, the proposed optimization allows not only defining the operating times of the heat pumps for each house ($y_{h,t}$), but also their optimal operating point through the decision of the incremental air mass flows $\Delta F_{h,m,t}$.

It is worth to underline that the proposed method works by filling the "energy boxes" with the additional flexible energy associated to the heat pumps. To guarantee that the approach is also correct from a physical point of view, it is important to ensure that the energy boxes are filled in the right order (starting from the bottom). Similarly to what was discussed for the continuous model of the heat pump, this is guaranteed if the weights α are increasing for increasing values of the associated power. In fact, since the optimization works to minimize the sum of the weighted energies, a lower level energy box will be always preferred to those placed on top.

4.2 DSM Framework

The proposed DSM works coordinating through a centralized algorithm the operation of all the heat pumps included in the scenario. The use of a centralized algorithm has been preferred because it guarantees the effective minimization of the power peaks, avoiding the risk to create new peaks at a different time (with respect to the original one) due to the similar response of local uncoordinated energy management systems.

Figure 4 provides a schematic view of the conceived DSM scheme highlighting the inputs required to the optimization algorithm. In particular, it is possible to observe that three main inputs are needed.

- Forecast of the inflexible load: this is an essential input, since the goal of the heat pumps scheduling is to allocate their operation in order to fill the valleys and to avoid the use at peak hours. Typical patterns of power consumption for residential customers are available for many DSOs and, in general, this type of information can be extracted also for the grid of interest through the analysis of historical data.
- Forecast of outdoor temperature: the outdoor temperature is essential in the computation of the house thermal losses and therefore for the calculation of the thermal balance (see Eqs. (1) and (5)); detailed forecasts, having a good time resolution, can be usually found for many locations.
- House characteristics: these data are essential to implement the thermal models of the houses; in many countries, due to the regulations on the energy efficiency of the buildings, the thermal characteristics of the houses can be known for the buildings of recent construction or for those that have been renovated; in case these data are not available, ad hoc measurement campaigns have to be conducted in order to extract the needed information. In addition to the thermal characteristics of the house, customers also need to send their daily preference for the indoor temperature with the associated boundaries. The operational characteristics of the heat pump are required as well.

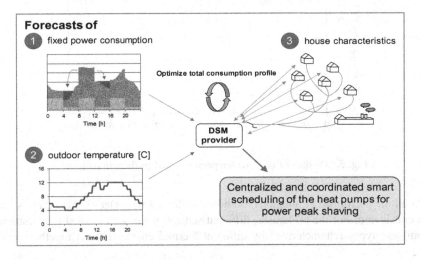

Fig. 4. Schematic view of the proposed DSM scheme.

Given the above set of inputs, the optimization algorithm is centrally executed by the DSM provider, who then sends the scheduling results back to the single customers.

5 DSM Results

5.1 Simulations Setup

DSM Inputs. To test the operation of the proposed DSM scheme and to assess the potential benefits deriving from its application, several simulations have been carried out. All the simulations have been performed by generating or retrieving the inputs needed for the optimization algorithm. Concerning the shape of the inflexible load, realistic profiles of aggregated daily consumption for residential customers have been taken into account. Needed data have been collected from a German database where different profiles, differentiated on a monthly basis and depending on the type of day (workday, Saturdays and Sundays), are made available [38]. In addition to the residential inflexible load, in the simulated scenarios, the presence of an additional inflexible industrial load has been also considered. This load has two levels of power consumption that are periodically alternated during the day. This particular shape has been considered to exacerbate some peaks over the daily profile and, consequently, to clearly evaluate and prove the proper operation of the proposed optimization.

Regarding the outdoor temperature profiles, real temperature data have been taken from historical data for the city of Aachen, Germany. In the tests presented in this paper, the temperatures of a day in May and a day in December have been considered. Figure 5 depicts the associated trend of the temperature profiles over the day.

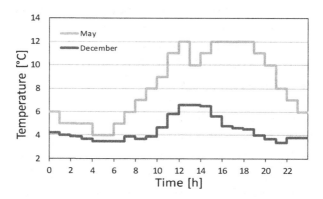

Fig. 5. Profiles of outdoor temperature used in the simulations.

In all the performed simulations, 60 house endowed with electric heat pumps have been considered. For the 60 houses, different settings have been created by combining 12 building types (characterized by different thermal characteristics) together with 5 different profiles of thermal comfort requirements for the customers. Figure 6 shows the thermal capacity and the heat loss factor of the 12 building types, whereas Fig. 7 depicts the 5 profiles of indoor temperature to be provided.

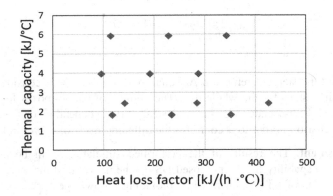

Fig. 6. Thermal characteristics of the simulated houses.

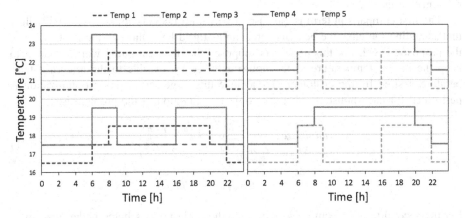

Fig. 7. Thermal comfort requirements for the final customers.

Regarding the heat pumps, the proposed approach obviously allows considering different heat pumps characteristics for each house. In the simulated scenarios, however, a unique heat pump characteristic curve (the one shown in Fig. 2) has been considered for the sake of simplicity. Table 1 shows the values associated to the operation curve. The air output temperature Γ^{HP} is 30 °C, whereas the considered minimum operating time τ_{min} is 30 min.

Table 1. Heat pump operational parameters.

Mode m	m = 0	m = 1	m = 2
$\beta_m \left[\dfrac{\text{Wh}}{\text{kg}}\right]$	0.939	1.86	3.70
$\widehat{\Phi}_m \left[\dfrac{\text{kg}}{\text{h}}\right]$	426	690	868
$\Delta\widehat{\Phi}_m \left[\dfrac{\text{kg}}{\text{h}}\right]$	426	264	178

Finally, a time step discretization Δt equal to 15 min has been considered, which leads to 96 time steps for the whole day. The number of energy boxes for time step, their height and the associated weights are instead varying depending on the considered simulation scenario in order to account for the different levels of power to be handled.

DSM Assessment. The results of the performed simulations will be analyzed to understand the capability for the proposed DSM scheme to fulfill the thermal comfort requirements of the final customers while providing a power peak shaving service to the DSO. Moreover, the benefits given by the continuous heat pump model presented in Sect. 3.2 will be also investigated. To this purpose, the results provided by the designed DSM scheme will be compared to those achievable through two different reference cases.

The first comparison term is the case in which no DSM is applied. This will allow understanding the effective benefits achievable at grid level through the application of the proposed DSM. For this case, it is supposed that the heat pump works in order to keep the indoor temperature of the house as close as possible to the reference temperature desired by the customer. To achieve this target, a quadratic optimization is performed for each house h, where the objective function of the problem is:

$$\underset{y_{h,t},\, \Delta F_{h,m,t}}{minimize} \sum_t \left(T_{h,t}^{IN} - \frac{\Gamma_{h,t}^{LB} + \Gamma_{h,t}^{UB}}{2} \right) \tag{21}$$

$$s.t. \quad \text{Eqs.}(1) - (17)$$

For the second term of comparison, instead, the same DSM scheme under analysis is used, but a binary heat pump model has been considered. The purpose of this comparison is to understand the additional value given by a more detailed modelling of the heat pump and by the inclusion of the choice of the heat pump operating point in the optimization. To this purpose, a unique operation point of the characteristic curve shown in Fig. 2 has been considered, which corresponds to the average value of air mass flow $\Phi = 647 \text{kg/h}$. The associated electric power consumption is $P = 809 \text{W}$.

5.2   Simulation Results

Simulation have been performed by referring to two different grid scenarios. In the first case, a feeder supplying 60 houses is considered, where all the 60 houses are assumed equipped with the electric heat pump. The objective in this case is to minimize the peak of the power flowing in the feeder. In the second scenario, a substation subtending four feeders is taken into account. The feeders supply energy to 240 customers and, among them, 60 are endowed with the heat pump. This case emulates a scenario with smaller penetration of heat pumps (25%); the goal of the DSM provider is here to minimize the overall power peak at the substation. For both the considered grid scenarios, simulations have been performed considering both the outdoor temperatures shown in Fig. 5, leading to the following set of simulations:

- Scenario 1: 60 houses; 100% heat pumps; outdoor temperature: May.
- Scenario 2: 240 houses; 25% heat pumps; outdoor temperature: May.
- Scenario 3: 60 houses; 100% heat pumps; outdoor temperature: December.
- Scenario 4: 240 houses; 25% heat pumps; outdoor temperature: December.

Figure 8 shows the results achieved at household level for the simulation performed with scenario 1. The upper part of the figure gives the obtained scheduling of the heat pump with the related power consumption levels, whereas the bottom part gives the evolution of the indoor temperature profile. The first outcome in Fig. 8 is that the conceived DSM is able to fulfill the requirements of the final customer in terms of thermal comfort. For both the DSM (binary and continuous model of the heat pump) the temperature can oscillate between the upper and the lower boundary but it is always within the allowed thresholds. Figure 8 also permits evaluating how the DSM optimization takes advantage of the temperature flexibility provided by the customer. In particular, it is possible to note that the heat pump mainly works during the night and the morning to store thermal energy in the house (the inflexible demand is low in this period), while its use is avoided during lunch time and in the evening when local power peaks occur in the grid. When no DSM is applied, instead, this flexibility cannot be exploited and the operation of the heat pump strictly follows the temperature

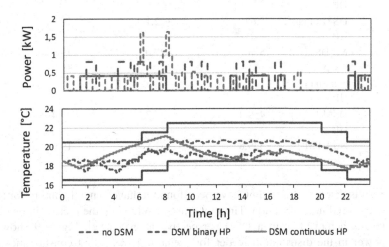

Fig. 8. Heat pump consumption and temperature profile for a sample house with May outdoor temperature [35].

requirements given by the customer. For this reason, large peaks of power consumption occur when a temperature increase is required, whereas with the DSM scheduling it is possible to modulate the operation of the heat pump in order to face such a requirement.

Regarding the DSM results obtained for the two different heat pump models, an important difference is from the energy consumption point of view. Table 2 shows the results in terms of heat pump power consumption and average indoor temperature for the same household represented in Fig. 8 and for the overall scenario 1. Looking at the single household, it is possible to notice that an increase of the energy consumption larger than 30% is obtained for the case with binary heat pump model, despite the fact that a lower average indoor temperature is provided to the house. The reason for such a bad energy performance is due to the worse efficiency of the operating point used to represent the binary heat pump model. When the continuous model is used, the advantage is that the optimization allows choosing the most favorable operating condition, thus fostering the operation in the points at better efficiency. This obviously translates into an overall reduction of the energy consumption. The same trend is also confirmed when looking at the results for the overall scenario. The growth of the energy consumption with respect to the DSM scheme with continuous heat pump is in this case even larger than in the single household example: this is partially due to the higher indoor temperature provided to the customers, but mostly due to the operation of the heat pumps at operation points with worse efficiency.

Table 2. Heat pump operational characteristic parameters.

Case		HP consumption [kWh]	Consumption increase [%]	Avg. indoor temperature [°C]
Sample household	DSM continuous HP	4.30	–	19.1
	DSM binary HP	5.68	+32.1	18.7
	No DSM	5.46	+27.0	19.7
Total	DSM continuous HP	333.6	–	18.9
	DSM binary HP	459.4	+37.7	19.4
	No DSM	439.8	+31.8	19.9

The results discussed until now have been found to have general validity for all the scenarios analyzed in this Section. Therefore, in the following, the focus will be mainly on the results achieved from a DSO perspective at grid level. Figure 9 shows the overall power in the distribution feeder for scenario 1. As a first consideration, it is possible to note the evident benefit given by the use of the DSM strategies. In fact, when no DSM is applied, the uncoordinated operation of the heat pumps leads to significant power peaks that could jeopardize the operation of the distribution grid. The reason for such large peaks is associated to the particular profiles of thermal comfort used for the customers. Looking also at Fig. 7, it is possible to see that the power peaks occur when an increase in the indoor temperature is required by the customers. Since no DSM is coordinating the heat pumps, many of them work simultaneously and with a significant level of power consumption to fulfill the temperature increase requirement. While the achieved result is clearly affected by the use of analogous thermal comfort requirements for several customers, it is important to remark that similar habits or needs are likely among residential customers (e.g. due to similar working hours and occupancy of the house) and, thus, issues similar to those in Fig. 9 are possible. Similar problems can also occur in case of distributed DSM like, for example, when using ToU tariffs: in this situation, the price-based scheme itself determines the synchronized reaction of the customers, leading to the possible creation of undesired power peaks.

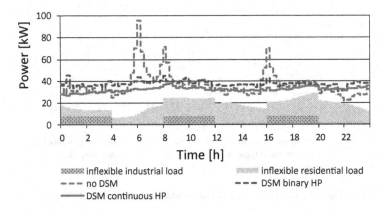

Fig. 9. Overall power at the distribution feeder for scenario 1 [35].

As shown in Fig. 9, the use of a centralized DSM allows achieving the goal of power peak shaving through the allocation of the heat pump operation during off-peak times and minimizing their use during peak times. As already mentioned, the main difference between the two DSM cases (binary and continuous heat pump model) is from an energy efficiency standpoint: in Fig. 9 it is possible to see again that the binary model requires more energy and this is clear from the larger level of power allocated on top of the inflexible load. Table 3 summarizes the results obtained for scenario 1 from a quantitative point of view. From the data, it can be observed that a large portion of the overall demand is associated to the heat pumps operation, emphasizing the importance to suitably manage this source of flexibility through ad hoc DSM policies. The DSM with the continuous heat pump model gives the best results in terms of power peak shaving, as it has the lowest peak of demand, and also has the best capabilities in flattening the daily power profile, as it can be deduced from the results of load factor (ratio between average power consumption over the day and power peak).

Table 3. Grid level results for scenario 1.

Case	Daily energy consumption [kWh]	Quote flexible energy [%]	Power peak [kW]	Load factor [%]
Inflexible load	442.2	–	28.9	63.6
No DSM	882.0	49.9	95.6	38.5
DSM binary HP	901.7	52.1	45.8	82.1
DSM continuous HP	778.9	37.8	37.8	85.6

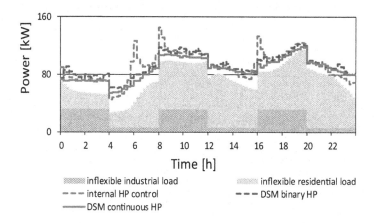

Fig. 10. Overall power at the substation for scenario 2 [35].

Table 4. Grid level results for scenario 2.

Case	Daily energy consumption [kWh]	Quote flexible energy [%]	Power peak [kW]	No. HP at peak time	Load factor [%]
Inflexible load	1768	–	115.7	–	63.6
No DSM	2209	19.9	145.1	40	63.4
DSM binary HP	2228	20.6	122.7	12	75.7
DSM continuous HP	2105	16.0	118.6	8	74.0

In scenario 2, the most important difference with respect to the previous scenario is that the flexible energy to allocate on top of the fixed load is relatively low. As a result, the operation of the heat pumps will not lead to a significant modification of the shape of the power consumption profile. Figure 10 shows the results concerning the aggregated power consumption at substation level. Despite the lower amount of flexible energy, even in this scenario it is possible to notice that additional power peaks arise for the case where no DSM is considered. The DSM strategies, instead, allow significantly flattening the power consumption profile. In particular, it is possible to see that only a minimum amount of power is added on top of the flexible load during local peaks, whereas a much larger number of heat pumps are working during off-peak times to fill the valleys. Similarly to the previous scenario, Table 4 shows the results from a numerical point of view. In general, the same considerations done for scenario 1 still hold. Looking at the total energy consumption, it can be seen that the DSM with the binary model of the heat pump is more inefficient than the one with the continuous model. Moreover, in this scenario, it is possible also to observe how the possibility to modulate the operating power of the heat pumps gives an additional degree of flexibility for the scheduling: during the peak time, at 20:00, 12 heat pumps work for the DSM with binary heat pump model, while only 8 are required for the DSM with the continuous heat pump. The best performance of the proposed DSM is also confirmed by the power peak values: the increase in the power peak is only 2.5% versus the 6% for the DSM with binary heat pump (more than 25% if no DSM is applied). The worse results for the load factor can be instead explained by the fact that the binary heat pump leads to consume more energy, which can be allocated in the valleys to flatten the profile. In general, however, both the DSM strategies lead a clear enhancement of the load factor with respect to the starting value associated to the inflexible load.

As last considerations, the results obtained when taking into account the colder outdoor temperature of a day in December are presented. The first outcome of these simulations is that, both in scenario 3 and 4, the DSM with binary heat pump model cannot find any scheduling solution. The reason for this is due to the impossibility to satisfy the dynamic variations in the thermal comfort requirements of some customers: using a fixed value of air mass flow from the heat pump, in fact, the indoor temperature of some households cannot remain within the allowed boundaries. When the DSM with the continuous heat pump model is used, instead, the possibility to modulate the heat flow allows finding feasible solutions. This aspect is shown in Fig. 11: it is possible to observe that, in the represented house, the heat pump needs to stay always on to deal with the colder outdoor temperature; in addition, its operation point has to be modulated in order to fulfill the thermal comfort requirements and to pursue the overall goal of power peak minimization.

Table 5 shows the results achieved at grid level for both scenario 3 and 4. Having also a look at Tables 3 and 4, for the corresponding results obtained with the May temperature, the following considerations can be drawn. First, it is possible to see that, as expected, the portion of energy consumption coming from the heat pumps increases. This is clearly due to the colder outdoor temperature that leads the heat pumps to work more often, and with higher values of power, to fulfill the thermal comfort demand of the customers (see also Fig. 11). Despite the larger amount of *flexible* energy consumption, the capabilities in terms of power peak shaving and flattening of the daily

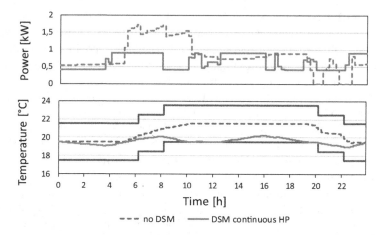

Fig. 11. Heat pump consumption and temperature profile for a sample house with December outdoor temperature.

power demand are reduced with respect to the corresponding scenarios for May. This outcome is due to the more severe impact brought by the thermal comfort requirements when the outdoor temperature decreases. As an example, in scenario 4, almost 50% of the heat pumps (28 out of 60) have to work during the peak time versus the only 8 scheduled to work in scenario 2. Regardless this reduction in the potential impact, Table 5 shows that the DSM strategy still has the capability to achieve the goal of power peak shaving and has the potential to significantly flatten the power profile. Obviously, this potential grows for increasing penetration of heat pumps in the grid.

Table 5. Grid level results for scenarios 3 and 4.

Case		Daily energy consumption [kWh]	Quote flexible energy [%]	Power peak [kW]	Load factor [%]
Scenario 3	Inflexible load	426	–	30.1	59.0
	No DSM	1045	59.2	95.5	45.6
	DSM continuous HP	916	53.4	47.4	80.4
Scenario 4	Inflexible load	1705	–	120.4	59.0
	No DSM	2324	26.6	149.8	64.6
	DSM continuous HP	2197	22.4	136.5	67.1

6 Conclusions

This paper presents a DSM scheme aimed at defining the optimal day ahead scheduling of electric heat pumps. The objective of the DSM program is twofold: providing a service to the DSO, by shaving the power peaks in the daily demand, and providing a service to the final customer, through a smart scheduling of the heat pump that guarantees the fulfillment of the end user thermal comfort requirements. An ad-hoc designed Mixed Integer Linear Programming formulation is proposed to tackle the above optimization problem, which relies on a continuous operation model of the heat pump. Presented results show the potential benefits arising from the application of the proposed centralized DSM scheme. The conceived optimization leads to clear benefits in terms of power peak shaving and of flattening of the daily curve of power demand, avoiding possible side effects usually present in price-based schemes.

References

1. Farhangi, H.: The path of the smart grid. IEEE Power Energy Mag. **8**(1), 18–28 (2010)
2. Barton, J., Huang, S., Infield, D., Leach, M., Ogunkunle, D., Torriti, J., Thomson, M.: The evolution of electricity demand and the role for demand side participation, in buildings and transport. Energy Policy **52**, 85–102 (2013)
3. Fan, J., Borlase, S.: The evolution of distribution. IEEE Power Energy Mag. **7**(2), 63–68 (2009)
4. Farhangi, H.: A road map to integration: perspectives on smart grid development. IEEE Power Energy Mag. **12**(3), 52–66 (2014)
5. Palensky, P., Dietrich, D.: Demand side management: demand response, intelligent energy systems, and smart loads. IEEE Trans. Ind. Inf. **7**(3), 381–388 (2011)
6. Murthy Balijepalli, V.S.K., Pradhan, V., Khaparde, S.A., Shereef, R.M.: Review of demand response under smart grid paradigm. In: Innovative Smart Grid Technologies – India (ISGT India), Kollam, pp. 236–243 (2011)
7. Logenthiran, T., Srinivasan, D., Shun, T.Z.: Demand side management in smart grid using heuristic optimization. IEEE Trans. Smart Grid **3**(3), 1244–1252 (2012)
8. Malík, O., Havel, P.: Active demand-side management system to facilitate integration of RES in low-voltage distribution networks. IEEE Trans. Sustain. Energy **5**(2), 673–681 (2014)
9. Caprino, D., Della Vedova, M.L., Facchinetti, T.: Peak shaving through real-time scheduling of household appliances. Energy Build. **75**, 133–148 (2014)
10. Graditi, G., Di Silvestre, M.L., Gallea, R., Riva Sanseverino, E.: Heuristic-based shiftable loads optimal management in smart micro-grids. IEEE Trans. Ind. Inf. **11**(1), 271–280 (2015)
11. Klaassen, E., Kobus, C., Frunt, J., Slootweg, H.: Load shifting potential of the washing machine and tumble dryer. In: 2016 IEEE International Energy Conference (ENERGYCON), Leuven, pp. 1–6 (2016)
12. Mattlet, B., Maun, J.C.: Assessing the benefits for the distribution system of a scheduling of flexible residential loads. In: 2016 IEEE International Energy Conference (ENERGYCON), Leuven, pp. 1–6 (2016)

13. Hayes, B., Melatti, I., Mancini, T., Prodanovic, M., Tronci, E.: Residential demand management using individualized demand aware price policies. IEEE Trans. Smart Grid **8** (3), 1284–1294 (2017)
14. Strbac, G.: Demand side management: benefits and challenges. Energy Policy **36**(12), 4419–4426 (2008)
15. Zhang, N., Ochoa, L. F., Kirschen, D. S.: Investigating the impact of demand side management on residential customers. In: 2nd IEEE PES International Conference and Exhibition on Innovative Smart Grid Technologies, Manchester, pp. 1–6 (2011)
16. US Federal Energy Regulatory Commission: Assessment of Demand Response and Advanced Metering. Technical report, Washington DC, USA (2011)
17. Smart Energy Demand Coalition: Mapping Demand Response in Europe Today. Technical report, Brussels (2015)
18. Ecofys – BUIDE12080: Heat Pump Implementation Scenarios Until 2030. Technical report, Cologne (2013)
19. Campillo, J., Wallin, F., Vassileva, I., Dahlquist, E.: Electricity demand impact from increased use of ground sourced heat pumps. In: 3rd IEEE PES Innovative Smart Grid Technologies Europe (ISGT Europe), Berlin, pp. 1–7 (2012)
20. Love, J., Smith, A.Z.P., Watson, S., Oikonomou, E., Summerfield, A., Gleeson, C., Biddulph, P., Chiu, L.F., Wingfield, J., Martin, C., Stone, A., Lowe, R.: The addition of heat pump electricity load profiles to GB electricity demand: evidence from a heat pump field trial. Appl. Energy **204**, 332–342 (2017)
21. Akmal, M., Fox, B.: Modelling and simulation of underfloor heating system supplied from heat pump. In: 18th International Conference on Computer Modelling and Simulation (UKSim), Cambridge, pp. 246–251 (2016)
22. Good, N., Zhang, L., Navarro-Espinosa, A., Mancarella, P.: Physical modeling of electro-thermal domestic heating systems with quantification of economic and environmental costs. In: Eurocon 2013, Zagreb, pp. 1164–1171 (2013)
23. Zhang, L., Good, N, Navarro-Espinosa, A., Mancarella, P.: Modelling of household electro-thermal technologies for demand response applications. In: IEEE PES Innovative Smart Grid Technologies Europe (ISGT Europe), Istanbul, pp. 1–6 (2014)
24. Arteconi, A., Hewitt, N.J., Polonara, F.: Domestic demand-side management (DSM): role of heat pumps and thermal energy storage (TES) systems. Appl. Therm. Eng. **51**(1), 155–165 (2013)
25. Arteconi, A., Hewitt, N.J., Polonara, F.: State of the art of thermal storage for demand-side management. Appl. Energy **93**, 371–389 (2012)
26. Diekerhof, M., Vorkampf, S., Monti, A.: Distributed optimization algorithm for heat pump scheduling. In: IEEE PES Innovative Smart Grid Technologies Europe (ISGT Europe), Istanbul, pp. 1–6 (2014)
27. Diekerhof, M., Schwarz, S., Monti, A.: Distributed optimization for electro-thermal heating units. In: IEEE PES Innovative Smart Grid Technologies Conference Europe (ISGT Europe), Ljubljana, pp. 1–6 (2016)
28. Molitor, C., Ponci, F., Monti, A., Cali, D., Müller, D.: Consumer benefits of electricity-price-driven heat pump operation in future smart grids. In: IEEE International Conference on Smart Measurements of Future Grids (SMFG), Bologna, pp. 75–78 (2011)
29. Loesch, M., Hufnagel, D., Steuer, S., Faßnacht, T., Schmeck, H.: Demand side management in smart buildings by intelligent scheduling of heat pumps. In: IEEE International Conference on Intelligent Energy and Power Systems (IEPS), Kyiv, pp. 1–6 (2014)
30. Bhattarai, B.P., Bak-Jensen, B., Pillai, J.R., Maier, M.: Demand flexibility from residential heat pump. In: IEEE PES General Meeting Conference & Exposition, National Harbor, MD, pp. 1–5 (2014)

31. Csetvei, Z., Østergaard, J., Nyeng, P.: Controlling price-responsive heat pumps for overload elimination in distribution systems. In: 2nd IEEE PES International Conference and Exhibition on Innovative Smart Grid Technologies, Manchester, pp. 1–8 (2011)
32. Zhang, L., Chapman, N., Good, N., Mancarella, P.: Exploiting electric heat pump flexibility for renewable generation matching. In: IEEE Manchester PowerTech, Manchester, pp. 1–6 (2017)
33. De Angelis, F., Boaro, M., Fuselli, D., Squartini, S., Piazza, F., Wei, Q.: Optimal home energy management under dynamic electrical and thermal constraints. IEEE Trans. Ind. Inf. **9**(3), 1518–1527 (2013)
34. Nielsen, K.M., Pedersen, T.S., Andersen, P.: Heat pumps in private residences used for grid balancing by demand response methods. In: IEEE PES Transmission and Distribution Conference and Exposition (T&D), Orlando, FL, pp. 1–6 (2012)
35. Cremer, J.L., Pau, M., Ponci, F., Monti, A.: Optimal scheduling of heat pumps for power peak shaving and customers thermal comfort. In: Proceeding of the 6th International Conference on Smart Cities and Green ICT Systems (SMARTGREENS), Porto, pp. 23–34 (2017)
36. Pau, M., Cremer J. L., Ponci, F., Monti, A.: Impact of customers flexibility in heat pumps scheduling for demand side management. In: 2017 IEEE International Conference on Environment and Electrical Engineering and 2017 IEEE Industrial and Commercial Power Systems Europe (EEEIC/I&CPS Europe), Milan, pp. 1–6 (2017)
37. Hedman, K.W., O'Neill, R.P., Oren, S.S.: Analyzing valid inequalities of the generation unit commitment problem. In: IEEE/PES Power Systems Conference and Exposition, Seattle, pp. 1–6 (2009)
38. BDEW. https://www.bdew.de/internet.nsf/id/DE_Standartlastprofile. Accessed 05 Aug 2017

The SAGITTA Approach for Optimizing Solar Energy Consumption in Distributed Clouds with Stochastic Modeling

Benjamin Camus[1], Fanny Dufossé[2], and Anne-Cécile Orgerie[3(✉)]

[1] Inria, IRISA, Rennes, France
`benjamin.camus@inria.fr`
[2] Inria, CRIStAL, Lille, France
`fanny.dufosse@inria.fr`
[3] CNRS, IRISA, Rennes, France
`anne-cecile.orgerie@irisa.fr`

Abstract. Facing the urgent need to decrease data centers' energy consumption, Cloud providers resort to on-site renewable energy production. Solar energy can thus be used to power data centers. Yet this energy production is intrinsically fluctuating over time and depending on the geographical location. In this paper, we propose a stochastic modeling for optimizing solar energy consumption in distributed clouds. Our approach, named SAGITTA (Stochastic Approach for Green consumption In disTributed daTA centers), is shown to produce a virtual machine scheduling close to the optimal algorithm in terms of energy savings and to outperform classical round-robin approaches over varying Cloud workloads and real solar energy generation traces.

Keywords: Data centers · Distributed clouds · Energy efficiency
Renewable energy · Scheduling · On/Off techniques

1 Introduction

The rapid increase of demand for Internet services leads Cloud providers to build more and more data centers for hosting these services. The data centers that constitute the Cloud infrastructures are usually geographically distributed for security reasons or to offer lower latency for their clients. This infrastructure increase comes with a dramatic growth of the power consumption globally drawn by data centers. As an example, in 2014, data centers in the U.S. consumed an estimated 70 billion kWh, representing about 1.8% of total U.S. electricity consumption [1].

To reduce this impact, Cloud providers resort to renewable energy sources which are either on-site or off-site [2]. Such energy sources are mostly intermittent by nature (wind, sun, etc.) with high variations, and periods of time without any production (during night for instance for photovoltaic panels). Energy storage

© Springer Nature Switzerland AG 2019
B. Donnellan et al. (Eds.): SMARTGREENS 2017/VEHITS 2017, CCIS 921, pp. 52–76, 2019.
https://doi.org/10.1007/978-3-030-02907-4_3

devices can help to overcome this issue. But, they constitute a costly investment and they intrinsically lose part of the energy stored [3]. Thus, without storage, renewable energy has to be consumed upon production or it is wasted. In this context, optimizing renewable energy consumption requires to know local availability for the distributed cloud infrastructure, in order to adequately allocate computing resources to incoming user requests. The goal is to geographically distribute the workload among the data centers so that, it fits at best the on-site renewable energy production that is variable and not known.

Here, we consider the problem of scheduling workload across multiple data centers for minimizing renewable energy loss. To solve this issue, we propose SAGITTA: a Stochastic Approach for Green consumption In disTributed daTA centers. SAGITTA uses a stochastic approach for estimating renewable energy production, and greedy heuristics for allocating resources to the incoming user requests and switching off unused servers. While SAGITTA was first introduced in [4], the original SAGITTA algorithm did not take into account the switching on and off energy costs. Here these costs are integrated in SAGITTA's algorithm and all the simulations have been redone. This chapter extends the first SAGITTA study [4] with the following contributions:

- modification of SAGITTA's original version to take into account the switching on and off energy costs;
- proof of local optimality of SAGITTA;
- proposition of an optimal algorithm based on dynamic programming to solve the problem;
- simulation results exploring the influence of green energy forecast and green energy production on SAGITTA and study of its scalability;
- performance comparison between SAGITTA and the optimal algorithm;
- study on the exactness of our green power production forecast.

Our simulation-based results show the efficiency of SAGITTA compared to classical allocation approaches. Indeed, compared to the optimal solution, SAGITTA consumes 5.2% more energy overall, while a classical round-robin solution consumes 12.9% more energy overall than optimum.

The remainder of the paper is structured as follows. Related work is presented in Sect. 2. A formal definition of the problem is given in Sect. 3. Section 4 details the SAGITTA approach. Section 5 exhibits an optimal algorithm for the considered problem. A simulation-based evaluation is conducted, simulation conditions are described in Sect. 6 and results are provided in Sect. 7. Future work is discussed in Sect. 8.

2 Related Work

Cloud infrastructures consist in geographically distributed data centers which are linked through communication networks [5]. With the emergence of the Future Internet and the dawning of new IT models such as cloud computing, the usage of data centers, and consequently their power consumption, increases dramatically.

As an example, for 2010, Google used 900,000 servers which consumed 1.9 billion kWh of electricity [6]. Other major Cloud companies present similar figures and similar issues [7].

Virtualization technology and its ability to pool resources through transparent sharing should have minimized worldwide data center consumption. But, the energy consumption of state-of-the-art servers grows inexorably as they embed more and more powerful cores and advanced features and technologies. Consequently, the global data center consumption keeps increasing rapidly [1]. This situation raises major environmental, economic and social concerns.

The first way to save energy at a data center level consists in locating it close to where the electricity is generated, hence minimizing transmission losses. For example, Western North Carolina, USA, attracts data centers with its low electricity prices due to abundant capacity of coal and nuclear power following the departure of the region's textile and furniture manufacturing [8]. This region has three super-size data centers from Google, Apple and Facebook with respective power demands of 60 to 100 MW, 100 MW and 40 MW [8].

Other companies opt for greener sources of energy. For example, Quincy (Washington, USA) supplies electricity to data facilities from Yahoo, Microsoft, Dell and Amazon with its low-cost hydro-electrics left behind following the shutting down of the region's aluminum industry [8]. Several renewable energy sources like wind power, solar energy, hydro-power, bio-energy, geothermal power and marine power can be considered to power up super-sized facilities.

In spite of these approaches, numerous data facilities have already been built and cannot be moved. Cloud infrastructures, on the other hand, can still take advantage of multiple locations to use green sources of energy with approaches such as follow-the-sun and follow-the-wind [9]. As sun and wind provide renewable sources of energy whose capacity fluctuates over time, the rationale is to place computing jobs on resources using renewable energy, and migrate jobs as renewable energy becomes available on resources in other locations. However, the migration cost, in terms of both energy and performance, may be prohibitive [10].

Within the data center itself, a range of technologies can be utilized to make cloud computing infrastructures more energy efficient, including better cooling technologies, temperature-aware scheduling [11], Dynamic Voltage and Frequency Scaling (DVFS) [12], and resource virtualization [13]. The use of Virtual Machines [14] brings several benefits including environment and performance isolation; improved resource utilization by enabling workload consolidation; and resource provisioning on demand. Nevertheless, such technologies should be analyzed and used carefully for actually improving the energy-efficiency of computing infrastructures [15].

One of the most efficient techniques for saving energy in not fully utilized data centers consists in shutting down unused resources as switched off resources consumes less power than idle ones [16]. The number of switched off resources can be increased by consolidation techniques [16]. However, switching on and off resources consumes time and energy and these costs need to be taken into account in order to effectively ensure energy savings [17].

Concerning green energy integration, Ren *et al.* have proposed an online scheduling algorithm which optimizes the energy cost and fairness among different data centers subject to queuing delay constraints [18]. While their work is based on a distributed Cloud model similar to ours, they aim at minimizing the cost of the consumed electricity, instead of the wasted renewable energy in our case. Tripathi *et al.* have presented a mixed integer linear programming formulation for capacity planning while minimizing the total cost of ownership [2]. Their model schedules demand considering the availability of green energy and its price variation to lower the total cost of ownership. Finally, a literature review of renewable energy integration in data centers can be found in [19].

3   Problem

First, the Cloud model and assumptions are described in Sect. 3.1 similarly to [4]. Then, Sect. 3.2 proposes the problem formulation.

3.1   Cloud Model

We consider a distributed Cloud infrastructure comprising several data centers geographically distributed and powered by the regular electrical grid on one side and on-site photovoltaic panels (PV) on the other side. The user management of the Cloud is assumed to be centralized. This Cloud model is shown on Fig. 1.

Incoming users requests can arrive at any time. Each request requires to be computed by a virtual machine (VM) located on any of the data centers. Each data center hosts a given amount of homogeneous servers.

3.2   Problem Formulation

We consider a system of M data centers spread over a large area. A data center DC_i is characterized by its number S_i of servers. Servers are considered homogeneous over the different data centers, in term of computing capabilities and energy consumption.

As for the application model, we consider identical VMs submitted at unpredictable rate. The VMs are supposed to be executable in less than one time slot. We can thus describe both computing and memory requirement of VMs by the number C of VMs that a server can complete in a single time slot. We consider that a server consumes at full capacity a power of P_s.

The main difference with previous paper [4] concern the energy consumption model. In [4], the model only considered the processing cost, that is the consumption of servers during the processing time. The switching ON/OFF costs were only used to evaluate the performance of resulting algorithms. In this paper, we take into account the switching ON/OFF costs in the basis model. More precisely, we consider that the switches are done at beginning of the time slots, and that the switching time is negligible compared to the time slot duration. For example, the duration to turn on will not impact the number of VM executed in

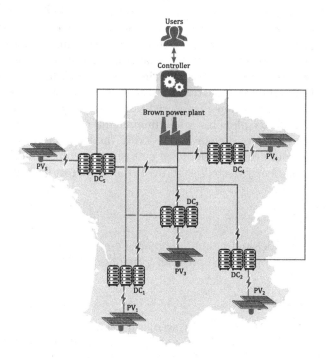

Fig. 1. Considered cloud model from [4]. (Color figure online)

the remaining time. We compute in this paper the energy at the scale of the time slots, and we do not evaluate if the energy is used mostly at beginning or at the end on a time slot. Therefore, in our model, the energy consumption, likewise the energy production, is smoothed on the time slot.

Consequently, the energy consumption of DC_i has three possible component. The first part is the power consumption of each server, and is proportional to the number of servers ON at current time slot t on DC_i, $U_i(t)$. The second part is the costs of switching on servers, proportional to the number of servers turned on $U_i^+(t) = \max(0, U_i(t) - U_i(t-1))$ and the last on the cost of shutting down servers, proportional to $U_i^-(t) = \max(0, U_i(t-1) - U_i(t))$. The total power consumed by the system is thus

$$\sum_{i=1}^{M} P_s \times U_i(t) + P_{ON} \times U_i^+(t) + P_{OFF} \times U_i^-(t).$$

This power requirement is to be compared with the green power produced at each data center. We model the green power available at time slot t in data center DC_i as a random variable $G_i(t)$ that follows a truncated normal distribution of mean $Eg_i(t)$ and standard deviation $p_i(t)$, with lower limit 0. Thus, the brown power consumed at time slot t in DC_i is equal to

$$max(0, P_s \times U_i(t) + P_{ON} \times U_i^+(t) + P_{OFF} \times U_i^-(t) - G_i(t)).$$

Our problem consists in allocating VMs to data centers, in order to minimize the consumption of brown energy. VMs are allocated by time slots. Then, our objective is to turn ON the adequate number of servers on the best locations for this criteria. We denote $N(t)$ the number of waiting VMs at time slot t. We thus need to have enough servers ON for all waiting VMs at time slot t: $\sum_{i=1}^{M} U_i(t) \geq N(t)/C$. All these notations are summarized in Table 1.

Table 1. Table of Notations.

Notation	Definition
Constants	
M	Number of data centers
DC_i	Data center number i
S_i	Number of servers in data center i
C	Maximum number of VMs in parallel on a server
P_s	Maximum power consumption of a server
Variables	
$N(t)$	Number of incoming VMs for time slot t (input)
$U_i(t)$	Number of machines ON at current time slot on data center i (output)
$G_i(t)$	Random variable of the green power produced at time slot t
$Eg_i(t)$	Expected green power generation at data center i during time slot t (input)
$p_i(t)$	Standard deviation of green power generation on data center i (input)
w	Workload portion (number of VM): $0 < w \leq N(t)$ (input)
$Ec_i(P, t)$	Expected brown consumption of data center i with power P at time slot t

4 SAGITTA

In this section, we present our approach named SAGITTA: a Stochastic Approach for Green consumption In disTributed daTA centers. The details for computing the expected green and brown consumption are provided in Sect. 4.1. SAGITTA's algorithms are presented in Sect. 4.2 and its local optimality is demonstrated in Sect. 4.3.

4.1 Expected Green and Brown Consumption

We now evaluate the expected brown power consumption of data center DC_i at time t, consuming power P, $Ec_i(P, t)$. We first evaluate the density function of the random variable of the green power generation of DC_i at time t $G_i(t)$.

Let X be a random variable following a normal distribution of parameters $Eg_i(t)$ and $p_i(t)$, density function

$$\phi(t) = \frac{1}{p_i(t)\sqrt{2\pi}} e^{-\frac{1}{2}\left(\frac{t - Eg_i(t)}{p_i(t)}\right)^2}$$

and distributive function

$$\Phi(t) = \frac{1}{2}\left(1 + \operatorname{erf}\left(\frac{t - Eg_i(t)}{p_i(t)\sqrt{2}}\right)\right).$$

Then, for $x > 0$,

$$P(G_i(t) < x) = P(X < x | X > 0) = \frac{P(0 < X < x)}{P(X > 0)}$$

and the density function of $G_i(t)$ equals $\phi_i(t) = \frac{\phi(t)}{P(X>0)}$.

Let $B_i(t)$ be the random variable of the brown consumption of DC_i at time slot t. Then, as shown in [4], we have:

$$Ec_i(P, t) = (P - Eg_i(t))\frac{\Phi(P) - \Phi(0)}{1 - \Phi(0)} - p_i(t)^2\frac{\phi(0) - \phi(P)}{1 - \Phi(0)},$$

with $\phi(x) = \frac{1}{p_i(t)\sqrt{2\pi}}e^{-\frac{1}{2}\left(\frac{x - Eg_i(t)}{p_i(t)}\right)^2}$ and $\Phi(x) = \frac{1}{2}\left(1 + \operatorname{erf}\left(\frac{x - Eg_i(t)}{p_i(t)\sqrt{2}}\right)\right).$

In the following, if the considered time slot is clear, we simply note $Ec_i(P)$ for the expected brown power consumption.

4.2 Algorithms Description

Our SAGITTA approach uses several algorithms to take decisions and allocate VMs to physical servers. These algorithms are designed to determine at any time slot on which data center to turn ON and OFF servers. At each time slot, our constraint is to turn ON the minimum number of servers that allows for executing all requested VMs, that is $\lceil N(t)/C \rceil$.

Algorithm 1. General algorithm.

---

    if $\sum\limits_{1 \leq i \leq M} U_i(t) < \lceil\frac{N(t)}{C}\rceil$ **then**

       Switch on decision;   (Algorithm 2)

    else if $\sum\limits_{1 \leq i \leq M} U_i(t) > \lceil\frac{N(t)}{C}\rceil$ **then**

       Switch off decision;   (Algorithm 3)

    end if

    Transfer decision;   (Algorithm 4)

    for $1 \leq i \leq M$ **do**

       Let $U_i(t)$ servers on and fill them, switch off the rest;

    end for

The general algorithm (Algorithm 1) is designed as follows. It first determines if the number of servers available is under or over the requested number. If there is not enough servers ON, Algorithm 2 determines the location of servers to switch on. If some servers are unnecessary, Algorithm 3 determines where servers should be shut down. These decisions are done regarding the expected green energy production in the different data centers. More precisely, Algorithm 2 compares the expected extra cost in brown energy consumption c_i induced by an additional server ON on any data centers, and selects the data center with minimum expected extra cost. The variable $U_i(t)$ is then incremented, but the servers are only switched on at end of Algorithm 1, when all decisions are taken on any data centers. The same way, Algorithm 3 selects one by one the servers to switch OFF to maximize the expected gain.

Finally, Algorithm 4 evaluates if the brown power consumption could be reduced by transferring the available processing power from one data center to another. More precisely, the algorithm determines some location where a fixed number of servers is turned off, and a new location where the same number of servers is turned on. One server is selected for switch OFF on the data center of maximum gain and another one

---

Algorithm 2. Switch on decision.

---

    for $1 \leq i \leq M$ **do**
        $P_i = U_i(t-1) \times P_s$
        $U_i(t) = U_i(t-1)$
        if $U_i(t) < S_i$ **then**
            Compute $c_i = Ec_i(P_i + P_{ON} + P_s) - Ec_i(P_i)$;
        else
            $c_i = +\infty$;
        end if
    end for
    while $\sum_{1 \leq i \leq M} U_i(t) < \lceil \frac{N(t)}{C} \rceil$ **do**
        Find j such that $c_j = \min_{1 \leq i \leq M} c_i$;
        $U_j(t) + +$;
        $P_j+ = P_{ON} + P_s$
        Recompute c_j;
    end while

---

to switch ON on the data center of minimum cost, if the gain on the first data center exceed the cost on the second one. The costs and gain are computing with respects to previous preallocation. For example, the cost of turning a server on is not the same if Algorithm 3 has decided to turn off a server at previous step, that is if $U_i(t) < U_i(t-1)$ or not. If no such decision was taken, the cost c_i is the same as in Algorithm 2, $c_i = Ec_i(P_i + P_{ON} + P_s) - Ec_i(P_i)$, that includes the cost to turn on and the cost of processing. In the other case, turning on a server correspond to cancel a decision to turn off a server, that is $c_i = Ec_i(P_i - P_{OFF} + P_s) - Ec_i(P_i)$. The same holds for gain g_i to turn off a server.

Algorithm 3. Switch off decision.

---

for $1 \leq i \leq M$ do
    $P_i = U_i(t-1) \times P_s$
    if $U_i(t) > 0$ then
        Compute $g_i = Ec_i(P_i) - Ec_i(P_i + P_{OFF} - P_s)$;
    else
        $g_i = -1$;
    end if
end for
while $\sum\limits_{1 \leq i \leq M} U_i(t) > \lceil \frac{N(t)}{C} \rceil$ do
    Find j such that $g_j = \max\limits_{1 \leq i \leq M} g_i$;
    $U_j(t) - -$;
    $P_j + = P_{OFF} - P_s$
    Recompute g_j;
end while

After running Algorithm 4, general Algorithm 1 applies all these decisions. The selected number of servers are turned ON and OFF and all VMs are allocated to available servers.

Algorithm 4. Transfer decision.

---

for $1 \leq i \leq M$ do
    if $U_i(t) > U_i(t-1)$ then
        $g_i = Ec_i(P_i) - Ec_i(P_i - P_s - P_{ON})$;
    else
        $g_i = Ec_i(P_i) - Ec_i(P_i + P_{OFF} - P_s)$;
    end if
    if $U_i(t) < U_i(t-1)$ then
        $c_i = Ec_i(P_i - P_{OFF} + P_s) - Ec_i(P_i)$;
    else
        $c_i = Ec_i(P_i + P_{ON} + P_s) - Ec_i(P_i)$;
    end if
end for
while $\max\limits_{1 \leq i \leq M} g_i > \min\limits_{1 \leq j \leq M} c_j$ do
    Find k such that $g_k = \max\limits_{1 \leq i \leq M} g_i$;
    Find l such that $c_l = \min\limits_{1 \leq j \leq M} c_j$;
    $U_k(t) - -$;
    $U_l(t) + +$;
    Recompute g_k and P_k;
    Recompute c_l and P_l;
end while

4.3 SAGITTA Local Optimality

We demonstrate in this section, that Algorithms 2, 3 and 4 are locally optimal. More precisely, at time slot t, with respect to the normal laws $(\mathcal{N}(Eg_i, p_i)^2))_{1 \leq i \leq M}$, these algorithms select the best servers to turn on or off, to minimize the expect brown power consumption. We first consider Algorithm 2.

Theorem 1. *Algorithm 2 is locally optimal with respect* $(\mathcal{N}(Eg_i, p_i)^2))_{1 \leq i \leq M}$.

Proof. We are studying the execution of SAGITTA during one unique time slot, so we do not precise t in this proof. For sake of simplicity, we denote in this proof $c_i(P) = Ec_i(P + P_{ON} + P_s) - Ec_i(P)$. We compare Algorithm 2 to any selection of servers to turn on. We do not compare to a configuration where some servers are turned on and others turned off.

We first prove that the function $c_i(P)$ is increasing, and then that this property induces the local optimality of Algorithm 2.

$$Ec_i(P) = P \times \frac{\Phi(P) - \Phi(0)}{1 - \Phi(0)} - \frac{\int_0^P x\phi(x)\mathrm{d}x}{1 - \Phi(0)}$$

with $\phi(x) = \frac{1}{p_i\sqrt{2\pi}} e^{-\frac{1}{2}\left(\frac{x - Eg_i}{p_i}\right)^2}$ and $\Phi(x) = \frac{1}{2}\left(1 + \mathrm{erf}\left(\frac{x - Eg_i(t)}{p_i(t)\sqrt{2}}\right)\right)$. Let first evaluate the derivative function of $Ec_i(P)$.

$$
\begin{aligned}
\tfrac{\mathrm{d}}{\mathrm{d}P} Ec_i(P) &= \frac{\Phi(P) - \Phi(0)}{1 - \Phi(0)} + P \times \frac{\frac{\mathrm{d}}{\mathrm{d}P}\Phi(P)}{1 - \Phi(0)} - \frac{\mathrm{d}}{\mathrm{d}P}\frac{\int_0^P x\phi(x)\mathrm{d}x}{1 - \Phi(0)} \\
&= \frac{\Phi(P) - \Phi(0)}{1 - \Phi(0)} + P \times \frac{\frac{\mathrm{d}}{\mathrm{d}P}\Phi(P)}{1 - \Phi(0)} - \frac{\mathrm{d}}{\mathrm{d}P}\frac{\int_0^P x\phi(x)\mathrm{d}x}{1 - \Phi(0)} \\
&= \frac{\Phi(P) - \Phi(0)}{1 - \Phi(0)} + P \times \frac{\phi(P)}{1 - \Phi(0)} - \frac{P\phi(P)}{1 - \Phi(0)} \\
&= \frac{\Phi(P) - \Phi(0)}{1 - \Phi(0)}
\end{aligned}
$$

Then,

$$\frac{\mathrm{d}}{\mathrm{d}P} c_i(P) = \frac{\Phi(P + P_{ON} + P_s) - \Phi(P)}{1 - \Phi(0)}.$$

The function Φ is strictly increasing, as it is the cumulative distributive function of law $\mathcal{N}(Eg_i, p_i)^2)$. Then, the derivative of function $c_i(U)$ is strictly positive, and $c_i(U)$ is strictly increasing. Intuitively, the more servers are turned on on a data center, the higher is the ratio of brown power consumption of this data center. For sake of simplicity, we denote $P_i(U)$, the power consumption corresponding to U servers on on data center DC_i. Function P_i is clearly strictly increasing.

We demonstrate now that the local optimality of Algorithm 2 can be deduced from this property. We consider here the location of servers to turn on, and not the possible transfers. Let $(U_i^{opt})_{1 \leq i \leq M}$ be the optimal choice for total brown power production, and $(U_i^{alg})_{1 \leq i \leq M}$ the decision of Algorithm 2.

First denote that by hypothesis, we have $\sum_{1 \leq i \leq M} U_i^{opt} = \sum_{1 \leq i \leq M} U_i^{alg}$. Suppose that for some i and j, $U_i^{opt} > U_i^{alg}$ and $U_j^{opt} < U_j^{alg}$.

Consider the step of Algorithm 2 at which the last server was decided to be turned on on DC_j. Let U_i be the number of servers this algorithm had decided to turn on on DC_i at this step. By definition of Algorithm 2, $c_j(P_j(U_j^{alg} - 1)) \leq c_i(P_i(U_i))$. As proven earlier, $c_i(P_i(U_i)) \leq c_i(P_i(U_i^{alg})) \leq c_i(P_i(U_i^{opt} - 1))$. As $(U_i^{opt})_{1 \leq i \leq M}$ is optimal, then $c_i(P_i(U_i)) = c_i(P_i(U_i^{alg})) = c_i(P_i(U_i^{opt} - 1))$ and as c_i and P_i are strictly increasing, $U_i^{alg} = U_i^{opt} - 1$. Thus, $c_j(P_j(U_j^{alg} - 1)) = c_i(P_i(U_i^{opt} - 1))$ and so, the decision to turn on a last server on DC_i or DC_j does not impact the expected brown power consumption.

As this property holds for all possible differences between $(U_i^{opt})_{1 \leq i \leq M}$ and $(U_i^{alg})_{1 \leq i \leq M}$, we can conclude that both selections have the same expected brown power consumption and that $(U_i^{alg})_{1 \leq i \leq M}$ is optimal.   □

Theorem 2. *Algorithm 3 is locally optimal with respect $(\mathcal{N}(Eg_i, p_i)^2))_{1 \leq i \leq M}$.*

Proof. This demonstration is very similar to the previous one, so we use similar intermediate results. Let $g_i(P) = Ec_i(P) - Ec_i(P + P_{OFF} - P_s)$. Then, $\frac{d}{dP} g_i(P) = \frac{\Phi(P) - \Phi(P + P_{OFF} - P_s)}{1 - \Phi(0)}$, $g_i(P)$ is strictly increasing.

Now, let $(U_i^{opt})_{1 \leq i \leq M}$ be the optimal choice for total brown power production, and $(U_i^{alg})_{1 \leq i \leq M}$ the decision of Algorithm 3. Suppose that for some i and j, $U_i^{opt} > U_i^{alg}$ and $U_j^{opt} < U_j^{alg}$. As in previous proof, we denote $P_i(U)$ the power consumption on data center DC_i corresponding to U servers on.

Consider the step of Algorithm 3 at which the last server was decided to be turned off on DC_i. Let U_j be the number of servers the algorithm had decided to turn on on DC_j at this step. By definition of Algorithm 3, $g_j(P_j(U_j^{alg} + 1)) \geq g_i(P_i(U_i))$. We have $g_i(P_i(U_i)) \geq g_i(P_i(U_i^{alg})) \geq g_i(P_i(U_i^{opt} + 1))$. By optimality, we obtain $g_i(P_i(U_i)) = g_i(P_i(U_i^{alg})) = g_i(P_i(U_i^{opt} + 1))$. As $g_i(P_i(P))$ is strictly increasing, $U_i^{alg} = U_i^{opt} + 1$ and the decision to turn off a last server on DC_i or DC_j does not impact the expected brown power consumption.

As in the previous proof, we obtain that $(U_i^{alg})_{1 \leq i \leq M}$ is optimal.   □

Theorem 3. *Algorithm 4 is locally optimal with respect $(\mathcal{N}(Eg_i, p_i)^2))_{1 \leq i \leq M}$.*

Proof. Let $(U_i^{opt})_{1 \leq i \leq M}$ be the optimal choice for total brown power production, and $(U_i^{alg})_{1 \leq i \leq M}$ the decision of Algorithm 4. Suppose that for some i and j, $U_i^{opt} > U_i^{alg}$ and $U_j^{opt} < U_j^{alg}$. As previously, $P_i(U)$ is the power consumption on data center DC_i for U servers on.

By definition of Algorithm 4, $c_i(P_i(U_i^{alg})) \geq g_j(P_j(U_j^{alg}))$. Moreover, we know $g_i(P_i(U_i^{opt})) = c_i(P_i(U_i^{opt} - 1)) \geq c_i(P_i(U_i^{alg}))$ and $g_j(P_j(U_j^{opt} + 1)) \leq g_j(P_j(U_j^{alg}))$. Thus, $g_i(P_i(U_i^{opt})) \geq c_j(P_j(U_j^{opt}))$, $U_i^{opt} - 1 = U_i^{alg}$ and $U_j^{opt} + 1 = U_j^{alg}$. The decision to turn on a last server on DC_i and DC_j does not impact the expected brown power consumption.

By induction, we obtain that $(U_i^{alg})_{1 \leq i \leq M}$ is optimal.   □

We have proven that our algorithms are locally optimal, it means that there are optimal only regarding the current time slot. It is clear that the only point in the model that impacts this locality is the switching ON/OFF costs. Without these costs, the power consumption on current time slot does not depend on previous time slot. Thus, we obtain the following theorems.

Theorem 4. *Algorithm 2 is optimal without switching ON/OFF costs, with respect* $(\mathcal{N}(Eg_i, p_i)^2))_{1 \leq i \leq M}$.

Theorem 5. *Algorithm 3 is optimal without switching ON/OFF costs, with respect* $(\mathcal{N}(Eg_i, p_i)^2))_{1 \leq i \leq M}$.

Theorem 6. *Algorithm 4 is optimal without switching ON/OFF costs, with respect* $(\mathcal{N}(Eg_i, p_i)^2))_{1 \leq i \leq M}$.

5 Computing the Optimal

We propose in this section an optimal algorithm that allows to compute the minimal power consumption for the whole experiment, knowing the workload trace and the green power production. This algorithm is then employed to evaluate the performance of our algorithms in the simulations. Obviously it cannot be used to allocate VMs, as it is based on complete data knowledge.

This algorithm (Algorithm 5) is a dynamic programming algorithm based on the concept of configurations. We consider as a configuration, a possible state of the platform, described by the number of servers on on each data center. More formally, configuration c at time slot t is defined as $c = (k_1, \ldots, k_M)$, where k_i is the number of servers on at time slot t on data center DC_i We defined as $P(c, t)$ the minimal power consumption of a schedule for the t first time slots, with k_i servers on on DC_i for each i at time slot t.

We denote $L(t)$ the set of pairs $(c, P(c, t))$, for all the possible configurations for time slot t. Notice that the total number of servers on during a time slot is directly related to the workload, thus $\sum_i k_i = \lceil N(t)/C \rceil$. As the complexity of this algorithm directly depends on the size of $L(t)$, this property permits to strongly reduce the execution time of experiments simulating this algorithm.

This algorithm computes recursively the set $L(t)$ for each time slot t, based on $L(t-1)$. More precisely, it first computes the possible configurations, that is the set of tuples $c = (k_1, \ldots, k_M)$ such that for all i, $k_i \leq S_i$ and $\sum_i k_i = \lceil N(t)/C \rceil$. Then, for each of these tuples, it computes $P(c, t)$ using $L(t-1)$. To that end, it considers each possible configuration c' at time slot $t-1$ and computes the power consumption at time slot t, based on the number of servers to turn on and off during this time slot and the power consumption for the servers on. When adding to $P(c', t-1)$, we obtain the minimum power consumption for t time slots with configuration c' at time $t-1$ and c at time t. We can then take the minimum for all $(c', P(c', t-1)) \in L(t-1)$ and we obtain $P(c, t)$. $L(0)$ is initialized with one unique tuple in the configuration c, $k_i = 0$ for all i and $P(c, 0) = 0$, this means all servers off, and no power consumed yet.

Algorithm 5. Optimal algorithm.

$L(t)$: list of values $(k_1, ..., k_M, P)$ where k_i is the number of servers ON on DC_i
at current time slot and P the lowest brown power consumed with this final
configuration;

for $1 \leq t \leq t_{max}$ **do**

    forall the $j_1, j_2, \cdots, j_M$ **such that** $\forall k, 0 \leq j_i \leq S_i$ **and** $\sum_i j_i = \lceil N_t/C \rceil$ **do**

        $P = +\infty$;

        forall the $(k_1, ..., k_M, P_{t-1}) \in L(t-1)$ **do**

            if $P_{t-1} \leq P'$ **then**

                $P' = 0$;

                for $1 \leq i \leq M$ **do**

                    $P_i = 0$;

                    if $k_i > j_i$ **then**

                        $P_i += (k_i - j_i) \times E_{OFF}$;

                    else

                      $P_i += (j_i - k_i) \times E_{ON}$;

                    end

                    $P_i += j_i \times P_S$;

                    $P' += max(0, P_i - GP(t))$

                end

                $P' += P_{t-1}$;

                if $P' < P$ **then**

                  $P = P'$;

                end

            end

        end

        Add $(j_1, ..., j_M, P)$ in $L(t)$;

    end

end

---

6   Validation Framework

We evaluate our algorithm through a modeling and simulation (M&S) process. In
the following, we first give an overview of the whole cloud implementation model
(Sect. 6.1). We then detail our implementation of the data centers (Sect. 6.2), of
the green power production (Sect. 6.3), of the cloud workload (Sect. 6.4), of the
algorithm implementation (Sect. 6.5), and the different simulations performed
(Sect. 6.6).

6.1   Simulation Overview

The whole cloud implementation model is described in Fig. 2. We simulate data
centers using the DCSim (Data Center Simulator) discrete-event M&S tool [20].
This simulator provides the power consumption of each data center as a function
of time.

We implement our algorithm in an ad-hoc way using the Java language into a simulated cloud controller. This simulator receives as inputs the green power production for each data center as well as the cloud workload (i.e. the number of VMs to deploy on the cloud for each time slot). Based on these inputs and on SAGITTA's algorithms, the controller generates for each server the VM allocation and the instructions which are directly sent to the simulated data center manager.

Note that we do not explicitly model the brown power production as we assume it to be infinite (at the scale of the cloud). We also ignore the telecommunication network as we assume it to have negligible impact on the system functioning (we assume network to be oversized for our scenario), and an almost constant power consumption over time if no energy-saving technique is applied [16]. Finally, we do not take into account here the energy consumed by the data centers' cooling systems.

In order to perform the simulations, we connect all these heterogeneous models using the MECSYCO (Multi-agent Environment for Complex-SYstem-CO-simulation) M&S platform [21,22] which is based on the DEVS (Discrete-EVent System specification) formalism [23]. We have defined a DEVS interface for DCSim, and implemented it in MECSYCO.

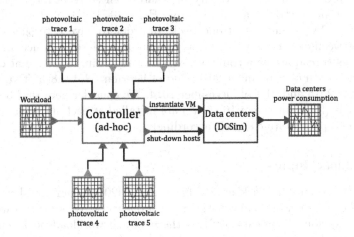

Fig. 2. Bloc diagram view of the cloud model from [4]. (Color figure online)

6.2   Data Center Simulation

Our cloud consists in five homogeneous data centers composed of five clusters. Each of these clusters contains 80 homogeneous nodes, so overall, the cloud comprises a total of 400 servers. The characteristics of each server are based on the Taurus servers of the French experimental testbed Grid'5000[1]. These

---

[1] https://www.grid5000.fr.

Taurus servers are equipped with 2 Intel Xeon E5-2630 CPU with 6 cores each, 32 GB memory, 598 GB storage and a 10 Gigabit Ethernet interface. In order to determine the power consumption of each node, we implement the power model of [24], which is based on real measurements made on Taurus nodes. These measurements notably state that a Taurus server consumes 8 W when powered OFF, 97W when idle, and 220 W at 100% CPU load (i.e. $P_s = 220W$ for our algorithm).

Within this cloud, we deploy homogeneous VMs that are equivalent to the Amazon EC2 "large" flavor[2] - i.e. each VM requires 4 CPU cores, 8 GB memory and 80 GB storage. Hence, three VMs can be simultaneously running on one node. For the sake of simplicity, we assume that, when deployed, a VM always works at full capacity. In the same way, we neglect the delays for the VM to start/stop. All the VMs are automatically deleted at the end of each time slot. A time slot lasts five minutes in our simulations.

6.3   Green Power Production

In order to feed the controller during the simulation, we use real recordings of green power production and real workload traces. We get the former from the Photovolta project[3] of the University of Nantes. These recordings correspond to the power produced by a single Sanyo HIP-240-HDE4 photovoltaic panel updated every five minutes over one week. In order to have heterogeneous trajectories between data centers (and thus to represent solar irradiance differences between sites spread across a country), we select recordings starting at different dates, namely: 4th of September 2016, 2nd of February 2014, 8th of June 2014, 22nd of June 2015 and 21st of December 2014. We consider here that 30 photovoltaic panels (for a surface of $165.6\,m^2$) are installed at each data center. Then we scale these photovoltaic signals accordingly.

6.4   Workload Input

We use the normalized ClarkNet HTTP trace of [20] for our cloud workload, shown in Fig. 3. This workload trace spans over one week. We scale this workload to 98% of the cloud total capacity (i.e. the maximal workload peak represents 98% of the total computing capacity of the cloud). The trace peaks are synchronized with the photovoltaic signal ones to have proper day-night cycles in our simulation.

6.5   Algorithm Implementation

The controller implementing our SAGITTA approach is run at each time slot (i.e. each five minutes). It saves all the data received from the green power sources during the current day. The controller computes at each time slot the standard

---

[2] https://aws.amazon.com/ec2/.

[3] http://photovolta2.univ-nantes.fr.

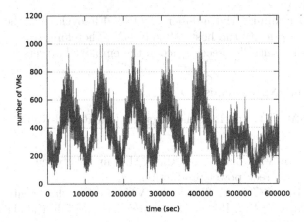

Fig. 3. The input workload used in the experiments from [4].

deviations $p_i(t)$ using this history. It computes each expected green power production $Eg_i(t)$ by averaging a reference green power production trajectory (the Photovolta project recording of the 20th of August 2013 in our case which is the day with the best yield) scaled according to the last green power production received from i. More precisely, we denote $P_{ref}(t)$ the green power production at corresponding hour the day of reference (see Fig. 4). We obtain the following formula:

$$Eg_i(t) = max\left(0, PV_i(t-1) + \frac{P_{ref}(t) - P_{ref}(t-1)}{2}\right).$$

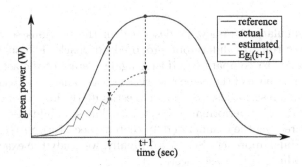

Fig. 4. Expected green power production computation for a time slot from t to $t+1$ from [4]. (Color figure online)

Note that we consider with this formula that $Eg_i(t)$ is equal to the average between the green power production received at $t-1$ and the one estimated at t. Thus, we take into account that the green power trajectory changes *during* the time slot, and not only at its beginning.

In order to minimize the number of ON/OFF cycles for the servers, the controller fills in priority the hosts already ON. Therefore, from a time slot to the next one, the controller keeps trace of the employed servers.

6.6 Simulated Approaches

We compare SAGITTA performance against two Round-Robin inspired algorithms:

- **Round-Robin-VM** distributes the VMs fairly between the data centers regardless their green power production.
- **Round-Robin-DC** starts filling with VMs the first data center (in an arbitrary predefined order). If this data center becomes full, the algorithm starts using the next one, and so on.

Like SAGITTA, these two algorithms employ in priority the nodes already ON.

As the performance of Round-Robin-DC strongly depends on the order of the data centers, we test two opposite configurations corresponding to the best and the worst possible contexts. To define these contexts, we sort the photovoltaic traces according to the total amount of green energy they provide. We assign then the traces to the data centers following this order. The best context corresponds to the case where the photovoltaic traces are sorted in a decreasing order. Thus, the first data center (i.e. the one filled in priority) will be supplied by the best photovoltaic power trajectory. The worst context corresponds then to the case where the traces are sorted in an increasing order (i.e. the data center with the worst green power supply will always be filled first).

7 Results

Based on the simulation framework described in the previous section, several experiments were run to validate our proposed approach. The simulations are performed in order to compare SAGITTA against state-of-the-art approaches (Sect. 7.1). The influence of the green energy forecast is analyzed (Sect. 7.2). Various green production scenarios are studied to estimate the impact of green energy location on SAGITTA's performance (Sect. 7.3). The scalability of SAGITTA is evaluated by increasing the number of data centers (Sect. 7.4). SAGITTA is compared to the optimal approach (Sect. 7.5). Finally, we study the exactness of the green power production forecast (Sect. 7.6).

7.1 Energy Consumption with SAGITTA vs. Round-Robin Approaches

The second set of simulation integrates the switching ON/OFF costs and estimates their impact on the algorithms' energy consumption to reflect this point. Following the data collected by [17] on the Taurus cluster, we add a static energy consumption penalty of 5.28 Wh (consumed in 150 s) for each switch-ON command, and 0.56 Wh (consumed in 10 s) for each switch-OFF command sent. As shown in Table 2, even when considering these penalties, simulations show that SAGITTA performs better than the other solutions with a difference of at least 10%.

Table 2. Total cumulative brown energy consumption.

	SAGITTA	Round-Robin-VM	Round-Robin-DC
Best	2.77 MWh	3.38 MWh	3.02 MWh
Worst	2.77 MWh	3.38 MWh	4.4 MWh

Figure 5 shows the power consumption over time of each data center in the simulated cloud using SAGITTA. This figure also shows the number of transfers made by Algorithm 4 - a negative (respectively positive) value meaning that the algorithm switches off (respectively on) hosts. This plot highlights the usefulness of the transfer algorithm. For instance, at time 173,700 s. which corresponds to early morning, DC 2 starts producing green energy slightly earlier than DC 0. SAGITTA takes then advantage of this situation by performing 19 transfers from DC 0 to DC 2. Transfers are highly correlated with discontinuities in the green power production trajectories. Thus, the transfer decision may enable adapting the VM allocation, and consequently the energy consumption, to unforeseen increases and decreases of the green power production. In the absence of transfer, the switch on and off decisions enable adapting the DC workload to their green power production - i.e. the data centers with higher power production are generally more used than the others.

For the sake of simplicity, in the following, we will consider the best case for the Round-Robin-DC algorithm (with data centers ranked by their overall green energy production). All the simulations in the next sections also include the switching ON/OFF costs.

7.2   Influence of the Green Energy Forecast

One basis of the SAGITTA approach is the green energy production forecast. The value $Eg_i(t)$, namely the expected PV production in DC_i at time slot t is computed regarding the electricity production at time slot $t - 1$. This approach permits a simple computation for the value $Eg_i(t)$ to parametrize the probability law of green energy production. However, this formula estimates the electricity production regarding only the previous time slot, despite of the high volatility of solar energy. We experiment in this section an evaluation of $Eg_i(t)$ on a sliding window of PV production values. We target here the optimal size of the window, and the weight to give to the values of the different time slots of the window.

We propose several solutions to determine $Eg_i(t)$ on a sliding window of size s. For the sake of simplicity, we denote $g_i(t) = PV_i(t) - P_{ref}(t)$, with $P_{ref}(t)$

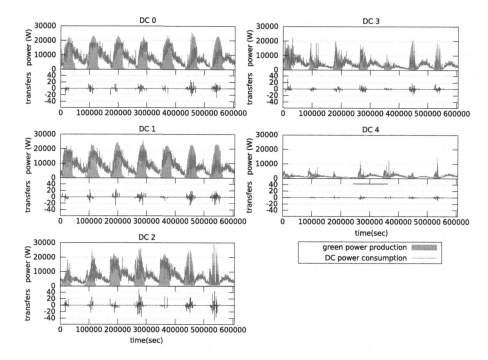

Fig. 5. Power consumption per data center with SAGITTA and transfer decisions. (Color figure online)

the daily production at same hour, the day of reference. We then make a weighted average value of values $g_i(t)$, with weight p_i:

$$Eg_i(t) = max\left(0, \frac{PV_i(t-1) + \frac{\sum_{k=1}^{s}(g_i(t-k) \times p_{s-k})}{\sum_{k=1}^{s} p_k} + P_{ref}(t)}{2}\right).$$

The first variant CST1 uses constant weigths $p_k = 1$ for recent and old values. In the second variant ADD1, the values of p_k increase linearly: $p_k = k + 1$. Finally, the values of p_k are multiplied by 2 at each step in PROD1: $p_k = 2^k$. In these variants, the computation includes values corresponding to the night, when $PV_i(t)$ and $P_{ref}(t)$ are both null. This impacts the estimation with useless values. Then, in variants CST2, ADD2 and PROD2, all values $g_i(t)$ corresponding to $P_{ref}(t) = 0$ are removed from the computation. Results of these computations are detailed in Fig. 6.

The first unexpected result is the very low values of the optimal size of the sliding window. Regardless of the variant, the best size of the window is always 2, with a slight reduction of the brown energy consumed. The good performance of algorithms PROD1 and PROD2 can be related to the large weight given to the earliest production values in the computation. The weight given to early values has indeed a large impact on the variants' performance.

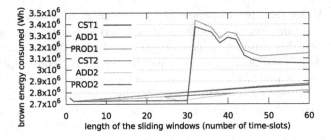

Fig. 6. Influence of Eg_i estimation from [4]. (Color figure online)

7.3   Influence of Green Energy Production

Cloud providers need to adequately dimension their on-site photovoltaic panels (PVs). This issue involves a trade-off between the financial cost of installing and operating PVs, and the financial gains they are bringing in terms of green energy produced and thus, electricity that has not to be bought from the regular grid.

We perform a set of experiments to determine the influence of green energy production on SAGITTA performance. As shown in Fig. 7, the number of PVs varies per data center and the total brown power consumption is recorded over one week. We can see that, as soon as green energy is available, SAGITTA consumes clearly less brown energy than the other approaches.

Figure 7 also shows that up to about 25 photovoltaic panels, the brown energy consumption curves have a steeper slope, leading to higher gains per photovoltaic panels. For more than 25 photovoltaic panels, the energy gains are lower per

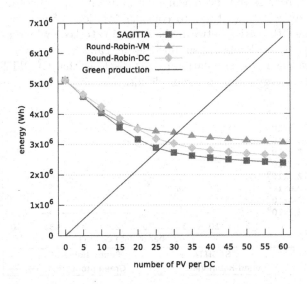

Fig. 7. Influence of green energy production on brown energy consumption from [4]. (Color figure online)

added panel. When reaching 45 panels, the green energy production exceeds the total energy consumption of the data center (represented by the case with 0 panel). However, this production is concentrated during the day (as shown in Fig. 5), whereas the workload, and consequently the energy consumption, spans over the day and the night. Thus, when reaching a number of photovoltaic panels whose production covers most of the Cloud energy consumption during daylight, adding panels can only save the energy consumption peaks at the beginning and the end of the day (when panels produce less energy), and their buying cost can thus exceed the monetary gains they generate.

Table 3. The considered cloud scenarios with increasing number of data centers.

Number of data centers	5	10	15	20	25	30	35	40
Total number of nodes	400	400	400	400	400	400	400	400
Number of PV per data centers	30	14	9	7	6	5	4	3

7.4   Scalability of SAGITTA

In order to check if the SAGITTA's energy savings scale up, we simulate the power consumption of distributed clouds with a larger number of data centers. For these different clouds, we progressively increase the number of data centers, and so the number of green power sources (still taken from the Photovolta project), while maintaining the same total number of nodes (and so an unchanged input workload). The total photovoltaic energy production is also kept as steady as possible by progressively decreasing the number of photovoltaic panels per data centers. Yet we decided not to consider fractions of panels, so the number of panels slightly varies between the scenarios to keep whole numbers. The compositions of these clouds are summed up in Table 3.

As shown in Fig. 8, the simulation results discloses that SAGITTA scales up: it maintains its energy gains in larger clouds, and always consumes less brown

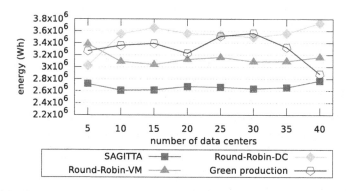

Fig. 8. Brown energy consumption of SAGITTA with increasing number of data centers. (Color figure online)

energy than the other approaches. From a computing time point of view, in our simulation environment, it takes 9 s to execute SAGITTA over the whole workload trace (representing one week) for the case with 5 data centers, and 28 s for the case with 40 data centers. While this computing time is increased by a factor of 3 (when increasing the data center number by a factor of 8), it still remains inconsequential for the scalability of SAGITTA.

7.5 Comparison with the Optimal

We compare SAGITTA with the optimal solution when considering ON/OFF switches penalties. In order to compute this optimal solution, we implement the algorithm 5 in python. Due to the high level of computing resources required, we parallelize and distribute the first **forall** loop of the algorithm. For each time slot, we use 30 hosts of the Grid'5000 platform to run in parallel the algorithm. Even with this optimization, we only were able to compute the optimal solution for a cloud composed of 5 data-centers of 20 hosts. The algorithm took about 2 weeks to perform 1 week of simulation.

The results are shown in Tables 4 and 5. We can see that:

- SAGITTA is very close to the optimal solution although it requires way lesser computing resources than the optimal algorithm.
- SAGITTA still performs better than the other two algorithms.

Table 4. Total cumulative energy consumptions over one week when considering ON/OFF penalties.

	Optimal	SAGITTA	Round-Robin-VM	Round-Robin-DC
Best	649,259Wh	666,238Wh	822,204Wh	733,304Wh
Worst	649,259Wh	666,238Wh	822,204Wh	1,086,626Wh

Table 5. Percentage of cumulative energy consumptions over the optimal when considering ON/OFF penalties.

	SAGITTA	Round-Robin-VM	Round-Robin-DC
Best	5.2%	26.6%	12.9%
Worst	5.2%	26.6%	67.4%

7.6 Exactness of the Green Power Production Forecast

The Table 6 shows the difference between the green power production predicted by SAGITTA and the actual ones. The Root-Mean-Square Deviation (RMSD) and the Normalized RMSD (NRMSD) are given in order to allow comparing SAGITTA with other future green power prediction models. We can see that SAGITTA makes an average error of only 311.3W, which corresponds by comparison to the power consumed by 1.4 working hosts. This is a relatively small error when considering that the green production of each DC ranges from approximately 0 W to 25,000W. This demonstrate that the prediction model of SAGITTA is accurate.

Table 6. Differences between the green power production predicted by SAGITTA and the actual ones.

	DC1	DC2	DC3	DC4	DC5	Average
Average	243.22W	332.96W	602.51W	303.71	74.44W	311.37W
Standard deviation	618.96W	884.74W	1456.11W	863.1W	270.21W	921.97W
RMSD	665.03W	945.32W	1575.84W	914.97W	280.27W	973.13W
NRMSD	2.66%	3.72%	6.22%	4.66%	2.14%	3.83%

8   Conclusion

In this chapter, we consider the problem of optimizing the green energy consumption of a geographically distributed cloud equipped with on-site photovoltaic panels. We tackle this challenge by distributing the cloud workload (composed of virtual machines) among the different data-centers.

We propose here a new version of the SAGITTA (a Stochastic Approach for Green consumption In disTributed daTA centers) approach which is based on this strategy. SAGITTA relies on a stochastic modeling of the expected green energy production in order to adequately allocate virtual machines to the data centers. The approach also switches off unused servers to save energy. In the new version, SAGITTA now natively takes into account of the energy costs arising from these on/off switches. This extension of SAGITTA is more robust and offers slightly better results.

We have proven in this paper that SAGITTA is locally optimal -i.e. that SAGITTA is optimal regarding the current time slot. In order to evaluate the performance of SAGITTA when considering the on/off switches costs, we also proposed a dynamic programming algorithm for computing the optimal energy consumption of the whole experiment. We conducted a simulation-based evaluation using real workload traces, wattmeter measurements on testbed servers, and real production traces from photovoltaic panels. We compared SAGITTA with two round-robin algorithms which do not consider the green energy production for allocating virtual machines in the cloud. The results show that SAGITTA can allocate virtual machines in a more energy-efficient way than these traditional approaches. Moreover, SAGITTA exhibits good results in term of brown energy consumption with a difference of only 5.2% with the optimal solution computed by our dynamic programming algorithm. The simulations also show that SAGITTA can adapt to different green energy production patterns as it outperforms traditional approaches in all these cases. Finally, we shown that SAGITTA can smoothly scale with the number of data centers belonging to the cloud.

In future work, we plan to extend SAGITTA by integrating the ability to dynamically migrate virtual machines and the energy production from one site to another. We also want to adapt SAGITTA to continuous (i.e. non time-slotted) workloads. Finally we want to integrate in our simulation the impact of network devices on the energy consumption.

Acknowledgments. This work has been supported by the Inria exploratory research project COSMIC (Coordinated Optimization of SMart grIds and Clouds).

References

1. Shehabi, A., et al.: United States Data Center Energy Usage Report. Technical report, Lawrence Berkeley National Laboratory (2016)
2. Tripathi, R., Vignesh, S., Tamarapalli, V.: Optimizing green energy, cost, and availability in distributed data centers. IEEE Commun. Lett. **21**(3), 500–503 (2017)
3. Wang, D., Ren, C., Sivasubramaniam, A., Urgaonkar, B., Fathy, H.: Energy storage in datacenters: What, where, and how much? In: ACM SIGMETRICS/PERFORMANCE, pp. 187–198 (2012)
4. Camus, B., Dufossé, F., Orgerie, A.C.: A stochastic approach for optimizing green energy consumption in distributed clouds. In: International Conference on Smart Cities and Green ICT Systems (SMARTGREENS), pp. 47–59 (2017)
5. Wang, L., Tao, J., Kunze, M., Castellanos, A., Kramer, D., Karl, W.: Scientific cloud computing: early definition and experience. In: IEEE International Conference on High Performance Computing and Communications (HPCC), pp. 825–830 (2008)
6. Koomey, J.: Growth in Data Center Electricity Use 2005 to 2010. Analytics Press, Berkeley (2011)
7. Katz, R.H.: Tech titans building boom. IEEE Spectr. **46**, 40–54 (2009)
8. : How dirty is your data? Greenpeace report (2011)
9. Figuerola, S., Lemay, M., Reijs, V., Savoie, M., St. Arnaud, B.: Converged optical network infrastructures in support of future internet and grid services using IaaS to reduce GHG emissions. J. Lightwave Technol. **27**, 1941–1946 (2009)
10. Callau-Zori, M., Samoila, L., Orgerie, A.C., Pierre, G.: An experiment-driven energy consumption model for virtual machine management systems. Technical Report 8844, Inria (2016)
11. Fan, X., Weber, W.D., Barroso, L.A.: Power provisioning for a warehouse-sized computer. In: ACM International Symposium on Computer Architecture (ISCA), pp. 13–23 (2007)
12. Snowdon, D., Ruocco, S., Heiser, G.: Power management and dynamic voltage scaling: myths and facts. In: Workshop on Power Aware Real-time Computing (2005)
13. Talaber, R., Brey, T., Lamers, L.: Using Virtualization to Improve Data Center Efficiency. Technical report, The Green Grid (2009)
14. Barham, P., et al.: Xen and the art of virtualization. In: ACM Symposium on Operating Systems Principles (SOSP), pp. 164–177 (2003)

15. Miyoshi, A., Lefurgy, C., Van Hensbergen, E., Rajamony, R., Rajkumar, R.: Critical power slope: understanding the runtime effects of frequency scaling. In: ACM International Conference on Supercomputing (ICS), pp. 35–44 (2002)
16. Orgerie, A.C., Dias de Assunção, M., Lefèvre, L.: A survey on techniques for improving the energy efficiency of large-scale distributed systems. ACM Comput. Surv. (CSUR) **46**, 47:1–47:31 (2014)
17. Raïs, I., Orgerie, A.-C., Quinson, M.: Impact of shutdown techniques for energy-efficient cloud data centers. In: Carretero, J., Garcia-Blas, J., Ko, R.K.L., Mueller, P., Nakano, K. (eds.) ICA3PP 2016. LNCS, vol. 10048, pp. 203–210. Springer, Cham (2016). https://doi.org/10.1007/978-3-319-49583-5_15
18. Ren, S., He, Y., Xu, F.: Provably-efficient job scheduling for energy and fairness in geographically distributed data centers. In: IEEE International Conference on Distributed Computing Systems (ICDCS), pp. 22–31 (2012)
19. Deng, W., Liu, F., Jin, H., Li, B., Li, D.: Harnessing renewable energy in cloud datacenters: opportunities and challenges. IEEE Netw. **28**, 48–55 (2014)
20. Tighe, M., Keller, G., Bauer, M., Lutfiyya, H.: DCSim: a data centre simulation tool for evaluating dynamic virtualized resource management. In: Workshop on Systems Virtualization Management, pp. 385–392 (2012)
21. Camus, B., et al.: Hybrid co-simulation of FMUs using DEV&DESS in MECSYCO. In: Symposium on Theory of Modeling & Simulation - DEVS Integrative M&S Symposium (2016)
22. Camus, B., et al.: MECSYCO: a multi-agent DEVS wrapping platform for the co-simulation of complex systems. Research report, LORIA (2016)
23. Zeigler, B., Praehofer, H., Kim, T.: Theory of modeling and simulation: integrating discrete event and continuous complex dynamic systems. Academic Press, Cambridge (2000)
24. Li, Y., Orgerie, A.C., Menaud, J.M.: Opportunistic scheduling in clouds partially powered by green energy. In: IEEE International Conference on Green Computing and Communications (GreenCom) (2015)

Using Energy Supply Scenarios
in an Interdisciplinary Research Process

Barbara S. Zaunbrecher[1](✉), Thomas Bexten[2], Jan Martin Specht[3],
Manfred Wirsum[2], Reinhard Madlener[3], and Martina Ziefle[1]

[1] Chair for Communication Science,
RWTH Aachen University, Campus Boulevard 57, 52074 Aachen, Germany
{zaunbrecher,ziefle}@comm.rwth-aachen.de
[2] Chair and Institute of Power Plant Technology, Steam and Gas Turbines,
RWTH Aachen University, Templergraben 55, 52062 Aachen, Germany
{bexten,wirsum}@ikdg.rwth-aachen.de
[3] Chair of Energy Economics and Management,
Institute for Future Energy Consumer Needs and Behavior (FCN),
E.ON Energy Research Center, RWTH Aachen University,
Mathieustr. 10, 52074 Aachen, Germany
{mspecht,rmadlener}@eonerc.rwth-aachen.de

Abstract. The sustainable energy transition (Energiewende) is a multidisciplinary challenge. While for technical disciplines, the focus is on the development of technologies which can supply, transmit and store energy in a sustainable way, economic research focuses for example on the analyses of costs and risks of different asset portfolios. Yet another perspective is taken by the social sciences who focus on social challenges associated with the implementation of measures for realizing the Energiewende (decarbonization, high energy efficiency, high shares of renewables, nuclear phaseout), for example their acceptability. A solution for energy supply and storage which is optimized only according to one of these perspectives will, however, fail to meet other essential criteria. To develop sustainable solutions for energy supply and storage, which are technically feasible, cost-effective, and supported by local residents, interdisciplinary cooperation of researchers is thus needed. Interdisciplinary research, however, is subject to many barriers, for example the need to agree on a common analytical framework. In this paper, a process model for interdisciplinary energy research is proposed, in which specific scenarios are used to aid interdisciplinary cooperation and reciprocal integration of results. Based on a current research project, the phases of the model and the use of the scenarios in disciplinary and interdisciplinary work packages are described, as well as challenges and shortcomings of the model.

Keywords: Renewable energy · Social acceptance · Economics
Technology · Interdisciplinarity · Electricity storage

© Springer Nature Switzerland AG 2019
B. Donnellan et al. (Eds.): SMARTGREENS 2017/VEHITS 2017, CCIS 921, pp. 77–93, 2019.
https://doi.org/10.1007/978-3-030-02907-4_4

1 Introduction

The inherent high volatility and limited predictability of the growing number of renewable power generation capacities in Europe and especially in Germany poses various challenges for a stable operation of the electricity grid (Holttinen [24]). This is a result of the requirement for an equilibrium between power generation and consumption at any time. In order to compensate for these characteristics of renewable power generation, flexible and efficient dispatchable energy conversion and storage units are needed today and in the future (Bouffard et al. [9]). Centralized large-scale units are one option to provide this required dispatchable capacity (e.g. coal fired power plants, combined cycle power plants, pumped storage power plants). An alternative is the deployment of smaller distributed units, e.g. in a municipal context. The utilization of these units leads to a better convergence of power production and consumption profiles on a local level and hence offers the potential of enhancing the electrical autarky of municipal energy supply systems while reducing the pressure on the higher voltage levels of the electric grid.

In recent years, the economic perspective on the energy transition in Germany seems to have changed to some extent. While the focus in the last decades was mainly set on getting the diffusion of renewable energy technologies started, the resulting increase of the electricity price imposed some pressure on policy makers to limit the rise. One measure was to switch from the promotion through guaranteed feed-in tariffs to an auctioning system for wind power and large PV plants, so that only the most competitive projects are realized (EEG 2016, §2 (3)). The first auctioning round conducted in spring 2017 resulted in citizens' energy initiatives receiving 93% of all awards (BMWi [15]). How this affects the diffusion of renewable energy projects and more specifically municipal energy systems remains to be seen. One option could be that local projects will forgo the strong competition for the declining national funding but turn towards business models that allow for local financing, e.g. with the support of a municipality or additional returns for local green electricity.

While the necessity of turning away from fossil fuels towards renewables is widely acknowledged and supported by the general public (Zoellner et al. [54]), specific energy projects have raised protests by (local) residents, especially large scale technologies and associated infrastructures (e.g., wind farms, transmission lines) (Wüstenhagen et al. [47]). While in the past, slow diffusion and a lack of social acceptance also occurred, the scope, pace and organization of protest has dramatically changed (Marg et al. [38]), delaying projects and leaving residents unsatisfied with the development process (Gross [20]). The reasons why local residents oppose energy infrastructure are manifold. Among other reasons, landscape impact of the energy infrastructure plays a role for its social acceptability (Wüstenhagen et al. [47], Hirsh and Sovacool [22], Johansson and Laike [27]), and, closely connected to this issue, environmental concerns (Krewitt and Nitsch [29]). Health concerns have also been shown to affect public attitudes toward energy infrastructure, for example, fear of infrasound in the case of wind power plants, or of electromagnetic fields in the case of transmission lines (Songsore and

Buzzelli [43], Baxter et al. [3], Wiedemann et al. [44]). Besides these concerns, the social setting of the planning process has an impact on local acceptance of energy infrastructure. It has been found that trust in the involved stakeholders (Huijts et al. [25], Bronfman et al. [11]) as well as perceived fairness of the decision process (Wolsink [46], Liebe et al. [35]) can also have an influence on the perception of local energy infrastructure. From a planning perspective, the claim has been made that participatory approaches, which value and integrate the parties concerned in early stages, are more likely to gain approval with local communities (Langer et al. [32], Raven et al. [40], Schweizer et al. [42]). This requires openness from the planners' perspective towards alternative options (Schweizer et al. [42]) and the acknowledgement that there is more to energy infrastructure planning than technical requirements.

An interdisciplinary approach, in which energy supply scenarios are not only evaluated from a technical, but also from an economic and social perspective, can help to develop solutions which take into account the technical, economic and social challenges associated with the changes to the energy supply system, and thus provide holistic solutions to a complex problem. Especially the early integration of social factors in early stages of the technology development process can help to overcome some of the above mentioned barriers (Zaunbrecher and Ziefle [48]). Interdisciplinarity, in this context, is understood as "a coordinated collaboration between researchers from at least two different disciplines, which can manifest itself in a simple exchange of ideas to the point of integration of methods, concepts and theories" is referred to (Hamann et al. [21]). Especially for global challenges like climate change or energy supply, interdisciplinary approaches are called for (Wilson [45]), because these complex topics cannot be answered by one discipline alone and "do not exist independently of their sociocultural, political, economic, or even psychological context" (Brewer [10]:329). While the methodological variety of interdisciplinary approaches offers the benefit of capturing a problem more holistically, the assembling of "their partial insights into something approximating a composite whole" (Brewer [10]:330) still presents a challenge. Barriers to interdisciplinary work are, e.g. different scientific cultures, thus also different frames of references and methods, with which problems are approached (Brewer [10]). Furthermore, the problem of communication, based on a different "language" of each discipline, can hinder successful interdisciplinary collaboration (Brewer [10], Holbrook [23], Jacobs and Frickel [26]). It requires the researchers involved to translate their concepts, approaches and ideas into terms that members of other disciplines can relate to (Holbrook [23]). Institutional barriers, such as incentives, funding, and the priority given to interdisciplinary over disciplinary work present a further challenge (Brewer [10]). It was in fact found that the more institutions are involved, the less knowledge outcomes are reported, due to higher coordination costs and more effort required to sustain strong working relationships (Cummings and Kiesler [14]). Distributed team members mostly do not know each other, and therefore have weaker ties, and, consequently, less communication (ibid.). Disciplinary structures, like specialized journals or conferences further hinder interdisciplinary exchange by supporting an inner-disciplinary communication rather than inter-disciplinary exchange (Jacobs and Frickel [26]). Moreover, there can be a lack of knowledge about possible contributions and opportunities for collaboration with other disciplines, due to "disciplinary assumptions about the "other" half of the system [based on] simplistic models" (Lélé and Norgaard [34]:968).

Although some of these issues might not be unique to interdisciplinary teams (Jacobs and Frickel [26]), the variety of challenges on a content and institutional level illustrates the complexity of interdisciplinary research.

In this paper, a process model for interdisciplinary research is presented, taking an energy-related project as an example, which seeks to overcome some of these challenges. It presents a specific application of an interdisciplinary research approach to questions of energy supply, and moreover, can serve as a guideline for other interdisciplinary projects in other contexts with regard to the various stages of cooperation. It is shown how in the different phases of the model, interdisciplinarity is achieved as a process from separated, multidisciplinary research to fully integrated interdisciplinary research. In particular, it is shown in detail how energy-supply scenarios were used in the process to facilitate communication and data exchange between the disciplines and how those scenarios were defined in a coordinated process between the disciplines, taking into account requirements on the one hand, and applications of the scenarios in disciplinary and interdisciplinary research on the other hand.

2   Interdisciplinary Process Model

The process model (Fig. 1) describes the research process in the research project KESS[1]. In this project, an interdisciplinary group of researchers develops energy supply scenarios for municipalities, including electricity production, transmission and storage. The energy supply scenarios are analyzed from a technical perspective by researchers from mechanical and electrical engineering, an economic perspective by researchers from energy economics, and a social perspective, by researchers from communication science and linguistics. The research process is illustrated in the following chapters by first referring to the process model which was followed and afterwards by a detailed description of the scenarios which were applied in key stages of the research process. Overall, the research model describes a continuum from an "informal communication of ideas" to "formal collaboration" (Lattuca [33]).

In the first stage, stage one, the disciplines are at the beginning of the interdisciplinary collaboration. In this stage, their collaboration is thus characterized by a multidisciplinary, not an interdisciplinary approach (Jungert [28]), as the connection is not yet established through collaboration, but -at least- through a shared research topic (energy supply for municipalities). Thus, the research style here is rather a "parallel play" (Aboelela et al. [1] than an integrated approach. During this phase, each discipline defines relevant research topics and thus lays

---

[1] For information on the project see http://www.comm.rwth-aachen.de/index.php?article_id=954&clang=1.

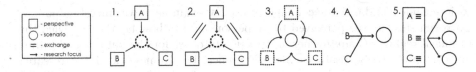

Fig. 1. Process model for Interdisciplinary Collaboration (Zaunbrecher et al. [53]).

the basis for later cooperations. The importance of this phase is underlined by the fact that disciplinarity is "considered [one of] the most important factors for successful interdisciplinary collaboration" (Hamann et al. [21]).

The research questions which are developed in this initial phase are thus also of uni-disciplinary nature, as "every component of [the] research problem calls for a different science" (Krohn [30]). Examples for uni-disciplinary research questions concern, e.g., the interconnectivity between different technical parameters from a technical perspective (Bexten et al. [5]), or the perception of single components of the system from a social perspective (Zaunbrecher et al. [52]). In order to align the results, the framing parameters of the energy supply scenario are loosely defined, e.g., which components define the energy supply system, how many inhabitants the municipality has, how large the annual electricity consumption is and how large the share of electricity produced by renewables is.

In the second stage, the multidisciplinary approach has progressed to a "multidisciplinary approach with exchange". Although all disciplines still approach the topic from their own perspective using their own methods, the process to interdisciplinarity is further progressed by an exchange between the three perspectives. This exchange includes communicating methods, approaches, terminology, ideas and requirements for further collaboration, in order to enhance mutual understanding (Armstrong and Jackson-Smith [2]). The mutual exchange can help to overcome misunderstandings between the disciplinary perspectives (Hamann et al. [21]), and, on a different note, enhance the understanding for possible contributions of the other disciplines and thus open the floor for further collaborations. It also creates the communicative basis which is needed for the negotiation of the energy supply scenarios used in later stages of the process model.

The first stage in which true interdisciplinarity is visible in the working process, in the research methodology applied, and the publications, is stage three. Central benefits of interdisciplinary collaboration can be achieved during this stage, e.g., the widening of the horizon of the researchers involved, and the innovative potential through the combination of knowledge (Hamann et al. [21]). It differs from stage two in the fact that now, research questions are formulated and approached which can no longer be solved by one discipline alone, thus requiring multiple disciplines to closely collaborate. In this stage, bilateral teams of two different disciplines approach a common research topic and align their methodological approaches. In the KESS project, which is referred to as an exemplary project, the research questions at this stage concerned socio-economic, socio-technical and techno-economic issues. For example, it was investigated how hydrogen storage was perceived by laypersons and in how far this matched the technical realities (Zaunbrecher et al. [50]). In order for the data acquired during this stage to be usable across disciplines, the framework of the research has to be more closely defined, to ensure transferability and comparability of the data. Therefore, scenarios are defined by boundary parameters which define the research context for all disciplines involved, comprising, e.g. obligatory and optional components of the system and the specific technologies involved. It is also negotiated on which level of detail the analyses will be conducted, in order to ensure the resulting data are comparable. The specific scenarios used in the exemplary project and how they were derived from the requirements of the different disciplines are presented in Sect. 3.

In stage four, mature interdisciplinarity and elaborate communication between the disciplines is achieved. The research topic is approached using a multi-method approach, combining viewpoints, methods and approaches from all perspectives. It is an advancement to stage three because instead of bilateral teams, all disciplines involved in the research project now collaborate on a single research question. These joint efforts are supported by the ever increasing trust of the partners into the potential of the collaboration in terms of working quality and scientific merit. In the KESS project, the energy scenarios (defined in stage three) are evaluated from all involved perspectives in a parallel working process, in order to achieve a multidimensional evaluation of the scenarios, in which the different properties of the scenarios (technical, social, economic) can, in a final step, be weighted against each other. For the social acceptability, the evaluation could refer to a relative preference value for each scenario which will be derived using conjoint analysis (for a similar procedure see, e.g., Zaunbrecher et al. [51]). The degree of self-sufficiency of the investigated energy supply system scenarios is one example for a core criterion from a technical point of view. Another possible candidate for a technical criterion, focusing on the ecological impact, is the total amount of CO_2 emissions during an analyzed time period. The economic assessment, in turn, tries to optimize the monetary value of a proposed scenario. The predefined scenarios function as a baseline for the level of detail for the analysis. By attributing one value per perspective to each scenario, an interdisciplinary exchange about the overall suitability of a scenario is possible. Furthermore, trade-offs between the perspectives can be discussed (e.g., in which context should a socially acceptable scenario be preferred over an economically efficient one?). According to the combined evaluation from three perspectives, the scenarios can then, as a final step, be qualitatively ranked according to suitability.

Stage five represents the transferral of the interdisciplinary evaluation of the scenarios into practice (transdisciplinarity). In this stage, the results from stage four are operationalized in a tool which allows decision makers to gain insights into different suitable energy supply scenarios according to his needs. The technical, economic and social requirements towards the energy supply can be predefined and, on this basis, possible suitable scenarios are suggested. As a prerequisite, the multidimensional evaluation from stage 4 is needed. The tool would allow decision makers to enter framework conditions according to their local requirements. The tool would then, based on the framework conditions, suggest potential energy supply scenarios, which are characterized by technical, economic and social attributes. Besides being useful for planners, such a tool could enhance understanding of laypersons of planning procedures and conditions which need to be taken into account for the planning of large infrastructure projects. Similar approaches can be found in the context of urban green space planning (Grêt-Regamey et al. [19]) or wind power planning (Cavallaro [12], Gamboa and Munda [18]).

3   Definition and Integration of Energy Scenarios

Beginning at stage 3, specific scenarios were defined in an interdisciplinary exchange to help coordinate research paths and align the depth of the analyzed data. This was considered an essential step in order to be able to compare data from the different disciplines. While the research scenario was only loosely defined in the first stages, specific scenarios were formulated for the final stages of the research process.

The scenarios used for the exemplary case presented in this paper are set up in order to represent a mid-sized municipality with a high share of volatile renewable power generation. It is assumed that there a roughly 10,000 inhabitants living within the municipality and that the associated households are the predominant consumers of power. The result of these assumptions is an annual power consumption of 20 GWh. Regarding the set up of the renewable power generation within the municipality, all scenarios follow the concept of "integral autarky". This means that the number of installed renewable power generation capacities (i.e. wind turbines and photovoltaic panels) is chosen in a way that the corresponding annual power generation is equal to the annual municipal power consumption. This approach is not comparable to full autarky of the municipality due to the inevitable temporal mismatches between volatile renewable power generation and power consumption. In order to maintain the balance between power generation and consumption, the municipality can interact with the grid. Renewable power is fed into the grid in times of excess generation while power is supplied by the grid in times of residual demand. In addition, energy storage and conversion units are integrated into the scenarios (i.e. battery storage and hydrogen storage). These units are used to store and produce electricity on a local level, thus reducing the need for grid interactions.

Regarding the renewable power generation capacities, a reference year in the region of Aachen, a mid-sized city in Western Germany, is chosen to provide data for solar irradiation and wind speeds. For solar power, installation on rooftops was assumed rather than a solar park. Furthermore, the types of components used (i.e. wind turbines, solar panels, battery storage, hydrogen storage) are technically specified for the scenarios (Bexten et al. [4]).

3.1 Disciplinary Parameters and Scenario Requirements

Apart from the reference framework described above, each perspective had specific requirements for the definition of the scenarios.

Technical: From a technical point of view, the main purpose of the scenarios is the analysis of the interaction between renewable power generation capacities, local consumers, and dispatchable energy storage and conversion units within a municipal energy supply system.

On the one hand, these investigations focus on overall system performance parameters that are influenced by the interaction between the renewable power generation portfolio and the configuration of the dispatchable energy storage and conversion units. Investigated performance parameters include the self-sufficiency of the municipal energy supply system, the power exchange with the upstream transmission grid and the total CO_2 emissions of the system. The scenarios used for these investigations have to incorporate a wide range of diverse renewable power generation portfolios in order to capture the individual characteristics of wind and solar based renewable power generation like seasonal and short-term volatility. In addition, the scenarios also have to incorporate an extensive set of energy storage and conversion unit configurations in order to highlight the individual capabilities of the investigated technologies (e.g. short-term battery storage vs. long-term hydrogen storage).

On the other hand, the scenarios are used for a detailed analysis of the dispatchable units operation regarding the degree of utilization and the flexibility requirements. These investigations require information on the time-dependent dispatch and performance of the individual system components within the scenarios. To be able to include these aspects into the scenarios, high fidelity models of the components, incorporating part-load characteristics and operational flexibility parameters, have to be integrated into the overall model of a municipal energy supply system. This approach subsequently enables the detailed time-dependent simulation of the energy supply system operation within a predefined scenario after a corresponding operational strategy is defined.

Economic: Scenarios help to estimate costs and benefits of different asset portfolios in the economic assessment. In early stages, they support decisions such as either to focus on a calculation with total values (e.g., a Net Present Value analysis) or to head for relative values such as levelized costs of electricity (LCOE). Especially for the optimization of scenarios that compare technologies with very different shares of capital and operational costs and different life expectancies, or to account for different operational strategies, LCOE are often preferable. However, a holistic economic analysis should not only account for the monetary cost benefit analysis but should also consider aspects such as portfolio optimization, investment risks and the capital structure. To preclude technically unfeasible constellations, predefined scenarios can narrow down the scope for an economic optimization, but still the scenario with the highest expected return does not necessarily have to be the best advice or preferred option for a planner. This is due to the investment risk and the fact that many, typically rather risk averse decision makers, should search for a trade-off between profitability and risk (Madlener [37]). Scenarios with a strong focus on only one source of uncertainty (e.g. "only wind plants") are often more vulnerable to external factors and errors in the assumptions, whereas the versatile scenarios (e.g. "wind power, PV and battery storage") provide more reliable estimates leading to a lower investment risk. Or, to put it differently, a scenario with reliable returns might still be preferred to a highly speculative scenario even with lower average returns.

Social: For the analysis of the social acceptability of energy scenarios, it is indispensable that the scenarios which are to be evaluated are technically feasible, in order to ensure technical relevance of the acquired results. Also, whenever users are included with the task to engage themselves with the scenarios and evaluate the social acceptability, it is mandatory that the scenarios are actually realistic. Therefore, the technical feasibility of the scenarios needs to be determined as a first step (cf. methodological considerations in Zaunbrecher et al. [49]). This included the number and combination of infrastructural elements, in this case, electricity production and storage infrastructure. Furthermore, information on the specific local impact of the technical infrastructure, for example in terms of size, was needed, in order to explore questions of local visual impact of the infrastructure (McNair et al. [39], Johansson and Laike [27], Devine-Wright and Batel [16]). Further important information included technical consequences of combinations of components, such as the degree of self-sufficiency of the municipality, determined by the type and number of PV panels, wind turbines and storage technologies. These technical consequences can serve as potential trade-offs for laypeople in their evaluation of the scenarios (e.g., more self-sufficiency means more local storage infrastructure). Despite the necessity of some technical framework conditions, it had to be taken into account that the participants in the socio-psychological studies should not be overstrained with too many technical details that are outside of their level of knowledge and not relevant for social acceptance on a broad level of scale. For example, although technically relevant, it was determined that the specific technical components, in terms of particular products with technical performance data, need not be determined in detail for the use in social-psychological analyses. This is justified by the explorative nature of the research: As literature on the social acceptability of electricity storage technologies is still scarce, the goal of the analyses within the scope of the project was to gain a general understanding of acceptance-relevant parameters of electricity storage in general, not with relation to one specific model of an electricity storage facility. In order to present systematically varying scenarios to the participants, it was also necessary to define attributes of the scenarios which could be implemented in different variations (e.g., attribute "storage" could be implemented as battery storage, hydrogen storage, no storage etc.).

3.2   Final Scenarios

The final specifications of the scenarios (Table 1) were the result of a balancing process between the three perspectives described in the previous section. The scenarios varied in the renewable power generation portfolio and the type of local storage technologies. The decision to define different shares of wind and solar based renewable power generation was mainly influenced by technical and social considerations. The shares should correspond to integer numbers of the same type of wind turbines and solar panels to make the scenario feasible from a technical point of view. At the same time, there should be substantial differences between the scenarios (e.g., not 33% vs. 35%), so that the differences are relevant to laypersons and the different technical characteristics of wind and solar based power generation are highlighted. According to these requirements, shares of around 30/70 and 50/50 were chosen. The impact of the integration of electricity storage into the scenarios was operationalized by the differentiation between battery and hydrogen storage systems. It was refrained from including different technical specifications of the battery or hydrogen storage systems, as these differentiations were considered to be too detailed information for laypersons. The combination of these two factors resulted in 12 scenarios (Table 1), which are used in subsequent stages for interdisciplinary research approaches.

Table 1. Energy supply scenarios (Zaunbrecher et al. [53]).

Scenario	Electricity mix	No. wind turbines	No. PV modules	Storage
A1	73% wind, 27% PV	3	1025	No storage
A2	73% wind, 27% PV	3	1025	Battery storage
A3	73% wind, 27% PV	3	1025	Hydrogen storage
A4	73% wind, 27% PV	3	1025	Hydrogen + battery storage
B1	49% wind, 51% PV	2	1960	No storage
B2	49% wind, 51% PV	2	1960	Battery storage
B3	49% wind, 51% PV	2	1960	Hydrogen storage
B4	49% wind, 51% PV	2	1960	Hydrogen + battery storage
C1	24% wind, 76% PV	1	2695	No storage
C2	24% wind, 76% PV	1	2695	Battery storage
C3	24% wind, 76% PV	1	2695	Hydrogen storage
C4	24% wind, 76% PV	1	2695	Hydrogen + battery storage

3.3 Integration of Scenarios in Disciplinary and Interdisciplinary Research

The scenarios defined in Table 1 were used in disciplinary and interdisciplinary research approaches.

Technical: In a first approach, the described scenarios were used as input parameters for the simulation of the municipal energy supply system operation. The subsequent analysis of the simulation results mainly focused on the impact of the different predefined dispatchable energy storage and conversion units on the self-sufficiency of the overall system and the remaining power exchange with the upstream transmission grid (Bexten et al. [6]). In addition to the analysis from a technical point of view, the main findings of this study also served as input parameters for subsequent studies focusing on the social acceptance of the scenarios. Due to the fact that the simulation results indicated a high operational flexibility requirement by the gas turbine as part of the hydrogen storage system, additional investigations were conducted. These investigations focused on options to reduce the number of start-ups and fast load changes of the gas turbine by using additional battery storage capacity (Bexten et al. [5]).

In a following step, the scenarios were used as a framework for more detailed investigations regarding the capability of individual dispatchable units to enable a more efficient integration of the volatile renewable power generation capacities into the overall energy supply system. An example for this kind of investigations is an analysis of the potential of wind farm forecast error compensation by the utilization of flexible combined heat and power units (Bexten et al. [7]).

In future studies, the scope of the scenarios and the associated simulations will be extended to the municipal heat demand and the potential to provide the required heat with dispatchable units. This allows for the conduction of a wide range of studies within the rapidly growing research field of "sector coupling". Besides a number of technical research questions that can be answered using the extended scenarios (e.g. optimal use of heat storage capacities), an additional dimension is added to the studies focusing on the social acceptance of municipal energy supply systems.

In addition to the extension of the technical scope of the scenarios, future work will also focus on a closer link between the technical and the economic aspects of the scenarios. Previous studies mainly used simplified operational strategies and predefined configurations of the renewable power generations capacities and the dispatchable energy storage and conversion units. The planned integration of economic parameters (e.g. investment costs, operational costs) and corresponding optimization algorithms will enable the determination of optimized forecast-based control strategies as well as the composition of economically optimized system configurations.

Economic: To enable potential decision makers to evaluate the trade-off between risk and value, a pre-simulator was programmed. The goal was to minimize computational time for this simulator to be able to use it live in discussions with decision makers or activists. As input to this simulator, the parameters and limitations from the technical perspective had to be taken into account. While less precise than the technical simulation, this pre-simulator allows for a quick overview of the economic viability and risks of different technical generation portfolios, which can subsequently be addressed from a social perspective. The results can help to keep risk at a socially acceptable level without losing too much of the economic value. Besides of the quick pre-simulation, the technical simulator was also taken up as base for an advanced economic simulation. Adding costs and using the Monte Carlo simulation technique to account for uncertainties, we currently investigate their impact on the value distribution. Since the acquisition of national funding became more complicated and competitive with the recently introduced auctioning scheme in Germany, a further focus was put on the evaluation of different financing schemes. Several alternatives, such as a focus on green electricity certifications or municipal funding, are discussed as an alternative, depending on the pursued energy solution. This again goes hand in hand with social acceptance, since it is unlikely that, for example, a municipality would support a project that lacks support by the residents.

Social: The scenarios were used as a basis for various socio-psychological studies in close cooperation with the researchers from technical disciplines. In a first, exploratory approach, the scenarios were used in focus groups. Focus groups are organized group discussions, which serve the purpose to gain broad insights into attitudes, experience and motives of participants regarding a specific topic (Krueger [31]). The scenarios were used as an anchor in the discussions, using a scenario builder (Fig. 2), while the participants discussed not only the single components of the scenarios, but also discussed the differences between the scenarios. The scenarios further helped to introduce participants to a the situation where they were asked to imagine the energy supply of their hometown would be renewed and different options were available, because the scenarios were sufficiently concrete. The results of this stage of research included general acceptance-relevant parameters for the single components of the energy supply system (battery and hydrogen storage, wind power and PV), as well as dimensions for trade-offs between the scenarios. In a next step, acceptance for the scenarios was quantified by means of a conjoint analysis (Luce and Tukey [36]), in which the scenarios were decomposed into their single components. Participants could then state their preferences for combinations of components (energy supply scenarios), so that the relative, quantified preference for each energy supply scenario (defined in Table 1) could be calculated. In this way, the preference for a scenario from a social point of view can, on the one hand, be integrated as a boundary parameter in technical simulations, and, on the other hand, be compared side-by-side to technical and economic evaluations of the scenarios (cf. Stage 4 of the Process model).

4   Discussion

While the interdisciplinary approach, for which a process model is proposed in this paper, has many advantages when complex topics such as energy supply are addressed, the intensive collaboration required on different levels also has its drawbacks. In order to align their research interests and to ensure comparability and the ability to integrate data from the different disciplines, the representatives of the disciplines involved have to agree on a certain level of detail of the analysis, as, in this example, was done when the scenarios were defined. From the (still) relatively broad definition of the scenarios (Table 1), it becomes obvious that the interdisciplinary approach bears the cost of disciplinary detail. From an economic, technical or social perspective alone, the scenarios would have been defined differently, with a different level of detail in certain aspects. It could thus be argued that this approach results in a lack of depth of analyses (Hamann et al. [21]). In order to counterbalance this caveat, continued disciplinary approaches, in addition to the more general, interdisciplinary analyses, are necessary. In this way, the level of detail which cannot be covered by interdisciplinary approaches can be tackled by more detailed, disciplinary approaches while at the same time providing a level of detail regarding the data which can still be used for integration with other disciplines.

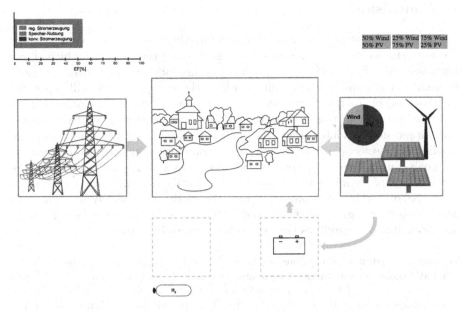

Fig. 2. Scenario builder for social acceptance studies (Zaunbrecher et al. [53]).

Regarding the transferability of the process model to other research projects, some limitations need to be mentioned. An advantage of the KESS project was the fact that all members of the research project were based at the same university, so that institutional barriers were probably lower as if different organizations had been involved (Cummings and Kiesler [13]). Moreover, while the model can provide some guidelines for interdisciplinary research projects, its application cannot guarantee the success of interdisciplinary collaborations. This might be subject to the research topic (content) of the project, the disciplines involved, or even the researchers themselves (Rhoten and Pfirman [41]).

While it is increasingly acknowledged that interdisciplinarity should be the methodological approach to address complex problems, and that it presents a core academic competence (Boddington et al. [8]), education at universities does not systematically incorporate interdisciplinarity as an inherent component of content-related questions across disciplines. It should therefore be an aim to "train future scholars and professionals to think way beyond the confines of their basic disciplines to attain the broadest perspectives so urgently needed for environmental protection." (Brewer [10]:333). Novel modules in different faculties, in which interdisciplinary methods are interlinked with content related questions to teach multiperspective problem solving, could address this need. First evaluations of such courses have shown the potential to change students' mindsets and promote openness towards interdisciplinary collaborations (Drezek et al. [17]).

5 Conclusions

The paper has presented an example of how interdisciplinary research in the field of energy supply can be achieved using a step-wise process model which shows how researchers from different disciplines can interact with each other to move on from multidisciplinary research, in which the disciplines are still separated from each other, to truly integrated, interdisciplinary research. Energy supply scenarios can help in this process to align research interests, and to provide a basis for mutual data integration. As an advantage of this research approach, the enabling of close cooperation as well as the communication about requirements and goals, and a common level of detail were achieved. As a disadvantage, the possible lack of detail of the analyses was identified, along with measures to counterbalance this development. While the model is generally applicable to other research projects, it should not be taken as a guarantee for successful interdisciplinary research, as this depends on multiple factors.

Acknowledgements. A previous version of this paper was presented at the 7th International Conference on Smart Cities and Green ICT Systems (SMARTGREENS) (Zaunbrecher et al. [53]). Special thanks go to the KESS project members for the fruitful discussions which have contributed and led to this paper. Thanks also to Dr. Klaus Baier, Julian Halbey, Iana Gorokhova and Saskia Ziegler for research support. The "KESS" project conducted at RWTH Aachen University is funded by the strategic funds of the Excellence Initiative of the German federal and state governments.

References

1. Aboelela, S.W., et al.: Defining interdisciplinary research: conclusions from a critical review of the literature. Health Serv. Res. **42**(1), 329–346 (2007)
2. Armstrong, A., Jackson-Smith, D.: Forms and levels of integration: evaluation of an interdisciplinary team-building project. J. Res. Pract. **9**(1), 1–20 (2013)
3. Baxter, J., Morzaria, R., Hirsch, R.: A case-control study of support/opposition to wind turbines: perceptions of health risk, economic benefits, and community conflict. Energy Policy **61**, 931–943 (2013)
4. Bexten, T., et al.: Modellbasierte Analyse der Auslegung und des Betriebs kommunaler Energieversorgungssysteme [Model-based analysis of the design and operation of municipal energy supply systems]. In: 14th Symposion für Energieinnovation, EnInnov 2016, Graz, Austria, 10–12 February 2016 (2016)

5. Bexten, T., et al.: Einfluss betrieblicher Flexibilitätsparameter einer Gasturbine auf Auslegung und Betrieb eines dezentralen Energiever- sorgungssystems [Influence of flexibility parameters of a gasturbine on design and operation fo a municipal energy supply system]. In: Kraftwerkstechnik 2016: Strategien, Anlagentechnik und Betrieb, Saxonia, pp. 541–552 (2016)

6. Bexten, T., et al.: Impact of dispatchable energy conversion and storage units on the electrical autarky of future municipal energy supply systems. In: Proceedings of the International Renewable Energy Storage Conference, IRES, Düsseldorf, Germany (2017)

7. Bexten, T., et al.: Techno-economic study of wind farm forecast error compensation by flexible heat-driven CHP units. In: Proceedings of ASME Turbo Expo 2017: Turbomachinery Technical Conference and Exposition GT2017, GT 2017-63555, Charlotte, NC, USA, 26–30 June 2017 (2017)

8. Boddington, A., Kermik, J., Ainsworth, T.: Interdisciplinary design in the college of arts and humanities at the university of Brighton. In: Banerjee, B., Ceri, S. (eds.) Creating Innovation Leaders. UI, pp. 239–254. Springer, Cham (2016). https://doi.org/10.1007/978-3-319-20520-5_15

9. Bouffard, F., Ortega-Vasquez, M.: The value of operational flexibility in power systems with significant wind power generation. In: Power and Energy Society General Meeting. IEEE (2001)

10. Brewer, G.D.: The challenges of interdisciplinarity. Policy Sci. **32**(4), 327–337 (1999)

11. Bronfman, N.C., Jiménez, R.B., Arévalo, P.C., Cifuentes, L.A.: Understanding social acceptance of electricity generation sources. Energy Policy **46**, 246–252 (2012)

12. Cavallaro, F., Ciraolo, L.: A multicriteria approach to evaluate wind energy plants on an Italian island. Energy Policy **33**(2), 235–244 (2005)

13. Cummings, J.N., Kiesler, S.: Collaborative research across disciplinary and organizational boundaries. Soc. Stud. Sci. **35**(5), 703–722 (2005)

14. Cummings, J.N., Kiesler, S.: Who collaborates successfully?: prior experience reduces collaboration barriers in distributed interdisciplinary research. In: Proceedings of the 2008 ACM Conference on Computer Supported Cooperative Work, pp. 437–446. ACM (2008)

15. Federal Ministry for Economic Affairs and Energy (BMWi): State Secretary Baake: "Citizens' energy" wins first onshore wind auction, 19 May 2017. https://www.bmwi.de/Redaktion/EN/Pressemitteilungen/2017/20170519-staatssekretaer-baake-buergerernergie-gro%C3%9Fe-gewinner-der-ersten-ausschreibungsrunde-wind-an-land.html. Accessed 10 July 2017

16. Devine-Wright, P., Batel, S.: Explaining public preferences for high voltage pylon designs: an empirical study of perceived fit in a rural landscape. Land Use Policy **31**, 640–649 (2013)

17. Drezek, K.M., Olsen, D., Borrego, M.: Crossing disciplinary borders: a new approach to preparing students for interdisciplinary research. In: Proceedings of the 38th ASEE/IEEE Frontiers in Education Conference, pp. F4F1–F4F6. IEEE (2008)

18. Gamboa, G., Munda, G.: The problem of windfarm location: a social multi-criteria evaluation framework. Energy Policy **35**(3), 1564–1583 (2007)

19. Grêt-Regamey, A., Celio, E., Klein, T.M., Hayek, U.W.: Understanding ecosystem services trade-offs with interactive procedural modeling for sustainable urban planning. Landsc. Urban Plan. **109**(1), 107–116 (2013)

20. Gross, C.: Community perspectives of wind energy in Australia: the application of a justice and community fairness framework to increase social acceptance. Energy Policy **35**(5), 2727–2736 (2007)

21. Hamann, T., Schaar, A.K., Calero Valdez, A., Ziefle, M.: Strategic knowledge management for interdisciplinary teams - overcoming barriers of interdisciplinary work via an online portal approach. In: Yamamoto, S. (ed.) HIMI 2016. LNCS, vol. 9735, pp. 402–413. Springer, Cham (2016). https://doi.org/10.1007/978-3-319-40397-7_38
22. Hirsh, R.F., Sovacool, B.K.: Wind turbines and invisible technology: unarticulated reasons for local opposition to wind energy. Technol. Cult. **54**(4), 705–734 (2013)
23. Holbrook, J.B.: What is interdisciplinary communication? Reflections on the very idea of disciplinary integration. Synthese **190**(11), 1865–1879 (2013)
24. Holttinen, H.: Impact of hourly wind power variations on the system operation in the Nordic countries. Wind Energy **8**(2), 197–218 (2005)
25. Huijts, N.M., Molin, E.J., Steg, L.: Psychological factors influencing sustainable energy technology acceptance: a review-based comprehensive framework. Renew. Sustain. Energy Rev. **16**(1), 525–531 (2012)
26. Jacobs, J.A., Frickel, S.: Interdisciplinarity: a critical assessment. Annu. Rev. Sociol. **35**, 43–65 (2009)
27. Johansson, M., Laike, T.: Intention to respond to local wind turbines: the role of attitudes and visual perception. Wind Energy **10**(5), 435–451 (2007)
28. Jungert, M.: Was zwischen wem und warum eigentlich? Grundsätzliche Fragen der Interdisziplinarität [What between whom and why? General questions of interdisciplinarity]. In: Interdisziplinarität, Theorie, Praxis, Probleme, pp. 1–12. Wissenschaftliche Buchgesellschaft, Bamberg (2011)
29. Krewitt, W., Nitsch, J.: The potential for electricity generation from on-shore wind energy under the constraints of nature conservation: a case study for two regions in Germany. Renew. Energy **28**(10), 1645–1655 (2003)
30. Krohn, W.: Interdisciplinary cases and disciplinary knowledge. In: The Oxford Handbook of Interdisciplinarity, pp. 31–49. Oxford University Press (2010)
31. Krueger, A.: Focus Groups: A Practical Guide for Applied Research. Sage, Thousand Oaks/London/New Delhi (1994)
32. Langer, K., Decker, T., Menrad, K.: Public participation in wind energy projects located in Germany: which form of participation is the key to acceptance? Renew. Energy **112**, 63–73 (2017)
33. Lattuca, L.R.: Learning interdisciplinarity: socio-cultural perspectives on academic work. J. High. Educ. **73**(6), 711–739 (2002)
34. Lélé, S., Norgaard, R.B.: Practicing interdisciplinarity. BioScience **55**(11), 967–975 (2005)
35. Liebe, U., Bartczak, A., Meyerhoff, J.: A turbine is not only a turbine: the role of social context and fairness characteristics for the local acceptance of wind power. Energy Policy **107**, 300–308 (2017)
36. Luce, R.D., Tukey, J.W.: Simultaneous conjoint measurement: a new type of fundamental measurement. J. Math. Psychol. **1**(1), 1–27 (1964)
37. Madlener, R.: Portfolio optimization of power generation assets. In: Zheng, Q., Rebennack, S., Pardalos, P., Pereira, M., Iliadis, N. (eds.) Handbook of CO_2 in Power Systems. ENERGY, pp. 275–296. Springer, Heidelberg (2012). https://doi.org/10.1007/978-3-642-27431-2_12
38. Marg, S., Hermann, C., Hambauer, V., Becke, A.B.: Wenn man was für die Natur machen will, stellt man da keine Masten hin [If you want to help nature, you don't build pylons in it]. In: Die neue Macht der Bürger: Was motiviert die Protestbewegungen? BP-Gesellschaftsstudie, pp. 94–138. Rowohlt (2013)
39. McNair, B.J., Bennett, J., Hensher, D.A., Rose, J.M.: Households' willingness to pay for overhead-to-underground conversion of electricity distribution networks. Energy Policy **39**(5), 2560–2567 (2011)

40. Raven, R.P., Mourik, R.M., Feenstra, C.F.J., Heiskanen, E.: Modulating societal acceptance in new energy projects: towards a toolkit methodology for project managers. Energy **34**(5), 564–574 (2009)
41. Rhoten, D., Pfirman, S.: Women in interdisciplinary science: exploring preferences and consequences. Res. Policy **36**(1), 56–75 (2007)
42. Schweizer, P.J., et al.: Public participation for infrastructure planning in the context of the German "Energiewende". Util. Policy **43**, 206–209 (2016)
43. Songsore, E., Buzzelli, M.: Social responses to wind energy development in Ontario: the influence of health risk perceptions and associated concerns. Energy Policy **69**, 285–296 (2014)
44. Wiedemann, P.M., Boerner, F., Claus, F.: How far is how far enough? Safety perception and acceptance of extra-high-voltage power lines in Germany. J. Risk Res. **21**(4), 463–479 (2018)
45. Wilson, G.: The world has problems while universities have disciplines: universities meeting the challenge of environment through interdisciplinary partnerships. J. World Univ. Forum **2**(2), 57–62 (2009)
46. Wolsink, M.: Wind power implementation: the nature of public attitudes: equity and fairness instead of 'backyard motives'. Renew. Sustain. Energy Rev. **11**(6), 1188–1207 (2007)
47. Wüstenhagen, R., Wolsink, M., Bürer, M.J.: Social acceptance of renewable energy innovation: an introduction to the concept. Energy Policy **35**(5), 2683–2691 (2007)
48. Zaunbrecher, B.S., Ziefle, M.: Integrating acceptance-relevant factors into wind power planning: a discussion. Sustain. Cities Soc. **27**, 307–314 (2016)
49. Zaunbrecher, B., Arning, K., Falke, T., Ziefle, M.: No pipes in my backyard? Preferences for local district heating network design in Germany. Energy Res. Soc. Sci. **14**, 90–101 (2016). https://doi.org/10.1016/j.erss.2016.01.008
50. Zaunbrecher, B.S., Bexten, T., Wirsum, M., Ziefle, M.: What is stored, why, and how? Mental models, knowledge, and public acceptance of hydrogen storage. Energy Proc. **99**, 108–119 (2016)
51. Zaunbrecher, B., Linzenich, A., Ziefle, M.: A mast is a mast is a mast...? Comparison of preferences for location-scenarios of electricity pylons and wind power plants using conjoint analysis. Energy Policy **105**, 429–439 (2017). https://doi.org/10.1016/j.enpol.2017.02.043
52. Zaunbrecher, B.S., Linzenich, A., Ziefle, M.: Experts and laypeople's evaluation of electricity storage facilities: implications for communication strategies. In: Proceedings of the International Renewable Energy Storage Conference, IRES, Düsseldorf, Germany (2017)
53. Zaunbrecher, B., Bexten, T., Specht, J.M., Wirsum, M., Madlener, R., Ziefle, M.: Using scenarios for interdisciplinary energy research. A process model. In: 6th International Conference on Smart Cities and Green ICT Systems, Smartgreens 2017, pp. 293–298. SCITEPRESS - Science and Technology Publications (2017). https://doi.org/10.5220/0006355702930298
54. Zoellner, J., Schweizer-Ries, P., Wemheuer, C.: Public acceptance of renewable energies: results from case studies in Germany. Energy Policy **36**(11), 4136–4141 (2008)

Integrated Electric Vehicle to Small-Scale Energy Management System

Muhammad Aziz$^{(\boxtimes)}$

Tokyo Institute of Technology, Tokyo 1528550, Japan
aziz.m.aa@m.titech.ac.jp

Abstract. The deployment of electric vehicle (EV) for transportation has received a massive intention due to its economic, environmental performance, and convenience. In addition, the controllable charging and discharging of EV lead to the potential of EV utilization to provide services to the electrical grid or energy management system (EMS). In this study, an integration of EV to support a small-scale EMS has been demonstrated and studied. This report covers the investigation of charging and discharging behavior of EV and the demonstration test of the developed integrated EV to small-scale EMS. Initially, charging behavior of EV under different ambient temperatures (seasons) were evaluated in order to clarify the impact of surrounding temperature to the charging rate. From the experimental test, it was found that charging in higher ambient temperature (during summer) results in a higher rate than charging in lower ambient temperature (during winter). Furthermore, the integration of EVs to small-scale EMS (such as office building) for peak-load shifting showed a very positive effect. Discharging of EVs during noon's peak load is able to cut and shift the peak load. Hence, high contracted capacity and large consumption of electricity with high price can be reduced leading to lower total electricity cost.

Keywords: Electric vehicle · Charging behavior · Grid integration
Peak-load cutting

1 Introduction

Electric vehicle (EV) has received an intensive attention due to its characteristics as vehicle having motor and battery. A massive deployment of EV, which is potential to replace the conventionally used internal combustion engine vehicle (ICEV), is considered potential to significantly increase the energy efficiency in transportation sector, and reduce both emission of greenhouse gases and consumption of fossil fuel [1]. Hence, lower environmental impacts can be realized. In addition, recent rapid development of EV was accelerated by some factors such as rising oil and gas prices, enhancement in battery technology, and policies related to environment and transportation [2, 3].

However, EV has also some challenging barriers in its dissemination due to high initial cost (high purchasing price), long charging time, and limited travelling distance. Although the running cost of EV is relatively lower than fossil fuels-based ICEV, the

B. Donnellan et al. (Eds.): SMARTGREENS 2017/VEHITS 2017, CCIS 921, pp. 94–107, 2019.
https://doi.org/10.1007/978-3-030-02907-4_5

manufacturing cost of EV is significantly higher [4, 5]. Hence, another value-added utilization of EV is essentially demanded to increase the economic performance of EV. Therefore, sustainable utilization of EV can be realized. However, a massive dissemination of EV is potential to give a significant impact to the grid, especially in case of massive uncontrolled charging and discharging. To minimize the impact of this problem, as well as increase the economic performance of EV, the concept of vehicle to grid (V2G) has been studied [6, 7].

EV utilization in supporting the grid has been evaluated by some researchers previously [8–12]. Kempton et al. [6, 8, 9] have proposed and performed theoretical study of V2G, and found that this kind of utilization is feasible both technically and economically. The integration of EV to grid (V2G) can be achieved due to the character of EV in both charging and discharging behaviors. As both of them can be fully controlled, scheduling and rate control become possible.

Being parked EVs can be considered as a potential battery as long as they are connected to the electrical grid or certain EMS. They are capable to absorb, store and deliver back the electricity from and to the grid responsively according to the given schedule and control value. V2G can be realized in case that minimally three essential requirements can be fully satisfied: (1) electricity connection between EV and grid, (2) communication which facilitates control flows between EV and electricity operator, and (3) accurate metering system which provides a fair measurement for the involved entities [13]. In V2G, operators of electrical grid or energy management system (EMS) are able to send a request to the parked and plugged EVs to discharge or absorb the electricity to and from the grid, respectively.

V2G leads to possibility of several ancillary services to grid such as load levelling, frequency regulation, spinning reserve, and electricity storage [14, 15]. The distributed EVs in large number and area are potential as massive energy storage that can be used to balance responsively the fluctuating supply such as PV and wind. In addition, because EVs are mobile, they can be used as energy carrier delivering the electricity from and to different places and time because of some factors including price difference and emergency condition. EV utilization in V2G is considered feasible because the estimated profit is higher than the current market price of EV batteries although with the consideration of the wear of battery [11].

On the other hand, charging and discharging of EVs are influenced by several factors including temperature. Therefore, measurement of the EV availability is not only limited to the available capacity for both charging and discharging, but also relates strongly to the possibility of both charging and discharging rates. Related to the latter, it is important to measure this kind of charging and discharging behaviors under different ambient temperature for the reference.

This paper reports mainly on two issues: (1) evaluation on charging behavior of EVs under different ambient temperature, and (2) impact of peak-load cutting and shifting in an integrated EVs and small-scale EMS based on demonstration test. The first study was conducted in different two seasons: summer and winter, while the second demonstration test was conducted for about 1 year of test.

2 Integration of EV to Grid

2.1 Involvement of EV in Community Energy Management System

Energy management is generally operated by an independent operator such as independent service operator (ISO) and regional transmission operator (RTO) in North America and ISO and independent transmission operator (ITO) in Europe. They are acting as the neutral organization which is responsible in monitoring, controlling, and coordinating the electric transmission throughout the region [16]. In case of Japan, a community EMS (CEMS) has been proposed previously with the main objective of integrating effectively the energy utilization, covering both demand and supply sides, in order to achieve highly efficient energy utilization and reduction of CO_2 emission. The motives behind the will to propose and adopt CEMS came from a strong intention to harmonize the energy services, minimize the environmental impacts, and maximize economic benefit throughout the community.

CEMS monitors and controls both demand and supply of energy, especially electricity, throughout the whole community. It has a responsibility to maintain the balance and harmonization throughout the community, therefore, the comfort, security, and safety of the community members can be improved. Furthermore, environmental parameters are also considered in parallel with the living quality.

Hence, in CEMS, both of information and energy are flowing simultaneously across the community. CEMS receives, processes, and manages the information and then delivers it to the community members according to their own function/position. Therefore, CEMS must be sufficiently robust and secure because it deals with personal and authentication information from the community.

Figure 1 shows the basic schematic structure of CEMS including electricity and information flows, especially the possible integration of EV inside CEMS. CEMS collects and manages the information from smaller EMS including building EMS (BEMS), house EMS (HEMS), and factory EMS (FEMS). In addition, CEMS also performs a communication with the outside of community and also upper energy supply such as electric utilities. Furthermore, CEMS predicts and maintains both energy supply and demand based on available previous historical data for certain time and forecasted weather information. Furthermore, CEMS also calculates and optimizes the energy balance with the objective of realizing the lowest energy cost throughout its community. In addition, in case of emergency such as disaster, CEMS evaluates and controls the energy conditions and communicates to its lower EMSs and negotiates with other CEMSs or electric utilities to cover its energy demand and recover the conditions.

On the other hand, charging and discharging of EVs are conducted through the EV charger. Currently, the chargers can be categorized into three main types according to its charging capacity: (a) slow charging (charging capacity is lower than 4 kW), (b) fast charging (charging capacity is in the range of 10–20 kW), and (c) ultrafast charging (the maximum charging capacity is about 50 kW or even higher). In addition, to facilitate charging and discharging to and from EV, respectively, bi-directional charger is demanded.

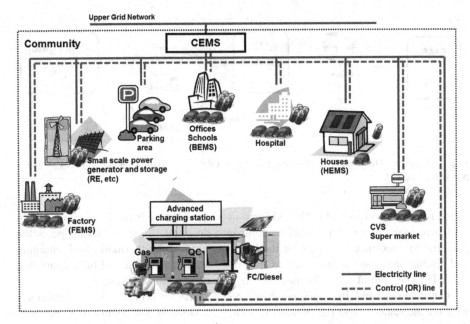

Fig. 1. Basic schematic structure of EV utilization in CEMS to provide ancillary services to the grid or certain EMSes [2].

2.2 Utilization of EV

Utilization of EVs to support the grid, including ancillary services and storage, is possible due to the characteristics of EV. In energy storage utilization, charging of EVs can be scheduled when the price of electricity drops because of electricity surplus in the grid (due to high supply and/or low demand). In addition, when the electricity price increases, electricity can be discharged from EVs and delivered back to the grid, leading to economic margin for the owner of EVs. The ancillary services from EV to grid includes frequency regulation (both up and down) and spinning reserve. Ancillary services are important to maintain the quality of the electricity. As the responses of EV in both charging and discharging are very fast, the ancillary service by EV is considered very feasible and it is potential to replace the current reserves such as thermal generators and pumped hydro.

Figure 2 shows the schematic utilization of EVs for grid support. In general, there are two possible schemes in EV utilization for grid support: direct and aggregator-based schemes. The collection of real time data in a certain interval from EVs includes battery state of charge (SOC), EVs position (GPS data) and predicted arrival time. This data collection is performed and managed further by a vehicle information system (VIS). Furthermore, in the real application, VIS can be owned and operated directly by EMS, aggregator or independent service operator.

In direct scheme, EV owners have the service contract directly with electricity-related entities, especially a small-scale EMS. In this scheme, both the electricity and information are handled privately and directly among two parties. This scheme is well

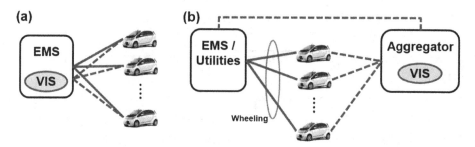

Fig. 2. Possible schemes in EV utilization to support the grid: (a) direct scheme, (b) aggregator-based scheme [2].

suited for a relatively small-scale EMS and in where EVs are parked and connected in relatively long time (such as office, company, hospital). Its main advantage is the potential to maximize the profit of the involved entities. Furthermore, both charging and discharging controls are easier as EVs are directly connected and fully controlled by EMS. The current study focuses on this type of utilization scheme.

On the other hand, in aggregator-based contract, EV owners have service contracts with the aggregator. The information including its position and battery SOC are handled by aggregator via VIS. Aggregator negotiates for electricity business with the electricity-related entities (EMS or electricity utilities). This kind of utilization scheme is prevalent for relatively large-scale EMS or electricity utilities. EVs may be distributed in different location, such as charging stations and parking areas. The electricity to and from EVs may be transferred via power wheeling system through the available grids. Aggregator offers some possible ancillary services to the EV owners. In turn, EV owners can choose them and receive their profit payment from aggregator.

Load levelling correlates strongly with the management of both demand and supply of electricity. Its aim is lowering the total power consumption in peak hours by shifting the load from peak to off-peak hours. Load levelling can be performed through peak-load shifting and peak-load cutting. The former is defined as moving the electricity load during peak time to off-peak time. It could be achieved through utilization of stationary battery or other storages. The latter deals with the effort to reduce the electricity purchased from the grid by generating or purchasing the electricity. In reality, it can be performed through harvesting the energy especially during peak hours, such as RE, or by purchasing the electricity from other entities including EVs. In this case, EVs are considered as energy storage and carrier storing and transporting the electricity from different time and place. Therefore, the economic performance of EV can be increased by joining this kind of ancillary program.

3 Charging Behavior of EV Under Different Ambient Temperatures

In general, EVs adopt li-ion batteries to store the electricity as power source due to their high energy density, stable electrochemical properties, longer lifetime, and low environmental impacts [17]. Temperature is considered as one factors influencing charging and discharging behaviors of li-ion batteries. Generally, lower temperature leads to poor charging and discharging performance because of electrolyte limitation [18] and changes in electrolyte/electrode interface properties including viscosity, density, electrolyte components, dielectric strength, and ion diffusion capability [19]. Liao et al. [20] found that as the temperature decreases, the charge transfer resistance increases significantly, higher than bulk resistance and solid-state interface resistance.

Unfortunately, lack of study deals with the effort to clarify the charging behaviors in different temperature or season. In this study, to clarify the effect of temperature (ambient temperature), to the charging behavior of EV, charging in different seasons: winter and summer, were conducted initially.

Table 1 shows the specifications of both EV and quick charger used for evaluating the charging rate of EV under different ambient temperature. Charging test was conducted in both summer and winter with ambient temperatures of about 30 and 10 °C, respectively.

Table 1. Specifications of EV and quick charger used for evaluating the charging rate of EV under different ambient temperature.

Component	Property	Value
EV	Car type	Nissan Leaf
	Battery type	Laminated lithium-ion battery
	Total battery capacity	24 kWh
	Maximum and nominal voltages	403.2 and 360 V
	Cell rated capacity	33.1 Ah (0.3 C)
	Cell average voltage	3.8 V
	Cell maximum voltage	4.2 V
DC quick charger	Standard	CHAdeMO
	Output voltage	DC 50–500 V
	Output current	0–125 A
	Rate power output	50 kW

Figure 3 shows the relation among charging rate, charging time, and SOC of EV battery both in winter (a) and summer (b). Generally, although the rated capacity of the charger is 50 kW, the charging power absorbed by EV battery is relatively lower, especially in winter. Compared to charging in winter, charging in summer leads to higher charging rate and shorter charging time. Numerically, to reach SOC of 80%, the required charging times in winter and summer were 35 and 20 min, respectively. In summer, higher charging rate (about 40 kW) could be achieved up to SOC of 50%. It decreased gradually following the increase of SOC and it showed the charging rate of 16 kW when SOC reached 80%. On the other hand, in winter, the charging rate reached about 35 kW instantaneously in very short time and decreased following the increase of SOC. The charging rate when SOC reached 80% was about 10 kW.

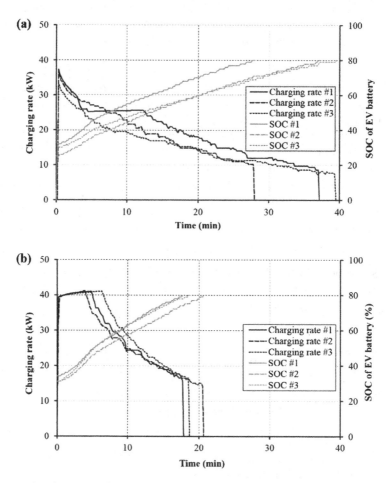

Fig. 3. Results of charging performance tests of EV in different seasons: (a) winter, (b) summer [21].

Figure 4 shows the correlation among charging current, voltage, and time in winter and summer, corresponding to the charging rate in Fig. 3. It can be observed that the curves of the charging current in Fig. 4 are nearly similar to those of the charging rate in Fig. 3. Lithium-ion batteries are charged in general using a constant current (CC)–constant voltage (CV) method [22]. Charging at lower temperature led to a gradual decrease in the charging current, especially at higher SOC, leading to longer charging time, and vice versa [23]. In case of summer, a higher CC of about 105 A could be achieved in the initial 5–10 min of charging (SOC up to about 50%). Although there was no significant difference in charging voltage, charging in a warmer temperature (summer) resulted in a slightly higher initial charging voltage before settling at a constant value. Hence, the CV condition could be achieved faster [24].

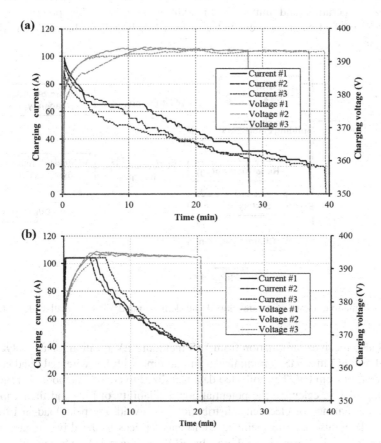

Fig. 4. Correlation among charging current, voltage and time in different seasons: (a) winter, (b) summer [21].

4 V2G Demonstration in Small-Scale EMS

Figure 5 shows the schematic diagram of EV integration test to support the electricity in a small-scale EMS. EMS is controlling all the electricity demand and supply throughout the certain building or area. EMS requests, manages, and integrates several information including electricity load, weather information from meteorological agency, EV information from vehicle information system (VIS), and electricity condition from CEMS and utilities. The meteorological agency sends periodically the weather information to EMS. Therefore, this information is used by EMS to calculate the electricity load and the possibly generated electricity from RE. In general, building load can be classified into base and fluctuating loads. The base load or fixed load is the minimum demand which is consumed by the system to operate for 24 h continuously. Hence, it is almost constant throughout the day and insignificantly affected by the weather condition. In addition, the fluctuating load depends strongly on the behavior of the residents. Human behavior is strongly influenced by the weather condition including temperature and humidity. In addition, the generable power from RE, including wind and solar, is predicted also by EMS based on the received weather information.

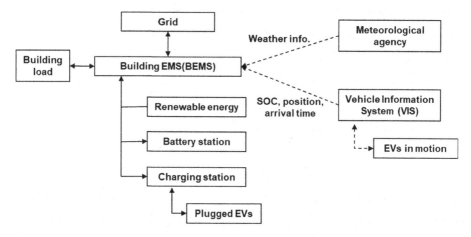

Fig. 5. Schematic diagram of the developed and demonstrated V2G in small-scale EMS [7].

EMS also receives information from VIS including EV position, battery SOC, and estimated arrival time. VIS communicates in real time with EV wirelessly and collects the information from EV. The collected data is utilized to coordinate both charging and discharging of EV, calculate the potential and availability of EVs and their capacity, and keep the balance of electricity distribution and avoid any peak-load in EMS. In addition, VIS is also able to facilitate additional services to the driver regarding the availability of ancillary services offered by EMS or aggregator. On the other hand, EMS also requests from electric utility the electricity condition and price information. These will be used to calculate the demand as well as the charging and discharging behaviors of both EVs and used EV batteries.

4.1   Developed V2G Concept

Figure 6 shows the concept of peak-load cutting and shifting (load leveling) developed for a small-scale EMS. It consists of four main subsequent steps: (1) forecast of load and power from RE, (2) forecast of the amount of load leveling, (3) correction of calculated value, and (4) charging and discharging controls of EVs. Forecasting of load and power from RE is basically conducted for 24 h-ahead. In addition, the load and control duration is set 30 min.

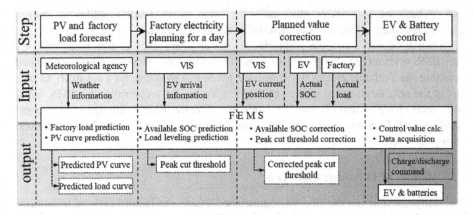

Fig. 6. Concept applied in load-levelling during the V2G demonstration [7].

During forecasting of load and generable electricity from RE, EMS initially requests the weather information from meteorological agency to calculate the generable electricity from RE including PV and wind. In this study, the generated electricity from RE is completely consumed, without being stored, and used for peak-load cutting. The weather information is also utilized to predict the fluctuating load mainly due to human behavior inside the building, especially related to air conditioning and lighting. The outputs from the first step are the forecasted RE generation and load curves for the next 24 h ahead.

Once both of the load and generated electricity from RE have been forecasted, EMS calculates the possible load leveling which can be achieved in the next 24 h. For this purpose, EMS communicates with VIS in estimating the number of EVs and their SOC states which are available to join the program. VIS initially receives the travelling schedule from the drivers and, subsequently, VIS transfers this information to EMS including EV's ID, planned departure time and estimated arrival time. Furthermore, the registration of travelling schedule by the driver should be conducted up to 24 h before the departure. As the available resources for peak-load cutting and shifting and load curves have been estimated, EMS is able to calculate the peak-load cutting threshold for the next day. Peak-load cutting threshold is defined as the maximum amount of electricity which is purchased from the grid by EMS. It is theoretically calculated based on some factors including electricity price, contracted capacity, available power generation, and storage. When the electricity consumed by the building increases and the purchased electricity from the grid reaches the peak-load cutting threshold, EMS sends the control command to EVs and used batteries to discharge their electricity. Therefore, the electricity purchased from the grid can be controlled/maintained to be the same to of lower than the calculated peak-load cutting threshold. In the real application, it may avoid a higher price of electricity during peak time.

When EVs are in motion, the information of EVs is transmitted to VIS. VIS sends the data to EMS which is used to recalculate the available electricity from EV. EMS also additionally recalculates the building load based on the real weather and the real load of building to achieve more accurate value. Next, EMS modifies its energy management plan, especially the peak-load cutting threshold.

When EVs arrive and are connected to the chargers, they communicate directly with EMS without transmitting it to VIS. Therefore, all the information is updated including the available electricity from EVs. From this moment, EVs are ready to take part in the ancillary service offered by EMS. EVs are fully controlled by EMS, especially their charging and discharging behaviors. Finally, EMS sends the control command to each EVs and used batteries to achieve its previously calculated peak-load cutting threshold.

4.2   Results of Demonstration Test

The demonstration test facility was constructed inside the factory area of Mitsubishi Motors Corporation in Okazaki, Japan. Table 2 shows the specifications of the developed demonstration test bed. It was connected and utilized to support the electricity of the main office building. As RE generator, 20 kW PV panels were installed on the roof top of the test bed. Five EVs, Mitsubishi i-Miev G, were taking part in the demonstration test and the drivers were also the employee. Therefore, EVs were mostly parked and plugged to the charging poles during working hours (from 09:00 to 17:00). In addition, five used EV batteries, used for about 1 year from the same type of EV, were also installed and directly owned by EMS. EMS are managing all the demand and supply sides of the test bed. On the other hand, VIS was also developed as a standalone system to facilitate a communication with EVs when EVs are in motion or unconnected to EMS and transfer the information to EMS.

Table 2. Specification of the demonstration test bed.

Component	Property	Value
EV and used batteries	Car type	i-Miev G
	Number of each EV and used battery	5
	Battery capacity	16 kWh
	Maximum charging capacity	DC 370 V, 15 A
	PCS capacity for each component	AC 200 V, 15 A
	Used battery condition	After 1 year of usage
PV	Type	Mono crystalline
	Capacity	20 kW
	Installation direction and angle	South, 30°
	PCS capacity	AC 200 V, 100 A

Used EV batteries were practically used for peak-load shifting. These batteries were charged during the night time (off-peak hours) when the price of electricity was relatively cheaper. In this study, charging for used EV batteries was conducted from 00:00 to 06:00. The SOC thresholds during charging and discharging of both EV and used EV battery were fixed at 90% and 40%, respectively. In addition, as the total capacity amount of EVs and used batteries was significantly smaller than the total load of the office building, peak-load cutting was designed to start from 12:00 until 18:00 (targeting mainly on the afternoon peak).

Figure 7 shows the results of load leveling test in a representative weekday consisting of the total grid load, building load, generated power from PV, and charged/discharged electricity to and from EVs and used EV batteries, respectively. The grid load is defined as the net electricity purchased from the electrical grid. Light green blocks in positive and negative sides represent discharge and charged electricity amount from and to EVs and used EV batteries, respectively. The main objective of peak-load cutting and shifting is reducing the grid load, especially during peak-load hours, in order to reduce the total electricity cost of the building.

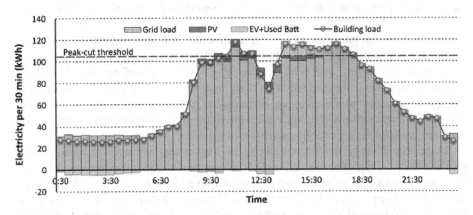

Fig. 7. Result of load leveling in a representative weekday [7]. (Color figure online)

PV generated the electricity during a day and it was consumed by the office building without being stored. The used EV batteries were charged during the night with a lower electricity price, starting from 00:00 to 06:00. As the result, the grid load during the night slightly increased. In the morning, around 08:00, EVs reached the office and they were plugged in the designated charging poles. From this time, both charging and discharging behaviors were fully controlled by EMS, as well as the information flows. Because the building load was smaller than the calculated peak-load cutting threshold, charging for EVs was performed until the building load reaches nearly the peak-load cutting threshold. Additional charging started again during noon break (12:00 to 13:00) because the building load dropped drastically.

Peak-load takes place twice in a weekday: before noon and afternoon peak-loads. Before noon peak load is lower than one in afternoon. The afternoon peak usually starts from 13:00 following the end of the noon break. As the peak-load is increasing significantly and it becomes higher than the calculated peak-load cutting threshold, EMS sends immediately the control command to both EVs and used EV batteries to discharge their electricity supporting the electricity of office building. As the results, the grid load can be reduced and maintained to be the same or lower than the peak-load cutting threshold. Unfortunately, as the total capacity of EVs and used EV batteries is significantly smaller compared to the total building load, peak-load cutting can only be achieved in a relatively short duration. It is believed that if the number of EVs participating in the load leveling program increases, the effect of load leveling becomes more significant. Hence, longer peak-load cutting and lower peak-load cutting threshold can be realized.

5 Conclusions

EV is believed able to give significant roles in our future transportation and energy systems, especially electrical grid. Integration of EVs with certain EMS, including small (HEMS, BEMS, FEMS) and large scales (EMS), is highly potential to improve the economic performance of EV, as well as supporting the electrical grid to balance both supply and demand.

In this study, an integration of EV to small-scale EMS was evaluated and demonstrated in order to measure the feasibility of ancillary services provided by EVs, especially peak-load cutting and peak-load leveling. Two stages of research have been conducted. The first is the evaluation on charging behavior of EV to clarify the effect of ambient temperature to the charging performance of EV. From the experiments conducted in different seasons of summer and winter, charging in higher ambient temperature (summer) showed higher charging rate than one in lower ambient temperature (winter).

In the second study, demonstration of peak-load cutting and leveling by integrating EVs and used EV batteries to small-scale EMS was performed and evaluated. Although the achieved capacity of peak-load cutting and leveling was relatively small, it was shown that the application of EVs and used EV batteries to support EMS is feasible and promising. Addition of the number of EVs and used EV batteries is believed able to increase the amount of peak-load cutting and leveling. Therefore, lower contracted capacity and lower electricity cost can be achieved.

Acknowledgments. The author expresses his deep thanks to Mitsubishi Corporation and Mitsubishi Motors Corporation for both financial support and cooperation during demonstration test.

References

1. Oda, T., Aziz, M., Mitani, T., Watanabe, Y., Kashiwagi, T.: Number of quick charging stations and their quick chargers for electric vehicles—a case study considering the traffic density of the road type. J. Jpn. Soc. Energy Resour. **37**(6), 7–12 (2016). (in Japanese)
2. Aziz, M., Oda, T., Mitani, T., Watanabe, Y., Kashiwagi, T.: Utilization of electric vehicles and their used batteries for peak-load shifting. Energies **8**(5), 3720–3738 (2015)
3. Oda, T., Aziz, M., Mitani, T., Watanabe, Y., Kashiwagi, T.: Actual congestion and effect of charger addition in the quick charger station—case study based on the records of expressway. IEEJ Trans. Power Energy **136**(2), 198–204 (2017)
4. Thiel, C., Perujo, A., Mercier, A.: Cost and CO_2 aspects of future vehicle options in Europe under new energy policy scenarios. Energy Policy **38**, 7142–7151 (2010)
5. Oda, T., Aziz, M., Mitani, T., Watanabe, Y., Kashiwagi, T.: Mitigation of congestion related to quick charging of electric vehicles based on waiting time and cost–benefit analyses: a Japanese case study. Sustain. Cities Soc. **36**, 99–106 (2018)

6. Kempton, W., Letendre, S.E.: Electric vehicles as a new power source for electric utilities. Transp. Res. Part D **2**, 157–175 (1997)
7. Aziz, M.: Charging/discharging behaviors and integration of electric vehicle to small-scale energy management system. In: Proceedings of the 6th International Conference on Smart Cities and Green ICT Systems - Volume 1: SMARTGREENS, pp. 346–351. SCITEPRESS, Porto (2017)
8. Kempton, W., Kubo, T.: Electric-drive vehicles for peak power in Japan. Energy Policy **28**, 9–18 (2000)
9. Tomic, J., Kempton, W.: Using fleets of electric-drive vehicles for grid support. J. Power Sources **168**, 459–468 (2007)
10. White, C.D., Zhang, K.M.: Using vehicle-to-grid technology for frequency regulation and peak-load reduction. J. Power Sources **196**, 3972–3980 (2011)
11. Aziz, M., Oda, T., Morihara, A., Murakami, T., Momose, N.: Utilization of EVs and their used batteries in factory load leveling. In: Proceedings of 2014 IEEE PES Innovative Smart Grid Technologies Conference (ISGT), pp. 1–5. IEEE, Washington (2014)
12. Aziz, M., Oda, T., Kashiwagi, T.: Extended utilization of electric vehicles and their re-used batteries to support the building energy management system. Energy Procedia **75**, 1938–1943 (2015)
13. Drude, L., Pereira, L.P., Ruther, R.: Photovoltaics (PV) and electric vehicle-to-grid (V2G) strategies for peak demand reduction in urban regions in Brail in a smart grid environment. Renew. Energy **68**, 443–451 (2014)
14. Kempton, W., Tomic, J.: Vehicle-to-grid power fundamentals: calculating capacity and net revenue. J. Power Sources **144**, 268–279 (2005)
15. Gao, B., Zhang, W., Tang, Y., Hu, M., Zhu, M., Zhan, H.: Game-theoretic energy management for residential users with dischargeable plug-in electric vehicles. Energies **7**, 7499–7518 (2014)
16. ISO/RTO Council, www.isorto.org/Pages/Home. Accessed 18 Aug 2017
17. Aziz, M., Oda, T.: Load levelling utilizing electric vehicles and their used batteries. In: Modeling and Simulation for Electric Vehicle Applications, pp. 125–147. InTech, Rijeka (2016)
18. Xiao, L.F., Cao, Y.L., Ai, X.P., Yang, H.X.: Optimization of EC-based multi-solvent electrolytes for low temperature applications of lithium-ion batteries. Electrochim. Acta **49**, 4857–4863 (2004)
19. Jansen, A.N., Dees, D.W., Abraham, D.P., Amine, K., Henriksen, G.L.: Low-temperature study of lithium-ion cells using a LiySn micro-reference electrode. J. Power Sources **174**, 373–379 (2007)
20. Liao, L., et al.: Effects of temperature on charge/discharge behaviors of $LiFePO_4$ cathode for Li-ion batteries. Electrochim. Acta **60**, 269–273 (2012)
21. Aziz, M., Oda, T., Ito, M.: Battery-assisted charging system for simultaneous charging of electric vehicles. Energy **100**, 82–90 (2016)
22. Seyama, Y., Shimozono, T., Nishiyama, K., Nakamaura, H., Sonoda, T.: Development of large-scale lithium ion batteries "LIM series" for industrial applications. GS News Tech. Rep. **62**, 76–81 (2003)
23. Aziz, M.: Advanced charging system for plug-in hybrid electric vehicles and battery electric vehicles. In: Hybrid Electric Vehicle, pp. 63–81. InTech, Rijeka (2017)
24. Aziz, M., Oda, T.: Simultaneous charging of electric vehicles using battery-assisted charger. In: Pistoia, G., Liaw, B. (eds.) Behaviour of Lithium-Ion Batteries in Electric Vehicles. Green Energy and Technology, pp. 201–224. Springer, Cham (2018). https://doi.org/10.1007/978-3-319-69950-9_9

Smart Cities

An Influence-Based Model for Smart City's Interdependent Infrastructures: Application in Pricing Design for EV Charging Infrastructures

Upama Nakarmi[1,2] and Mahshid Rahnamay-Naeini[1,2(✉)]

[1] Electrical Engineering Department, University of South Florida, Tampa, FL, USA
mahshidr@usf.edu
[2] Computer Science Department, Texas Tech University, Lubbock, TX, USA

Abstract. While smart critical infrastructures are principle components of smart cities, isolated and individually optimized infrastructures will not necessarily realize the vision of smart city in providing efficient solutions and services to citizens. This is mainly due to the interdependencies among critical infrastructures, which suggest the need for collaborative solutions and synergistic modeling and analysis of these systems. In this paper, an integrated framework based on influence model, a networked Markov chain framework, is proposed for modeling interdependent infrastructures and capturing their interactions based on the rules and policies governing their internal and interaction dynamics. To demonstrate the benefits of synergistic approaches, in this paper, the interdependencies between transportation networks and power infrastructures, through electric-vehicle (EV) charging infrastructures, are considered. Particularly, the proposed integrated framework is used to design an algorithm for assigning dynamic charging prices for the EV charging infrastructure with the goal of increasing the likelihood of having balanced charging and electric infrastructures. The proposed scheme for charging prices is traffic and power aware as the states and interactions of transportation and power infrastructures are captured in the integrated framework. Finally, the cyber infrastructure plays a critical role in enabling such collaborative solutions and their role is also discussed.

Keywords: Smart city · Integrated framework
Synergistic approach · Stochastic modeling · Electric vehicles
Charging infrastructure · Power grid · Interdependency · Pricing
Load distribution

1 Introduction and Motivation

Dependable, resilient, and efficient critical infrastructures are the center piece of sustainable societies. Critical infrastructures such as power, energy (gas and oil), communication, transportation, emergency services, water and food supply

© Springer Nature Switzerland AG 2019
B. Donnellan et al. (Eds.): SMARTGREENS 2017/VEHITS 2017, CCIS 921, pp. 111–130, 2019.
https://doi.org/10.1007/978-3-030-02907-4_6

are interdependent systems due to services and influences they receive from one another. The complex mesh of interdependencies among these infrastructures could be both problematic, by introducing vulnerabilities, and beneficial, by introducing opportunities to more effective operation of these systems.

As the urban population is expected to grow by 72% by 2050, according to recent studies [14], it is essential to develop smart-city solutions and services, which lead to optimal utilization of cities' limited resources and their reliability and sustainablity enhancement. The latter requires collaboration of currently vertical and isolated city infrastructures. However, the general mechanism of interactions among infrastructures and the opportunities of exploiting the inter-dependencies to enhance the design and operation of critical infrastructures are yet to be explored and understood. In this paper, we propose a synergistic app-roach for modeling and analyzing interdependent infrastructures. We specifically propose an integrated mathematical framework based on influence framework [2], which is a mathematically tractable probabilistic framework based on a network of Markov chains. The integrated framework allows for modeling interdependent infrastructures and capturing their interactions based on the rules and policies governing their internal and interaction dynamics.

We apply the proposed synergistic approach to power (electricity) and trans-portation critical infrastructures and discuss how their cooperation can lead to a more reliable operation of the power system and improve certain aspects of transportation systems through one source of their interdependency: the electric vehicles (EVs) charging infrastructure. Particularly, the increase in the num-ber of hybrid electric transportation systems, including plugin hybrid EVs and hybrid electric trains have introduced new interdependencies between the power and transportation infrastructures [4,12,13,17,20,26,28]. For instance, vehicle-to-grid (V2G) technology allow EVs to discharge their energy to the power grid using bi-directional power electronic DC/AC interfaces, which can help in sta-bilizing the power grid during disturbance and power shortage [21,24]. Another source of interdependency between power and transportation infrastructures comes from the EV charging infrastructure. The EV charging infrastructures are emerging in cities [33] similar to the traditional gas stations. On one hand, in the charging infrastructure, traffic patterns and population distribution can affect the power demand in the electric grid at various times and locations. On the other hand, the demand on the power grid can affect the charging price and consequently affect the traffic pattern in the transportation system. Such interde-pendencies are important as, for instance, during the peak-energy-consumption hours, inappropriate energy pricing signals at charging stations that motivate EV users to use specific charging stations, along with other factors, can lead to power demand profiles that result in instability of the electric grid and in worse cases power outages [37]. As such, it is essential to design and operate these charging infrastructures while considering the interdependency between power and transportation systems and the state of these systems. In particular, design-ing pricing incentives can provide a controlling mechanism for interdependency and reliable operation of these systems. The incentives will be communicated to the users through the cyber infrastructure. In general, the cyber infrastructure plays a key role in enabling such collaboration and cooperation among infrastructures, while also explicitly benefit from the reliable power system as the source of electricity.

During the operation of infrastructures, each component evolves through various states. The dynamics of components are governed by processes and the collective interaction of the processes among and across infrastructures, which define the behavior of the whole system. An important aspect of such dynamics is that although they are based on engineered processes, they are stochastic due to various random events that can affect the dynamics. For instance, in the power and transportation systems the load on different components of the system is stochastic. To address this aspect, the proposed integrated framework in this paper is built on an abstract probabilistic framework, which can also capture the role of structure of the systems on the dynamics to model the dynamics and specifically, the demand and traffic distribution in EV charging infrastructures. The goal of the model applied to EV charging infrastructures is to identify incentives, when and where they are needed, to design dynamic energy pricing signals based on the state of both the power and transportation systems, such that the incentives help in appropriate distribution of load in both systems and orchestrating their operation. Based on the proposed integrated model, we identify incentives in terms of charging prices using an algorithm based on *topological sort* on the active influence graph of the charging infrastructure. We will show that the identified incentives based on this model lead to higher probabilities of stable and balanced systems both for the power and transportation systems. This work extends our analyses presented in [23] by capturing the more detailed dynamics of the power systems. This work is an effort toward promoting the need and importance of integrated frameworks for modeling and analysis of smart cities.

2 Related Work

We review the related work in two main categories. First, we briefly review recent efforts on modeling, simulation, operation and design of integrated and interdependent infrastructure frameworks for smart cities. Second, as the focus of this paper is on charging infrastructures, we review the related work on different aspects of design, operation and optimization of charging infrastructures.

2.1 Integrated and Interdependent Infrastructures for Smart Cities

Critical infrastructures and particularly their reliability have been the focus of many recent research efforts. The role of integrated, interdependent infrastructures in designing efficient smart cities has been emphasized by smart-city research community [7]. Moreover, the vision of smart cities has been described in different ways among practitioners and academia [7]. Hall [5] visions the smart city as a city that monitors and integrates conditions of all of its critical infrastructures to optimize its resources and services to its citizens. Similar smart city visions has been described in [8,11]. In the last decade a large body of work has emerged in modeling and understanding interdependent infrastructures. The general concepts of interdependencies among critical infrastructures, challenges in modeling interdependent systems and their control and recovery mechanisms have been intensively discussed in [1,19,22,29]. These works mainly discuss the intrinsic difficulties in modeling interdependent systems and suggest new methodologies for their modeling and simulation as a single coupled system.

The majority of such integrated, theoretical frameworks has been focused on analyzing the reliability of coupled systems and the negative aspects of the interdependencies among critical infrastructures [9,30,31]. For instance, one of the problems of concern in interdependent infrastructures is their reliability to cascading failures and propagation of faults. Recently, many researchers have studied cascading failures in interdependent systems (see for example, [27,30] and references therein).

The work presented in the current paper is an effort to present an abstract and unified framework to model interactions among infrastructures, which can be used to design various smart-city solutions based on the state of interacting systems, for instance, the pricing mechanism based on the state of the EV charging infrastructure and the electric grid.

2.2   Charging Infrastructures

In recent years, a large body of work is focused on optimal placement of EV charging stations [6,10,13,15,18,34,36,37,40]. In particular, optimization formulations with various criteria have been used for addressing this problem [10,15,18,37]. Examples of such criteria include, maximizing sustainability from the environment, economics and society perspective [10], minimizing the trip time of EVs to access charging stations [15], minimizing trip and queuing time [18], and maximizing the coverage of charging stations [37]. In the work presented in [6,36], the set cover algorithm is used to optimize the location of charging stations from a set of possible locations. In addition, agent-based [34] and game-theoretic approaches [13,40] have also been adopted in characterizing optimal deployment of charging infrastructures. Reference [16] presents a more detailed review of various approaches used for the optimal deployment of EV charging stations.

Another research aspect of charging infrastructures is their pricing mechanisms. Studies of traditional fueling infrastructures [38,39] show that the price of fuel impact the behavior of drivers, which suggests that the charging price for EVs can also impact the users' choice and behavior. Specifically, authors in [41], discuss that the optimal placement of charging stations will be insufficient to handle rapid changes in traffic patterns and urbanization, hence an efficient pricing model that also minimize the social cost of traffic congestion and congestion at EV charging stations is needed. As another example, the impact of energy price and the interplay between the price and other factors, such as cost and emissions, on the charging decisions have been studied in [33]. Besides the studies on the impact of price on charging decisions and traffic patterns, some efforts are focused on designing and optimizing pricing and analyzing their impact on the users' behavior and the system operation. Examples of such efforts include the work presented in [17], which uses a game theoretical approach to study the price competition among EV charging stations with renewable power generators and also discusses the benefits of having renewable resources at charging stations. Similarly, game-theoretic approaches that model a game between the electric grid and their users, specifically for EV charging, in order to design pricing schemes, have been studied, for example in [35]. The model in [35] provides strategies to EV chargers to choose the amount of energy to buy based on a pricing scheme to operate the charging infrastructures at their optimal levels.

The work presented in this paper is closest to the studies on pricing mechanism design and also the interplay between the electric and EV charging infrastructures. At the same time, it is different in the approach as it considers the stochastic dynamics of the interdependent EV charging infrastructures and the power grid and their interactions in designing the charging prices at stations.

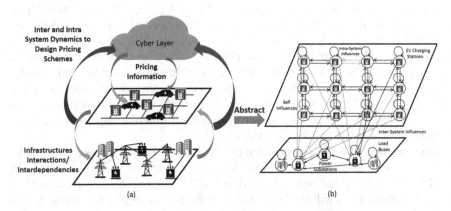

Fig. 1. Interdependent networks of power and transportation through charging infrastructure and the role of cyber infrastructure.

3   System Model for Interdependent EV Charging and Power Infrastructures

In this section, we describe our system model for the interdependent EV charging and power infrastructures; however, the model is adequately general to be applied to any interdependent infrastructure with interacting components. The schematics of the system under study is depicted in Fig. 1. As the figure shows, our study considers three layers in the system: (1) the power/electric grid layer, (2) the EV charging infrastructure layer, and (3) the cyber layer, which enables the collaborative solution for the pricing between layer 1 and 2. Our modeling is mainly focused on the power and the EV charging infrastructures. While the cyber layer is not a part of the theoretical model, we will discuss its key role in Sect. 4. The key interactions among the layers of this system can be summarized as following. The EV charging infrastructure receives energy from the power grid and thus the load on charging stations may affect the load on power substations. The pricing scheme, which depends on the state of both power and EV charging infrastructures, will be communicated through the cyber layer to the users. Finally, the communicated price will affect the load distribution over the charging infrastructure and subsequently the load on power substations. Note that the power infrastructure includes other types of loads (residential or commercial), which they can also affect the load and thus the pricing policy.

First, let us present the system model for the charging infrastructure. We denote the set of charging stations in a region in the smart city by $\mathcal{C} = \{C_1, C_2, ..., C_k\}$. For simplicity, we assume that the charging stations are distributed over a grid region such that each cell in the grid holds one charging station as shown in Fig. 1-a). The charging stations are connected over a directed graph $G = (\mathcal{C}, E_c)$, where E_c represents the set of directed links specifying the possibility of travel between charging stations for the users. For instance, $e_{ij} \in E_c$ implies that users in the cell containing the station C_i can travel to station C_j for charging. These links help specifying the constraints on the travel for charging, for instance, based on the distance that the users are willing to travel and the distance that a EV with the need for charging can travel before it runs out of energy. We will explain later that when the right incentives are applied, there is a likelihood for each user to travel to other stations with direct links (a probabilistic behavior). In this paper, we focus on a graph, in which charging stations in adjacent cells are connected. Other graphs with different topologies can also be considered and will not change the model.

Next, we describe the power infrastructure layer. In this paper, we consider the power grid loads (e.g., residential or commercial loads) denoted by $\mathcal{B} = \{B_1, B_2, ..., B_k\}$ and substations denoted by $\mathcal{S} = \{S_1, S_2, ..., S_m\}$ and their internal dynamics (as will be explained in Sect. 3.1) as well as the intra-system interactions (based on the physics of the electricity that can lead to propagation of voltage or current stresses and instabilities among substations). This represents a simplified model for a power system that is an extension to [23], which only considered individual substations with no interactions. Although the presented model for power system still does not capture the complete power grid model and dynamics with generators and power lines; it enables an abstraction for points of contact with the EV charging infrastructure and components which will directly impact it. The links among substations and loads are defined as following. The loads and substations are connected over a directed graph $G = (\mathcal{B} \cup \mathcal{S}, E_p)$, where link $e_{ij} \in E_p$ from load node to substation implies that the load is receiving energy from that substation and a $e_{ij} \in E_p$ between two substations imply that they can affect each others stress level and stability. In this model, we assume there will not be any link from substation to loads.

To model the inter-system interactions between the power and charging infrastructures, we assume that multiple charging stations belong to the distribution network of one substation, as such we consider a set of inter-system links denoted by $\mathcal{L}$, where $L_{ij} \in \mathcal{L}$ specifies that charging station C_i affect the load of substation S_j. In this model, $C_i \in \mathcal{C}$ should have a link to one specific $S_j \in \mathcal{S}$ while each S_j can have multiple incoming links from different geographically co-located charging station. Also, note that there will be no links from S_j to any node in $\mathcal{C}$. Such interactions and the effects of power substations on charging stations will be indirectly through the incentives communicated by the cyber layer. Based on the above discussion, the total integrated system can be denoted by a graph as $G_u = (\mathcal{C} \cup \mathcal{S} \cup \mathcal{B}, E_c \cup E_p \cup \mathcal{L})$. However, the model for the system is not simply a graph. Next, we will explain how each component in this graph stochastically and dynamically evolves and interacts with other components. We will specifically present a model to capture such dynamics. We have chosen a probabilistic approach for the modeling as various aspects of this system is stochastic. For instance, the state of a charging station (e.g., being busy or not) varies probabilistically at different times of the day and week and due to EV users mobility pattern and behavior. The state of the load in a substation also varies due to stochastic nature of the demand. The interactions among components are also stochastic and as components influence each other depending on their state. For instance, if charging stations, which have a link to substation S_j, become busy and overloaded with lots of demand then this increased demand will increase the likelihood of S_j to become overloaded and hinder the stability of the power grid. In such cases, we would like to distribute the load in the system using pricing incentives to increase the willingness of EV users to travel to other charging stations. These stochastic interactions and dynamics will be modeled in an influence framework as explained next.

3.1   Influence Model for Integrated Infrastructures

Here, we briefly review the Influence Model (IM) as first introduced in [2,3] and present an IM-based framework for modeling the integrated charging and power infrastructures. This review is borrowed from [23].

The IM is a framework consisting of a weighted and directed graph of interconnected nodes, in which, the internal stochastic dynamics of each node is represented by a Markov Chain (MC) and the states of the nodes varies in time due to the internal transitions of MCs as well as the external transitional influences from other nodes. The weights on the directed links represent the strength of influences that nodes receive from one another. In the following, we put the IM model in perspective with respect to the integrated charging and power infrastructures. In our model, graph G_u with different types of links and nodes (as introduced in Sect. 3) will serve as the underlying graph for the IM. To represent the internal dynamics of nodes, we consider that the state of the charging stations can be abstracted to three levels: (1) underloaded, (2) normal, and (3) overloaded. As such, we define a MC with state space of size three for each $C_i \in \mathcal{C}$. These states help describing the load (in terms of power demand) on a charging station at each time. In general, the state of a C_i may change due to departure or arrival of EV users. On the other hand, we model a substation S_j with an internal MC, which has two possible states: normal and stressed. These states specify if a power substation is overloaded and stressed or it is working under normal conditions. We model the load B_j with an internal MC, which has three possible states including underloaded, normal and overloaded, similar to the C_i states. The transition probability matrix of the internal MC for a node, say node $i \in \mathcal{C} \cup \mathcal{S} \cup \mathcal{B}$, is denoted by $\mathbf{A}_{ii}$, which is an $m \times m$ row stochastic matrix, where m is the size of the state space. We use a data driven approach to characterize the transition probabilities of these internal MCs based on datasets of system dynamics and simulations as will be explained later. The links in graph G_u specify the influence relation among the nodes. In particular, there are four types of influences in our model: (1) when a charging station influences another charging station, then it means there is a likelihood that it will send users (using proper incentives) to the influenced station; (2) when a charging station influences a power substation, then it means that there is a likelihood that the charging station increases the load on the power substation to a level that could change the state of the power substations (e.g., from normal to stressed); (3) when a substation influences another substation, then it means that there is a likelihood that the substation stresses the other substation if it is stressed or help a stressed substation to stabilize if it is normal; and (4) when a load node influences a substation then it means that it increases the load on the power substation to a level that could change the state of the power substations. The weights on the links also specify the strength of the influence. Specifically, the influences among the nodes of the network is captured by the *influence matrix* denoted by $\mathbf{D}$, where each element d_{ij} is a number between 0 and 1 representing the amount of influence that node i receives from node j. The larger the d_{ij} is the more influence the node i receives from node j; with the two extreme cases being $d_{ij} = 0$ meaning that node i does not receive any influence from node j and $d_{ij} = 1$ meaning that the next state of node i deterministically depends on the state of node j. Note that receiving influence from a node itself (self influence), i.e., d_{ii}, specifies how much the state evolution of a node depends on its internal MC. The total influence that a node receives should add up to unity i.e., $\sum_{j=1}^{n} d_{ij} = 1$, and therefore, matrix $\mathbf{D}$ is a row stochastic matrix too.

In IM, the status of a node, say node i, at time t is denoted by $s_i[t]$, a vector of length m, where m is the number of possible states for the node. At each time, all the elements of $s_i[t]$ are 0 except for the one which corresponds to the current state of the node (with value 1). In our model, $s_{i1}[t]$, $s_{i2}[t]$, and $s_{i3}[t]$ correspond to overloaded, normal and underloaded states, respectively, for charging stations and loads. Similarly, $s_{i1}[t]$ and $s_{i2}[t]$ correspond to normal and stressed states for power substations, respectively. The statuses of all the nodes concatenated together as $\mathbf{S}[t] = (s_1[t]s_2[t]...s_n[t])$ described the state of the whole system in time t, where $n = |\mathcal{C} \cup \mathcal{S} \cup \mathcal{B}|$ and $|.|$ denotes the cardinality of the set.

The influence matrix $\mathbf{D}$ specifies how much two nodes influence each other. In order to specify how the states of the nodes will change due to the influences, we also need state-transition matrices $\mathbf{A}_{ij}$, which capture the probabilities of transiting to various states due to the state of the influencing node. Matrix $\mathbf{A}_{ii}$ represents the special case of self-influence, which is described by the internal MC of the node. Note that the $\mathbf{A}_{ij}$ matrices are row stochastic. In the general IM [2], the collective influences among the nodes in the network is summarized in the total influence matrix $\mathbf{H}$ defined as:

$$\mathbf{H} = \mathbf{D}' \otimes \{\mathbf{A}_{ij}\} = \begin{pmatrix} d'_{11}A_{11} & \cdots & d'_{1n}A_{1n} \\ \vdots & \ddots & \vdots \\ d'_{n1}A_{n1} & \cdots & d'_{nn}A_{nn} \end{pmatrix}, \tag{1}$$

where $\mathbf{D}'$ is the transpose of the matrix $\mathbf{D}$ and $\otimes$ is the generalized Kronecker multiplication of matrices [2]. Finally, based on the total influence matrix $\mathbf{H}$ the evolution equation of the model is defined as

$$\mathbf{p}[t+1] = \mathbf{S}[t]\mathbf{H}, \tag{2}$$

where vector $\mathbf{p}[t+1]$ describes the probability of various states for all the nodes in the network in the next time step. Steady state analysis of IM has some similarities with that of MCs and has been discussed for various scenarios in [2,3]. For a more detailed discussion on the IM please refer to [2,3].

The work in [32] extends the original IM to a constraint or rule-based influence framework such that the influences among the nodes can dynamically get activated and deactivated depending on the state of the system. Also, as explained in [32], influencing can change the state of the influencer as well (transiting from overload to normal due to sending load to another station). [32] specifically defined a constraint matrix $\mathbf{C}$, where the entry c_{ij} for $i, j \in \mathcal{C} \cup \mathcal{S}$ is a binary variable specifying whether node i gets influenced by node j or not. In particular, $c_{ij} = 1$ indicates that node i gets influenced by node j and $c_{ij} = 0$ indicates otherwise. Moreover, each node always influences itself based on its internal MC (i.e., $c_{ii} = 1$ for all $i \in \mathcal{C} \cup \mathcal{S} \cup \mathcal{B}$). As explained in [32], one can define the value of c_{ij} according to boolean logic to capture the rules of interactions in the network. In other words, c_{ij}s are functions of the state of the nodes. For instance, when a charging station in the EV charging infrastructure is in overloaded state and based on G_u it has a link to another station, which is underloaded, the influence over that link should get activated to motivate the EV users to travel from the overloaded state to underload state. These types of rules can be specified using boolean functions such as the following examples. Function $c_{ij} = s_{i3}s_{j1} + s_{i2}s_{j1}$, where $i, j \in \mathcal{C}$ specifies the rules that can be applied to the transport layer of the model to show influences from charging station j to charging station i. Specifically, a transport node i will receive influence from transport node j if node i is underloaded and node j is overloaded or if node i is normal and node j is overloaded. Also, the power substations receive influences from the charging stations because overloaded charging stations can cause a power substation to go to overloaded state. Example of boolean function describing this rule is $c_{k\ell} = s_{k1} \prod_{j \in C_{S_k}} s_{ji} + s_{k2} \prod_{j \in C_{S_k}} s_{ji}$, where $k \in \mathcal{S}$

and $\ell \in \mathcal{C}$ and $C_{S_k} \subseteq \mathcal{C}$ is the set of charging stations connected to the power substation k. Specifically, a power substation, say k will receive influence from charging station ℓ when all the charging stations connected to the power substation are overloaded. As a power station is generally built with a capacity to accommodate large demand, the power substation will go to a stressed state provided that all the influence links connected to it are activated. This is just one example of influence rule and other conditions to specify the rules are also possible.

Note that as the goal of the integrated study of these two systems is to increase the probability of having power substations in normal conditions and charging stations not overloaded, the interaction rules defined in $\mathbf{C}$ should support this goal. In order to achieve this goal the influences among the charging stations should be engineered such that it forces the whole system toward desirable states. The second type of influence, which is from the charging station to power substations cannot be engineered and we assume that when the charging stations, which are receiving power from substations, are overloaded they influence (increase the likelihood) the substation to transit to a stressed state (similarly for the load buses influencing substations).

In [32], the constraint matrix $\mathbf{C}$ and the influence matrix $\mathbf{D}$ are used to define the constraint-based influence matrix denoted by $\mathbf{E}$, as

$$\mathbf{E} = \mathbf{D} \circ \mathbf{C} + \mathbf{I} \circ (\mathbf{D} \times (\mathbf{1} - \mathbf{C}')), \tag{3}$$

where $\circ$ is the Hadamard product (aka entrywise product), $\mathbf{1}$ is the matrix with all elements equal to 1 and $\mathbf{C}'$ is the transpose of matrix $\mathbf{C}$. Using $\mathbf{E}$, the IM-based state evolutions can be summarized as

$$\mathbf{H} = \mathbf{E}' \otimes \{\mathbf{A}_{ij}\}, \tag{4}$$

and $\mathbf{p}[t+1] = \mathbf{S}[t]\mathbf{H}$.

As discussed in [32], this formulation may or may not allow the asymptotic analysis of the behavior of the system. However, no matter if the analytical solution of the model exists or not, this model can be used for Monte-Carlo simulation of the behavior of the system in order to study how influences and interactions affect the state of the whole system. Based on this formulation, as the state of the system varies in time, various sets of influences get activated. Note that in IM, when a node influences another node, it may result in state change for the influenced node based on the adjusted transition probabilities that are captured through $\mathbf{H}$ and the formulation of $\mathbf{p}[t+1]$. As such, an activated influence in our model increases the probability of transitioning to a normal state for an underloaded charging station due to receiving load from the influencer (based on our definition of influence). In real-world, proper incentives for the users are needed to make that influence occur (transfer of load from one charging station to another). As such, to achieve the goal of the system which is increasing the probability of normal states, we use the status of the influence links (active or inactive) to guide the charging price design as discussed next.

Algorithm 1. Algorithm for Price Assignment to Charging Stations [23].

1: **Input**   ▷ Graph of active influences, $G_t(\mathcal{C}, E_a(t))$. A maximum electricity price limit **A** and a reduction factor in price, α.
2: **Output** ▷ Charging price in each charging station in $\mathcal{C}$ such that the price of the influencer station is higher than the influenced station.
3: Calculate the topological sort **T** for G_t.
4: **for** i=1 to $|\mathcal{C}|$ **do**
5:     **if** $|I(\mathbf{T}(i))| = 0$ **then**
6:         Price($\mathbf{T}(i)$) = **A**
7:     **else**
8:         Price($\mathbf{T}(i)$)=$\sum_{j \in I(\mathbf{T}(i))}$ Price(j)$/|I(\mathbf{T}(i))|$
9:     **end if**
10: **end for**
11: Return Price.

---

4   Designing Charging Prices

The model described in the previous section needs an external factor in real-world scenarios to provoke an EV user to travel from one charging station to another for charging (i.e., activating the described influence between charging stations in real-world). This external factor can be in terms of incentives or hampers that an EV user may get if they move from one cell to another. A good incentive would be lower charging prices (whenever the influence should be active) in the station, which should receive some load. The lower prices can motivate the EV users to move from their currently occupied cell to the other station. However, not every EV user will respond to such incentives in the same way and thus not every user will travel from the initial cell. Particularly, the probabilistic nature of the IM helps in capturing the random behavior of the users. Intuitively, the higher the influence strength the more we expect that the users travel to the other station, which can help in characterizing the price reduction that is needed. A key point to notice is that the cyber layer plays a key role in letting the desired influences to occur and to let the system identify its next states based on IM. Specifically, the cyber layer should communicate the lower charging price only to the users in the initial cell of concern. Otherwise, if the reduced price is communicated in the system globally and all the EV users in the city know about the reduced price in a station, this will activate influences among neighbor stations (neighbors are defined as according to G_u) that should not be activated according to the IM model. Thus, in order to only activate the influences that the IM model identifies for leading the system to a more balanced system in each step, the cyber layer plays a key role in communicating the prices to the right EV users based on their location.

In our model based on IM, whenever the set of activated influence links varies, we need to identify new set of prices for each station such that if station say i has an active influence link to station j, then the price at station i should be higher than that of station j. As discussed in [23], to identify the set of prices that satisfy this condition in the whole system, we apply Algorithm 1. This algorithm is similar to a constrained graph coloring problem. However, the problem of price assignment to the stations based on the above constraint is solvable with complexity $O(|\mathcal{C}| + |E|)$, which is because the graph of active influences denoted by $G_t(\mathcal{C}, E_a(t))$ and obtained from simulation of IM at step t is a directed and acyclic graph (DAG) (note that $E_a(t) \subseteq E$, also note that $E_a(t)$ does not include self-influences as they do not affect the pricing). This property is due to the rule set with the goal of balancing the load in the system, which never result in a cycle in the graph of active influences. In other words, the rule set in the model is very important to ensure that the load is not circulating in the system and purposely directed to the proper charging stations. Algorithm 1 for price assignment uses a topological sort of the graph and then assigns the prices based on the identified order such that the prices ensure that the stations appearing later in the topological sort have lower prices (as they should receive influences or loads). In our algorithm, we consider a maximum price limit of A and each price reduction occurs by a constant α. The values of A and α are considered fixed for simplicity, but can be variable and adjusted based on other factors in the system. In this algorithm, function $I(.)$ receives a node and returns the set of nodes, which influences the input node.

5   Evaluation and Results

In order to demonstrate the process of assigning prices to the charging stations dynamically as the system evolves in time while trying to lead both systems to more balanced states, we use an example network as shown in Fig. 2 with 12 charging stations, three substations, which two of them directly deliver electricity to the charging stations, and two load buses. We will compare the results with our previous model presented in [23]. To extract some of the parameters of the IM, we use a data driven approach based on data sets of traffic information. Specifically, we used the taxi data in [25], which contains GPS trajectories of 536 taxis in San Francisco, California from May 17, 2009-July 10, 2009 specifically to estimate $\mathbf{A}_{ii}$s. An example of $\mathbf{A}_{ii}$ based on the dataset is as following, where rows and columns are ordered from overload to normal and then underload:

$$\mathbf{A}_{ii} = \begin{pmatrix} 0.89473684 & 0.1052632 & 0.00000000 \\ 0.07262570 & 0.8770950 & 0.05027933 \\ 0.07142857 & 0.2142857 & 0.71428571 \end{pmatrix}. \tag{5}$$

In addition to $\mathbf{A}_{ii}$s, which characterize the internal dynamics of each charging station, we also need to consider $\mathbf{A}_{ij}$s to specify how the influences between two stations result in state transitions. An example of $\mathbf{A}_{ij}$ is shown in (6) in which each column specifies the probability of transition to overload, normal, and underload, respectively, depending on each row, which specifies the state of the influenced node. For simplicity and due to lack of detailed information in the datasets to characterize this matrix for all cells, we have simplified this matrix to have equal transition probabilities independent of the state of the influenced node (i.e., the same rows). Based on our model and the rules of influences, in order to lead the systems to balanced states a station only tries to send load to another station if the other station is not overloaded. As such the last row of the matrix in (6) does not play a role in the analysis.

$$\mathbf{A}_{ij} = \begin{pmatrix} 0.2 & 0.5 & 0.3 \\ 0.2 & 0.5 & 0.3 \\ 0.2 & 0.5 & 0.3 \end{pmatrix}. \tag{6}$$

Similarly, an example of $\mathbf{A}_{ij}$ for power substations is as following where rows and columns are ordered from normal to stressed to capture the influence among substations depending on their normal or stressed states:

$$\mathbf{A}_{ij} = \begin{pmatrix} 0.4 & 0.6 \\ 0.2 & 0.8 \end{pmatrix}. \tag{7}$$

We assume that the internal dynamics of each substation, i.e., $\mathbf{A}_{ii}$ can be captured by an identity matrix implying that the internal state of the substations can only change due to influences (i.e., without influences the probability of transition to the other state is zero). In our earlier work in [23], we did not consider the intra-power interactions, as such $\mathbf{A}_{ii}$ was considered to be the following matrix to compensate for the missing information on the intra system interactions that can cause state changes.

$$\mathbf{A}_{ii} = \begin{pmatrix} 0.8 & 0.2 \\ 0.5 & 0.5 \end{pmatrix}. \tag{8}$$

According to (8), when the system is stressed (i.e., the second row on the matrix in (8)), we assumed that there is an equal chance to get into normal state or stay stressed based on internal dynamics. However, as a part of influences in our IM-based model whenever the charging stations go back to normal or underloaded states then they can externally help the power substation to transit back to the normal state. To model the stochastic residential/commercial loads, we consider the following $\mathbf{A}_{ii}$ to consider the stochastic dynamics of loads (which can also be derived using a data-driven approach using load profile datasets).

$$\mathbf{A}_{ii} = \begin{pmatrix} 0.5 & 0.3 & 0.2 \\ 0.25 & 0.5 & 0.25 \\ 0.1 & 0.4 & 0.5 \end{pmatrix}. \tag{9}$$

Specifically, the set of rules for this study can be described as: (1) node i gets influenced by node j if and only if node i is underloaded and node j is overloaded or node i is in normal state and node j is overloaded for charging stations, and (2) for the influences between the power substations and the charging stations, the power substation gets influenced by a charging station or load bus if the power station is normal and the charging stations or load buses receiving power service from the substation are overloaded or if the power substation is stressed and the charging stations are normal or underloaded.

As mentioned earlier, based on the state of the components in the system, the influences among nodes may get activated and deactivated. In Fig. 3, we show two samples of active influence graphs for the network shown in Fig. 2. The activated links between charging stations suggest that the load should be transferred from one station to the station on the end of the directed link. Also, Fig. 4 shows the system model and the activated links for the setting defined in (8) and our previous analyses from [23].

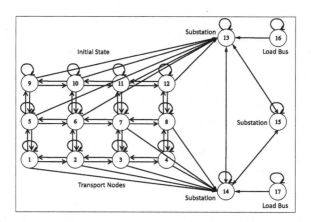

Fig. 2. The integrated charging and power infrastructures model with 12 charging stations and three substations and two load buses.(i.e., graph G_u).

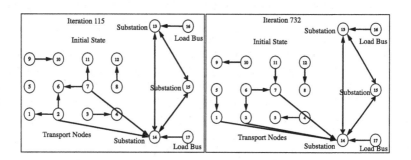

Fig. 3. Two samples of activated influence links for the model in Fig. 2.

The set of activated influences in each iteration prompts a change of state in the charging stations and power substations as shown in Figs. 3 and 4. It can be observed from Figs. 5 and 6 that the aggregated behavior of the system is helping the system towards being balanced (e.g., the likelihood of normally loaded charging stations and normal power substations is higher than other states). The results in Figs. 5, 6, 7 and 8 are obtained over 1000 steps of the IM simulation. Figure 7 shows the state distribution of the charging stations and power substations with various initial states.

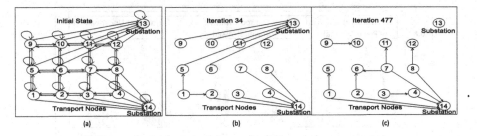

Fig. 4. (a) The initial system model without intra-power system dynamics (i.e., the model from [23]); (b) an example of activated influences at iteration 34; and (c) an example of activated influences at iteration 477.

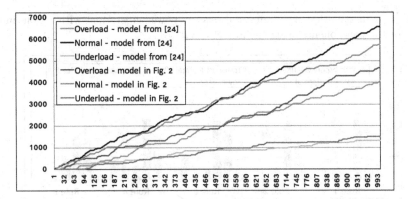

Fig. 5. Aggregated states distribution for overloaded, normal and underloaded states for charging stations for the system model in Fig. 2 and the model from [23].

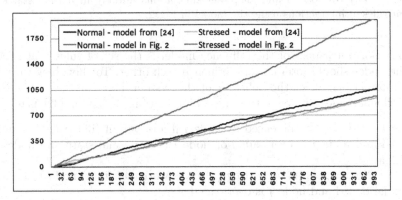

Fig. 6. Aggregated state distribution for normal and stressed states for the power substations for the system model in Fig. 2 and the model from [23].

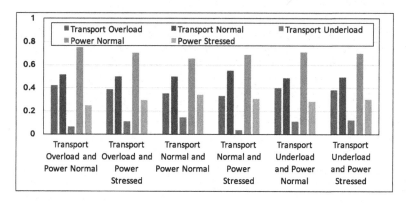

Fig. 7. State distribution of charging stations and power substations with various initial states for the components using the model in Fig. 2.

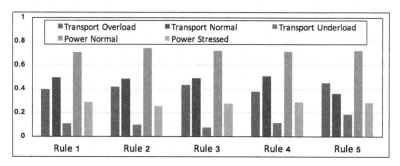

Fig. 8. State distribution of charging stations and power substations with all charging stations initially overloaded and all power substations initially in normal states for different rules of interactions using the model in Fig. 2.

An important aspect of the influence model is the set of rules that specify how the nodes should interact and influence each other. To show how the rules of the interactions affect the behavior of the system, here, we have considered other influence rules similar to the rules of interactions defined in [32] as follows:

- *Rule 1:* Node i gets influenced by node j if and only if (iff) node i is underloaded and node j is overloaded or node i is in normal state and node j is overloaded.
- *Rule 2:* Node i gets influenced by (receives workload from) node j iff node i is underloaded and node j is overloaded.
- *Rule 3:* Node i gets influenced by node j iff node i is underloaded and node j is overloaded or node i is underloaded and node j is in normal state.
- *Rule 4:* Node i gets influenced by node j iff either node i is underloaded and node j is overloaded, node i is underloaded and node j is in normal state or node i is in normal state and node j is overloaded.
- *Rule 5:* Node i gets influenced by node j iff either node i is underloaded and node j is overloaded, node i is underloaded and node j is in normal state,

node i is in normal state and node j is overloaded or node i is in normal state and node j is in normal state too.

Note that these rules only focus on the interactions/influences among the charging stations. Figure 8 shows the state distribution of nodes with all charging stations initially overloaded and all power substations initially normal for different rules applied to the model in Fig. 2. It can be seen that rule 5 performs the worst among all as the number of overloaded charging stations are higher compared to other cases. We observed similar performance for the model in [23].

Table 1. Charging prices in each EV charging station over various iterations.

Charging stations												
Iteration	1	2	3	4	5	6	7	8	9	10	11	12
1	A	A	A	A	A	A	A	A	A	A	A	A
99	A	A	A	A-α	A-α	A-α	A	A	A	A-α	A-α	A-α
368	A-α	A	A	A	A	A	A	A	A	A	A	A
562	A-α	A	A	A	A	A	A	A	A	A	A	A
600	A	A-α	A-α	A	A	A	A-α	A-α	A-α	A	A	A
653	A-α	A	A	A-α	A	A-α	A	A	A	A-α	A-α	A-α
779	A	A-α	A-α	A	A	A	A-α	A-α	A-α	A	A	A
987	A	A	A	A-α	A-α	A-α	A	A	A	A-α	A-α	A-α
993	A	A-α	A	A	A-α	A	A-α	A	A	A	A	A-α

To design the incentives that enable influences and lead to the shown results, we need to design the prices for each charging station. To do so, we have used Algorithm 1 over the active influence graph obtained at each step of the simulation whenever there was a change in the active influence graphs. Note that Algorithm CHalgorithm receives graphs similar to the ones shown in Fig. 3 where the self-edges are omitted. The price assignment based on this algorithm at each station is shown in Table 1 for sample steps of our simulation (with Rule 1). As it can be observed from the table, initially all the twelve stations have the same price of A but the prices vary over the network as the stochastic dynamics of the system change the states of the nodes. The set of the prices is different from Table 1 in [23], which did not consider the intra-power system dynamics.

In this section, we showed our study of collaborative pricing solution between the EV charging and electric infrastructures based on our IM-based model. We observed similar behavior between the models in Fig. 2 and [23]; however, the rate of changes in the state of the systems are different as seen from Figs. 5 and 6. Key takeaways from our results include: (1) by designing proper rules of interactions among the integrated systems, the load distribution can be improved in both systems, (2) the pricing assignment based on the obtained active influence graph enables the implementation of appropriate influences and (3) the intra- and inter-system influences and dynamics both affect the overall behavior and thus change the obtained pricing schemes.

6 Conclusions

In this paper, we discussed that critical infrastructures of smart cities are inter-dependent and thus it is important to consider collaborative solutions among them for designing services. To do so, we proposed a synergistic approach toward modeling and analysis of critical interdependent infrastructures that enable capturing the state and stochastic dynamics of inter and intra-system interactions. To demonstrate the benefit of collaborative solutions, in this paper, we focused on interdependent EV charging and power infrastructures and developed an integrated framework for modeling their interactions based on influence model, which is a networked Markov chain framework. We also proposed an algorithm, which assigns prices to charging stations based on the set of active links that can lead to more balanced systems. We discussed the role of the cyber infrastructure in enabling this pricing scheme, which considers the state of both of the systems. The work presented in this paper is an effort toward stimulating collaboration among various critical infrastructures and analyzing them using integrated models to develop collaborative solutions for smart cities. In future, we will study, both analytically and using simulations, the role of various parameters of the model in the behavior of the system. We also hope that this study and modeling approach can be extended to other smart city solutions and infrastructures.

Acknowledgement. This material is based upon work supported by the National Science Foundation under Grant No. 1541018.

References

1. Amin, M.: Toward secure and resilient interdependent infrastructures. J. Infrastr. Syst. **8**(3), 67–75 (2002)
2. Asavathiratham, C.: The influence model: a tractable representation for the dynamics of networked markov chains. Ph.D. thesis, Citeseer (2000)
3. Asavathiratham, C., Roy, S., Lesieutre, B., Verghese, G.: The influence model. IEEE Control Syst. **21**(6), 52–64 (2001)
4. Bass, R., Zimmerman, N.: Impacts of electric vehicle charging on electric power distribution systems (2013)
5. Bowerman, B., Braverman, J., Taylor, J., Todosow, H., Von Wimmersperg, U.: The vision of a smart city. In: 2nd International Life Extension Technology Workshop, Paris, vol. 28 (2000)
6. Chen, C., Hua, G.: A new model for optimal deployment of electric vehicle charging and battery swapping stations. Int. J. Control Autom. **7**(5), 247–258 (2014)
7. Chourabi, H., et al.: Understanding smart cities: an integrative framework. In: 2012 45th Hawaii International Conference on System Science (HICSS), pp. 2289–2297. IEEE (2012)

8. International Electrotechnical Commission, et al.: Orchestrating infrastructure for sustainable smartcities. Published in Geneva, Switzerland (2014)
9. Das, A., Banerjee, J., Sen, A.: Root cause analysis of failures in interdependent power-communication networks. In: 2014 IEEE Military Communications Conference, pp. 910–915. IEEE (2014)
10. Guo, S., Zhao, H.: Optimal site selection of electric vehicle charging station by using fuzzy topsis based on sustainability perspective. Appl. Energy **158**, 390–402 (2015)
11. Harrison, C., et al.: Foundations for smarter cities. IBM J. Res. Dev. **54**(4), 1–16 (2010)
12. Hatton, C., Beella, S., Brezet, J., Wijnia, Y.: Charging stations for urban settings the design of a product platform for electric vehicle infrastructure in Dutch cities. In: Towards zero emission: EVS 24 International Battery, Hybrid and Fuel Cell Electric Vehicle Symposium and Exhibition, 13–16 May 2009, Stavanger, Norway. European Association of Electric Road Vehicles (2009)
13. He, F., Wu, D., Yin, Y., Guan, Y.: Optimal deployment of public charging stations for plug-in hybrid electric vehicles. Transp. Res. Part B: Methodol. **47**, 87–101 (2013)
14. Heilig, G.: World urbanization prospects: the 2011 revision. United nations, department of economic and social affairs (DESA), population division. Population Estimates and Projections Section, New York (2012)
15. Hess, A., Malandrino, F., Reinhardt, M.B., Casetti, C., Hummel, K.A., Barceló-Ordinas, J.M.: Optimal deployment of charging stations for electric vehicular networks. In: Proceedings of the First Workshop on Urban Networking, pp. 1–6. ACM (2012)
16. Islam, M.M., Shareef, H., Mohamed, A.: A review of techniques for optimal placement and sizing of electric vehicle charging stations. Przeglad Elektrotechniczny **91**(8), 122–126 (2015)
17. Lee, W., Xiang, L., Schober, R., Wong, V.W.: Electric vehicle charging stations with renewable power generators: a game theoretical analysis. IEEE Trans. Smart Grid **6**(2), 608–617 (2015)
18. Li, Y., Luo, J., Chow, C.Y., Chan, K.L., Ding, Y., Zhang, F.: Growing the charging station network for electric vehicles with trajectory data analytics. In: 2015 IEEE 31st International Conference on Data Engineering, pp. 1376–1387. IEEE (2015)
19. Little, R.G.: Controlling cascading failure: understanding the vulnerabilities of interconnected infrastructures. J. Urban Technol. **9**(1), 109–123 (2002)
20. Liu, J.: Electric vehicle charging infrastructure assignment and power grid impacts assessment in Beijing. Energy Policy **51**, 544–557 (2012)
21. Liu, R., Dow, L., Liu, E.: A survey of pev impacts on electric utilities. In: 2011 IEEE PES Innovative Smart Grid Technologies (ISGT), pp. 1–8. IEEE (2011)
22. Min, H.S.J., Beyeler, W., Brown, T., Son, Y.J., Jones, A.T.: Toward modeling and simulation of critical national infrastructure interdependencies. Iie Trans. **39**(1), 57–71 (2007)
23. Nakarmi, U., Rahnamay-Naeini, M.: Towards integrated infrastructures for smart city services: a story of traffic and energy aware pricing policy for charging infrastructures. In: 6th International Conference on Smart Cities and Green ICT Systems, SmartGreens. IEEE (2017)
24. Pillai, J.R., Bak-Jensen, B.: Impacts of electric vehicle loads on power distribution systems. In: 2010 IEEE Vehicle Power and Propulsion Conference, pp. 1–6. IEEE (2010)

25. Piorkowski, M., Sarafijanovic-Djukic, N., Grossglauser, M.: Crawdad data set epfl/mobility (v. 2009-02-24) (2009)
26. Rahman, I., Vasant, P.M., Singh, B.S.M., Abdullah-Al-Wadud, M.: Intelligent energy allocation strategy for PHEV charging station using gravitational search algorithm. In: AIP Conference Proceedings, vol. 1621, pp. 52–59 (2014)
27. Rahnamay-Naeini, M., Hayat, M.: Cascading failures in interdependent infrastructures: an interdependent markov-chain approach. IEEE Trans. Smart Grid **7**(4), 1997–2006 (2016)
28. Recker, W.W., Kang, J.E.: An activity-based assessment of the potential impacts of plug-in hybrid electric vehicles on energy and emissions using one-day travel data. University of California Transportation Center (2010)
29. Rinaldi, S.M.: Modeling and simulating critical infrastructures and their interdependencies. In: Proceedings of the 37th Annual Hawaii International Conference on System Sciences, pp. 8-pp. IEEE (2004)
30. Shao, J., Buldyrev, S.V., Havlin, S., Stanley, H.E.: Cascade of failures in coupled network systems with multiple support-dependence relations. Phys. Rev. E **83**(3), 036116 (2011)
31. Shin, D.H., Qian, D., Zhang, J.: Cascading effects in interdependent networks. IEEE Netw. **28**(4), 82–87 (2014)
32. Siavashi, E.: Stochastic modeling of network interactions: conditional influence model. Texas Tech Master Thesis. https://ttu-ir.tdl.org/ttuir/handle/2346/67107
33. Sioshansi, R.: Or forum-modeling the impacts of electricity tariffs on plug-in hybrid electric vehicle charging, costs, and emissions. Oper. Res. **60**(3), 506–516 (2012)
34. Sweda, T.M., Klabjan, D.: Agent-based information system for electric vehicle charging infrastructure deployment. J. Infrastr. Syst. **21**(2), 04014043 (2014)
35. Tushar, W., Saad, W., Poor, H.V., Smith, D.B.: Economics of electric vehicle charging: a game theoretic approach. IEEE Trans. Smart Grid **3**(4), 1767–1778 (2012)
36. Vazifeh, M.M., Zhang, H., Santi, P., Ratti, C.: Optimizing the deployment of electric vehicle charging stations using pervasive mobility data. arXiv preprint arXiv:1511.00615 (2015)
37. Wagner, S., Götzinger, M., Neumann, D.: Optimal location of charging stations in smart cities: a points of interest based approach (2013)
38. Walsh, K., Enz, C.A., Canina, L.: The impact of gasoline price fluctuations on lodging demand for us brand hotels. Int. J. Hospit. Manage. **23**(5), 505–521 (2004)
39. Weis, C., Axhausen, K., Schlich, R., Zbinden, R.: Models of mode choice and mobility tool ownership beyond 2008 fuel prices. Transp. Res. Rec.: J. Transp. Res. Board **2157**, 86–94 (2010)
40. Xiong, Y., Gan, J., An, B., Miao, C., Bazzan, A.L.: Optimal electric vehicle charging station placement. In: Proceedings of the 24th International Joint Conference on Artificial Intelligence (IJCAI), pp. 2662–2668 (2015)
41. Xiong, Y., Gan, J., An, B., Miao, C., Soh, Y.C.: Optimal pricing for efficient electric vehicle charging station management. In: Proceedings of the 2016 International Conference on Autonomous Agents and Multiagent Systems, pp. 749–757. International Foundation for Autonomous Agents and Multiagent Systems (2016)

All Eyes on You! Impact of Location, Camera Type, and Privacy-Security-Trade-off on the Acceptance of Surveillance Technologies

Julia Offermann-van Heek[✉], Katrin Arning, and Martina Ziefle

Human-Computer Interaction Center, RWTH Aachen University,
Campus-Boulevard 57, 52074 Aachen, Germany
vanheek@comm.rwth-aachen.de

Abstract. While surveillance technologies are increasingly used to prevent or detect crimes and to improve security, critics perceive recording and storage of data as a violation of individual privacy. Thus, it has to be analyzed empirically where and to what extent the use of surveillance technologies is accepted and whether the needs for privacy and security differ depending on the location of surveillance, the type of technology, or the individual characteristics of city residents. By applying a conjoint analysis, our study investigated the relationship between different locations of surveillance, different types of cameras, increase of safety implemented by reduction of crime and intrusion of privacy operationalized as different ways of handling the recorded footage. Findings show that locations are the most important factor for crime surveillance scenario preferences, followed by increase of security, and intrusion of privacy. In the decision scenarios, the type of camera played only a minor role. Sensitivity analyzes enabled detailed examinations of the trade-off between privacy and security and a segmentation of different respondent profiles led to an identification of influencing characteristics on the acceptance of crime surveillance. Outcomes show the importance of integrating city residents' preferences into the design of infrastructural city concepts.

Keywords: Surveillance technologies · Technology acceptance
Conjoint analysis · Privacy-Security-Trade-off

1 Introduction

In the last years, the use of crime surveillance technologies (CST) in private and public environments has increased significantly, mainly due to terrorist attacks and rising crime rates [1]. Considering urbanization processes and the demographic change, the majority of people will live in cities rather than other regions by 2030 [2]. Therefore, it will be of great importance to what extent city residents accept surveillance technologies, on the one hand, and what information about benefits and caveats needs to be provided so residents are adequately informed.

At the present time, several hundred million cameras are installed in public and private environments around the world [3, 4]. CST, e.g., cameras, microphones, detection and localization technologies, are used increasingly to enhance security.

© Springer Nature Switzerland AG 2019
B. Donnellan et al. (Eds.): SMARTGREENS 2017/VEHITS 2017, CCIS 921, pp. 131–149, 2019.
https://doi.org/10.1007/978-3-030-02907-4_7

However, critics fear an invasion of privacy by recording and storage of data material [5]. Hence, research concerning CST acceptance is inevitable to determine at what locations and on what terms it is accepted and which circumstances may lead to changes in the need for privacy and security.

1.1 Usage of Surveillance Technologies in Urban Areas

Aiming at investigation or detection of crime and an increase of security, a rising number of CST is currently used in almost every city in the world [3, 6, 7]. This development is heavily critized, especially by data protection specialists who see the recording and storage of data as violation of a human's privacy and personal rights [8, 9]. In particular, the usage or processing purpose of recorded data material in urban areas is raising concerns: By whom, where, and for how long is the recorded data stored? Is the recorded data used for localization or recognition purposes, who has access to it, and who benefits from it?

Thus, the conflict between security and privacy is increasingly gaining importance that leads to central questions regarding the implementation of CST in urban areas: At which locations and on what terms is privacy or security more important? Are people willing to sacrifice their privacy to gain perceived safety? More and more CST are used without considering the requirements and needs of city residents or involving them into decision-making processes on the installation of CST in cities [10, 11]. In order to reach an accepted and positively perceived urban design, it is necessary to include residents into the implementation process and take their wishes, fears, and needs into account. Only this way, a long-term acceptance and adoption of CST in urban environments can be achieved.

Previous studies primarily focus on innovative technical and functional features of CST such as camera, microphone, localization, and detection technologies [12, 13]. In part, some politically or police motivated studies examine the effectiveness of CST, usually in terms of a decrease in crime rates at monitored locations or increased rates of revealed offenses [14, 15]. Furthermore, the regulation of the use and proliferation of surveillance technologies and associated legal, juristic and ethical aspects are discussed [16]. However, only little knowledge exists about the acceptance of CST at private and public locations because CST are usually integrated into urban environments without considering opinions and needs of urban residents. CST acceptance is, if at all, only superficially addressed: Attempts were made to understand acceptance of CST by means of theoretical models (e.g., [17]), in which city residents are not directly integrated though. In most of the previous studies of CST, it was determined whether crime surveillance is generally accepted or rejected (e.g., [18]), without detailing the underlying reasons. Thus, no consideration of potential impact factors has been conducted, e.g., different types of CST or locations of crime surveillance. Therefore, an empirical study is necessary that investigates the acceptance of CST as a function of several impact factors such as locations, type of technology, and different needs for privacy and security.

1.2 Acceptance of Surveillance Technologies

Perceived safety and protection of one's own privacy are important determinants of CST acceptance [9]. Understanding factors that influence technology acceptance is essential for a successful adoption and integration of innovative technologies [19].

The Technology Acceptance Model (TAM) is a well-established theoretical approach to explain and predict the adoption of technologies [20]. It provides a basis for many other acceptance models and has been adapted for a variety of contexts (e.g., [21]). However, for the purpose of CST acceptance, there is no established acceptance model yet. Previous acceptance models might not be transferrable to the context of CST for several reasons. Most acceptance models are limited to two key factors: ease of use and perceived usefulness of a technology or application [20]. Thus, other acceptance determinants are disregarded. The TAM was originally developed for end-user computing, directed on ICT usage in a job-related context. Since CST are implemented in a completely different and much more complex usage context, we assume that relevant factors of CST acceptance are not yet adequately considered. Previous research shows that even the same technology applied to different usage contexts evokes different acceptance patterns and underlying motives and barriers [22]. Further, technology acceptance models focus on an evaluation of *complete* technical systems. As a result, only general benefits and barriers of a technology or the generic intention to use can be assessed while insights into the importance of single technical characteristics and practical design guidelines are not possible. More specifically, nothing can be said about which type of surveillance technology is most accepted at which location and, depending on this, where needs for either privacy or safety predominate.

Even though the TAM successors (e.g., [21] integrated individual factors into the acceptance model, the effect of these factors, such as attitudinal variables beyond demographic characteristics (age, gender), has not been investigated yet. Moreover, existing models do not allow to derive user profiles, which allow for a more target-group-specific formulation of design guidelines in city planning processes. Finally, coming to methodological issues, questionnaires designed on the basis of TAM, do not allow to holistically portray complex decision scenarios, in which several decision criteria are weighted against each other. More specifically, it is not possible to draw conclusions about relative importance, relationships, and interactions of factors concerning CST acceptance, e.g., locations of CST implementation, type of technology as well as needs for safety and privacy. By combining a conjoint analysis with a traditional questionnaire, more information can be obtained and different attributes of CST acceptance as well as their interrelations can be analyzed in detail.

1.3 Research Questions and Aim of the Study

So far, the acceptance of CST has been investigated by considering potential influencing factors separately. Thus, it was the aim of the present study to explore factors that have been proven to be relevant for the acceptance of CST by a direct weighting of these factors against each other. In doing so, the following research questions were investigated:

- Does the location, the applied technology, safety, or privacy influence the decisions in crime surveillance scenarios the most? (RQ1)
- At which turning points tendencies of acceptance do convert in rejection? (RQ2)
- How is the trade-off between safety and privacy evaluated in detail? (RQ3)
- Which user profiles can be derived that evaluate the crime surveillance scenarios differently? (RQ4).

The paper is an extension to previous work [23], in which general findings have been reported differing between diverse surveillance contexts. Here, the specific context of crime surveillance is focused in detail, as crime surveillance is an essential part of public life in smart cities.

2   Methodology

Based on the findings of a preceding study in which CST acceptance was examined by means of focus groups and questionnaires, it was revealed that CST acceptance is influenced by several key characteristics and attributes that do play important roles for residents [24]. The present study is a follow up of this previous work, using a conjoint measurement approach. The aim of this study was to assess preferences for video-based crime surveillance scenarios, considering different locations of surveillance, different camera types, security as benefit in terms of a reduction in crime rate, and privacy as barrier, operationalized with different ways of handling recorded data.

2.1   Conjoint Analysis

Conjoint analyses combine a measurement model with a statistical estimation algorithm and were developed by Luce and Tukey [25]. In contrast to conventional survey-based acceptance research, conjoint analyses enable a holistic and more valid examination of decision scenarios, an exposure of single attributes against each other, and direct simulations of relationships and interactions. Within a conjoint analysis, specific product configurations but also different scenarios are assessed by the respondents. These products or scenarios consist of multiple attributes and differ from each other in the attribute levels. Conjoint analyses allow simulations of decision processes and fragmentations of product or scenario preferences into separate part-worth utilities of the attributes and their levels [26]. As a result, the relative importance of attributes deliver information about which attribute influences the respondents' choice the most. Part-worth utilities label which attribute level is valued the highest. Preference ratings and resulting preference shares can be interpreted as indicator of acceptance. A choice-based-conjoint analysis approach (CBC) was chosen, because it imitates complex decision processes, in which more than one attribute influences the final decision [27].

2.2   Selection of Attributes

One of the first and most important steps in the conceptualization stage of conjoint analyses is the identification and selection of relevant attributes and the number of levels, because this affects the significance and generalizability of findings [28]. The designer of a conjoint study must ensure that all relevant attributes that determine respondents' preference, on the one hand, and are relevant for policy-makers or city-planers, on the other hand, are considered. Beyond an extensive literature analysis, we used the outcomes of the prestudy [24], in which relevant criteria for the acceptance of CST were identified and selected for the subsequent conjoint study. The following attributes were evaluated in the conjoint study: locations, camera types, crime reduction (indicating safety), and handling the recorded footage (indicating privacy). In the next sections, it is explained how and why these attributes have been selected.

Location of CST Implementation. Acceptance of CST is higher for public than for private locations [24, 29–31]. However, the question arises how CST are perceived at semi-public locations, especially in direct decision situations. The findings of our prestudy showed that perceived crime threat and CST acceptance varied among different public and semi-public places. Therefore, not only the contrast between private and public locations but also two semi-public locations were chosen as attribute levels. Based on the ratings of perceived crime threat at different locations [24], a representative prototype was selected:

- own house (private)
- store (semi-private)
- market (semi-public)
- train station (public).

Type of Surveillance Technologies. Concerning different types of surveillance technologies, we focused on video-based surveillance because it is the most established and used type of surveillance technology (e.g., [9]). It was also favored in the prestudy in contrast to other technologies, e.g., microphones. To examine attitudes towards video-based technologies in detail, we distinguished between four types of cameras differing primarily in their size and visibility:

- big, tracking camera (large & visible)
- dome camera (smaller & less visible)
- mini-dome camera (small & partly hidden)
- integrated camera (very small & hidden).

This distinction among the dimensions "size" and "visibility" was made because large and visible cameras are mainly used for surveillance in public space, but seamless camera integration (e.g., in objects or clothing) is gaining popularity across different application areas [32].

Security. The most important benefit of CST is an increase in security, because crimes are detected or can be prevented through deterrence [1]. In our prestudy, crime reduction was also perceived as one major benefit of CST. Therefore, the attribute *crime reduction* was chosen as third attribute for the conjoint analysis. The definition of crime reduction levels was based on findings regarding the effectiveness of video-based crime surveillance (e.g., [33]):

- 0%
- 5%
- 10%
- 20%

Privacy. Privacy concerns are an important public concern in the context of CST implementation [8]. More specifically, violation of privacy due to storage of data material and unwanted release of personal information is perceived as critical [5, 9]. In our prestudy, storage and re-use of personal data was evaluated as barrier of CST acceptance. Thus, the category handling of recorded footage was selected as fourth conjoint attribute. To further investigate the evaluation of different categories of "privacy intrusion", the following four levels were chosen:

- archiving of data by police
- storage in profile databases
- face recognition
- location determination.

2.3 Experimental Design

Based on these considerations, four attributes and relevant levels were chosen for the CBC studies. No prohibitions for level combinations were included because all chosen attribute levels were combinable. In every CBC task, four sets of scenario configurations were presented without the opportunity to choose a "none-option". In order to improve comprehensibility, the attribute levels were presented partly in visual and partly in written form (Fig. 1). Overall, participants had to evaluate ten choice tasks, each consisting of four different combinations of the attributes *locations, crime reduction, handling of recorded footage,* and *camera types.*

Because a combination of all corresponding levels would have led to 256 ($4 \times 4 \times 4 \times 4$) possible combinations, the number of tasks was reduced [27] and, overall, 10 random and one fixed task were presented to the respondents. A test of design efficiency was used to consider whether the design is comparable to the hypothetical orthogonal design [34]. The result confirmed that the design was sufficient regarding a total of at least 100 respondents.

2.4 Questionnaire

The questionnaire of the online survey was created using the SSI Web Software. Participants started the survey by opening a link which was sent via e-mail or advertised on public internet sites. Prior to the choice tasks, demographic data and information on type of residence and residential location were collected. This was followed by querying perceived needs for security and privacy. To gather data concerning security and privacy needs, for each four statements had to be evaluated on a six-point

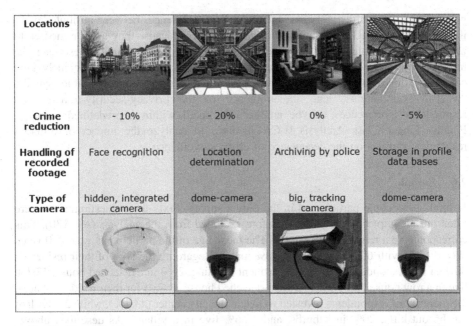

Locations				
Crime reduction	- 10%	- 20%	0%	- 5%
Handling of recorded footage	Face recognition	Location determination	Archiving by police	Storage in profile data bases
Type of camera	hidden, integrated camera	dome-camera	big, tracking camera	dome-camera

Fig. 1. Example of scenario decision with four attributes and their attribute levels [23].

Likert scale (1 = strongly disagree; 6 = strongly agree). In addition, perceived crime threat (PCT) was evaluated using a six-point Likert scale (1 = very low; 6 = very strong) adapted from a preceding study [31]. Based on ratings for different security needs, privacy needs and PCT, sum scores were calculated. Concluding this questionnaire, the participants were asked for previous experiences with crime. It was distinguished between slight crime (e.g., bicycle theft) and serious crime (e.g., robbery). This was followed by the introduction of attributes and their characteristics, including their visual representation. Afterwards, the respondents were prompted to select the preferred scenario in every scenario decision of the conjoint tasks. As a control, they were to imagine that they are alone at the respective locations during the day. Finally, the respondents were instructed to select a scenario in each choice task, that met their individual needs for security and privacy most closely.

2.5 Data Analysis

Data analysis (i.e., estimation of part-worth utilities, preference simulations) was conducted using Sawtooth Software: SSI Web and SMRT. In a first step, part-worth utilities were computed on the basis of hierarchical bayes estimation and part-worth utili-ties scores were deduced [27]. Relative importance of attributes deliver the information how important an attribute is relative to all other attributes for the scenario selection. The relative importance of an attribute was calculated by taking range of part-worth utility values for each factor and dividing it by the sum of the utility ranges for all factors. However, part-worth utility scores are interval-scaled within each attribute and, thus, a comparison of utility scores between different attributes is not possible [26]. In contrast, it is possible to compare differences between attribute levels, if using zero-centered differentials part-worth utilities, because they are summed up to zero within each attribute. In a second step, preference simulations estimate the influence on preferences if certain attribute levels change or are consciously kept constant within a specific scenario [26]. The simulation of preferences enables specific "what-if"-examinations, e.g., the influence of the privacy-security-trade-off on respondents' preferences can be analyzed in detail within a predefined scenario. Finally, Latent Class Analysis (LCA) is used to analyze the impact of different respondent profiles on the acceptance of crime surveillance.

2.6 Sample

In total, 193 people participated in the online survey. Sixty-three questionnaires were filled out incompletely and were therefore excluded from the analysis ($n = 130$). This corresponds to a return rate of 67.4%. The mean age of the participants was 32.0 years ($SD = 12.2$) with 60% females and 40% males. Regarding the type of their residence, 60% of the respondents live in an apartment building, 20% in a detached house, 13.1% live in a row house and 6.9% in a semi-detached house. Asked for their residential area, the majority of respondents revealed to live in the city center (43.1%), while 22.3% live on the outskirts, 20% in suburbs, and 14.6% live in a village. As described above, different needs for privacy and security as well as PCT were calculated as sum scores ($min = 4$; $max = 24$). Overall, an average need for security ($M = 12.4$, SD = 4.7) and an average perceived crime threat ($M = 10.0$, $SD = 3.7$) were existent. Needs for privacy were generally very high ($M = 22.2$ (out of 24 points max), SD = 2.3).

3 Results

First, the relative importance of attributes are presented, followed by part-worth-utility estimation results for all attribute levels. Afterwards, the results of different preference simulations and segmentation analyzes are described.

3.1   Relative Importance Scores (RQ1)

Hierarchical Bayes analysis was used to determine the importance scores of attributes and, thus, to discover main factors influencing the acceptance of crime surveillance.

As can be seen in Fig. 2, locations had the highest importance score in the scenario selection (42.4%) and, therefore, it is the most important determinant concerning crime surveillance acceptance. Crime reduction (23.1%) and handling of recorded footage (20.0%) gained similar importance, while crime reduction has a slightly more important meaning for the scenario decisions. Types of camera (14.6%) had the lowest influence on the scenario selection and, thus, it presents the attribute with the lowest impact on the acceptance of crime surveillance.

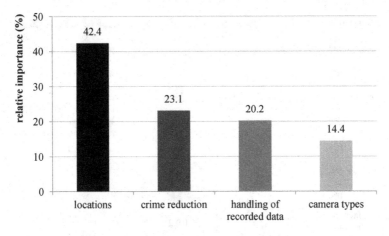

Fig. 2. Relative importance scores of attributes.

3.2   Part-Worth-Utility Estimation (RQ2)

Following, the part-worth utilities are presented for all attribute levels (Fig. 3). This way, the attribute levels with the highest utility values were identified in order to provide recommendations to city planners of future urban quarters about which scenario would, hypothetically, reach the highest acceptance in the context of crime surveillance. The best rated scenario was surveillance at a "train station", with a "crime reduction of 20%", "archiving" of data material, and using the "large, tracking camera".

Regarding absolute utility values, the attribute levels of *locations* reached the highest and lowest utility values, explained by the high relative importance score of this attribute. "Train station" received the highest utility value (44.0) and was consequently most accept-ed. "Market place" (20.9) and "department store" (10.6) reached lower scores, while the "own home" was strongly rejected with the lowest utility score (−75.5).

The second largest span and an almost linear behavior of the utility function belonged to the attribute *crime reduction*: Here, the "crime reduction of 20%" (33.7) received the highest value and was most accepted, while "crime reduction of 0%" (−41.2) reached the lowest utility value and was rejected. "Crime reduction of 10%" (14.9) was rated higher than a "crime reduction of 5%" (−7.4).

Within the attribute *handling of recorded footage,* "archiving by police" (21.7) obtained the highest utility value while "face recognition" (−19.2) clearly received the lowest utility value. "Location determination" (1.9) was evaluated slightly higher than "storage in a profile database" (−4.4).

The utility function of the attribute *camera types* shows a nearly linear relation-ship. The highest value achieved the "large moving camera" (15.2), while the "inte-grated camera" (−14.6) got the lowest score. The "dome camera" (3.9) reached a slightly higher utility value than the "mini-dome camera" (−4.5).

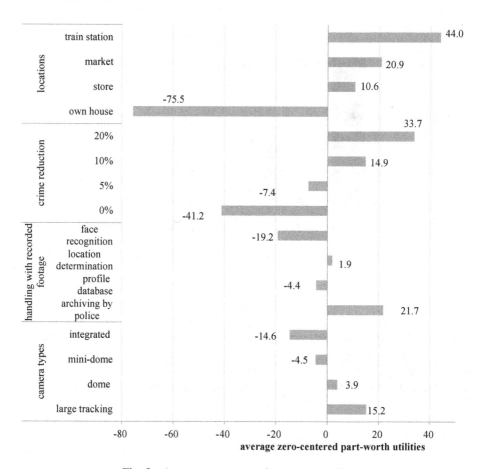

Fig. 3. Average zero-centered part-worth utilities.

3.3 Trade-off Between Security and Privacy (RQ3)

Using the market simulator of SSI web software, the trade-off between security and privacy was analyzed in so-called sensitivity analyzes. For that, the calculated values of the Hierarchical Bayes estimation were imported into the SMRT software. Preference simulations examine the extent to which relative preferences of a respondent vary, if single levels of an attribute change while other levels of attributes are kept constant [35].

As the relative importance of the attributes "crime reduction" (indicating security) and "handling of recorded footage" (indicating privacy) was very similar, the sensitivity analysis was regarded for constant privacy and security attributes to analyze the trade-off in detail. Based on the findings in previously reported part-worth utility analyses, three scenarios of attribute levels settings were constructed:

1. *high security* and *low privacy* with the levels "crime reduction of 20%" and "face recognition";

2. *high privacy* and *low security* with the levels "archiving by police" and "crime reduction of 0%".

3. *average security and privacy* with the levels "crime reduction of 10%" and "storage in a profile database".

These levels were kept constant in the preference simulation while all the other levels of attributes changed (locations and camera types). Outcomes are pictured in Fig. 4.

A clear preference for the scenario *high security* was found, especially in public places. This scenario received ratings three to four times higher than the other scenarios (e.g., "train station": *high security*: 66.5%, *high privacy*: 20.0%, *average*: 27.1%). Interestingly, the security scenario was rejected for private environments, since the attribute level "own home" was rated very low in general (min = 9.0%; max = 15.3%). The scenario *high privacy* received the lowest ratings - in private as well as in public spaces and for all camera types. The *average* scenario is rated a little bit higher than the *high privacy* scenario, except for the private environment (own home), which was the least preferable scenario. Concerning the camera type, the ratings were not widely dispersed inside the scenarios. In all scenarios, the maximal rating belonged to the large tracking camera with a decreasing tendency towards the hidden, integrated camera. Here, high ratings regarding the *high security* scenario were also striking (e.g., "large tracking camera": *high security*: 66.5%; *high privacy*: 14.8%; *average*: 24.9%).

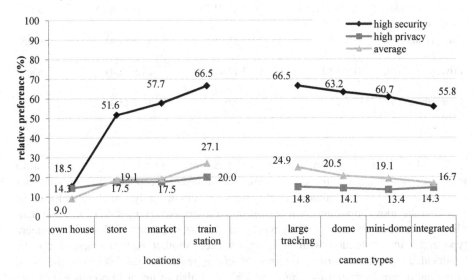

Fig. 4. Results of sensitivity analysis concerning the trade-off between security and privacy.

3.4 User Profiles (RQ4)

So far, the acceptance of crime surveillance technologies was reported for the whole group of respondents. However, residents in urban environments are highly hetero-geneous, which suggests the existence of group-specific acceptance patterns. In order to detect groups of respondents with similar preferences based on their choices in the CBC questionnaires, latent class segmentation analysis (LCA) was applied [36]. A three-group solution (Table 1) showed a sufficient data fit according to the criteria percentage certainty, consistent Akaike information criterion (CAIC), and relative Chi square.

- *Group 1,* in the following simplified designated as the "worried" (N = 22), mainly consisted of people with a stronger need for security and a higher PCT.
- *Group 2* (N = 68), simplified designated as the "unworried", represented participants with a rather low security need and PCT.
- *Group 3* (N = 40), simplified designated as the "undecided", showed a balanced relationship concerning security need as well as PCT.

Further group differences in demographic characteristics missed statistical significance in ANOVAs. Group comparisons revealed different importance patterns (see Fig. 5).

Table 1. Group segmentation (based on LCA results)

	Group 1 (n = 22; "worried")	Group 2 (n = 68; "unworried")	Group 3 (n = 40; "undecided")	P
Age (M; SD)	31.2(13.0)	33.1(13.0)	30.7(10.2)	n.s.
Gender (female, male in %)	68.2%, 31.8%	58.8%, 41.2%	57.5%, 42.5%	n.s.
Security need (M; SD)	14.5 (3.3)	11.5 (5.0)	12.8 (4.4)	<.05
Privacy need (M; SD)	21.8 (2.3)	22.3 (2.4)	21.9 (2.4)	n.s.
Crime threat (M; SD)	11.6 (2.9)	9.2 (3.9)	10.7 (3.4)	<.05

For the "worried" group, the attribute locations (55.6%) clearly was the most important determinant influencing crime surveillance acceptance, while all other attributes were of less but almost equal importance. Interestingly, the attribute locations was also the most important attribute for the "unworried" group (50.3%), while handling of recorded footage (23.2%) was second most important, followed by camera type and crime reduction with least importance. Another pattern emerged for the "undecided" group: security in terms of crime reduction (50.7%) was the most important determinant and locations (37.4%) were also of major importance. For this group, camera type and especially handling of recorded footage (privacy) were not important at all.

According to an extremely different relative importance of all attributes depending on the three groups, the part-worth utilities for all attribute levels were also extremely diverse (see Fig. 6).

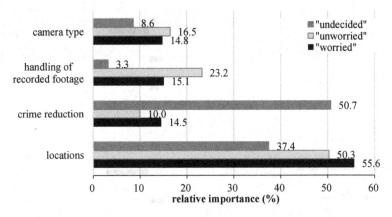

Fig. 5. Relative importance scores for segmented groups.

Concerning *locations*, very contrasting patterns were found. While the "unworried" and "undecided" groups preferred public locations for surveillance (e.g., train station: 78.8; 57.3), this was rejected by the "worried" group (−72.1) that preferred and accepted surveillance only at home (150.3).

The patterns of the attribute *crime reduction* were more similar. In all groups, a crime reduction of 20% was rated best and acceptance rose with increasing crime reduction. Crime reduction of 5% was an exception, because it was rated lowest by the "worried" group (−32.8). According to the high relative importance of the attribute, there were the lowest (crime reduction 0%: −112.1; crime reduction 20%: 90.6) and highest ratings in the "undecided" group.

Privacy in terms of handling of recorded footage was evaluated more differently. For the "undecided" group it was completely unimportant. The "unworried" group, preferred archiving of recorded data by the police (55.7) and strongly rejected the possibility of face recognition (−37.3). In contrast, the "worried" group preferred location determination (42.9) and rejected all other options of handling of recorded footage.

The attribute camera type also showed contrasting patterns. The "worried" group only accepted the integrated, hidden camera (41.4) and rejected all other types. However, the large tracking camera was preferred by the "unworried" (36.7) and the "undecided" (13.0) groups, while all other camera types were rejected or merely tolerated

4   Discussion

The present study evaluated preferences for different video-based crime surveillance scenarios, consisting of the attributes locations of surveillance, security (crime reduction), privacy (handling of recorded video material), and type of camera in a conjoint analysis.

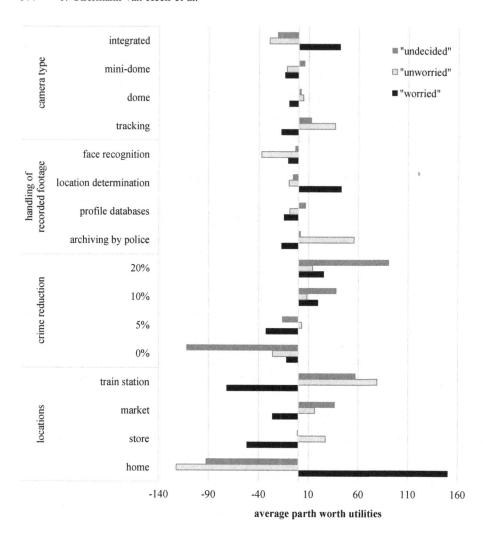

Fig. 6. Part-worth utilities (zero-centered diffs) for all attributes and levels for the groups "undecided", "unworried", "worried".

4.1   Acceptance of Surveillance Technologies in Urban Areas

Confirming previous research results (e.g., [18, 37]), the use of surveillance technologies is generally accepted in public locations in urban areas. Going beyond public spaces, our findings demonstrate that crime surveillance is not accepted at all in private spaces such as the own home. Interestingly, for medical surveillance applications, similar results were found for public and private locations [38]. In this study, the location of surveillance technology implementation was the most important factor in respondents' decisions for or against video-based crime surveillance scenarios. However, the importance of locations for surveillance acceptance is highly context-sensitive. For example, an equivalent conjoint study on medical surveillance showed a higher importance of security and privacy issues compared to locations [39].

In contrast to previous studies, which proved security and protection of privacy as important factors for acceptance without weighting them in direct decision situations [9, 37, 40–42], this study revealed that acceptance depends on perceived benefits in terms of increasing security and to a lesser extent on privacy-related issues. Technology-related aspects (such as the type of camera) played only a minor role for the acceptance of surveillance technologies.

However, the acceptance of surveillance technologies is no homogenous phenomenon but strongly shaped by individual factors of residents.

4.2   Specific Impacts on the Acceptance of Crime Surveillance

In this study, surveillance was not accepted at all in private spaces, but its acceptance rose with increasing publicity of locations. Thus, in public places, video-based surveillance systems are comparably well-accepted even in case of a rather low increase in security, i.e., low crime reduction.

Increase of security was another important factor in respondents' decisions, that increased linearly with the amount of crime reduction. Declines in the crime rate of "0%" and "5%" were perceived as insufficient, while a decline of "10%" was desired and preferences even redoubled for a crime reduction of "20%". Thus, the effectiveness of crime surveillance in terms of crime reduction is crucial for acceptance of crime surveillance.

In terms of privacy, archiving of data by police was most accepted, followed by location determination and storage in profile databases; face recognition was rejected. This result is especially important because face recognition technologies are increasingly used in public areas to fight and detect crime such as terrorist attacks [43].

Concerning camera types, acceptance decreases with increasing invisibility of cameras. The most visible large tracking camera is most accepted, while the hidden, integrated cam-era was rejected. This is especially surprising, because current technological developments – especially in the field of ambient assisted living environments – aim for designing smaller, less visible and seamlessly integrated technologies, e.g., in street or traffic lights, smoke detectors, smart homes [32, 44].

4.3   Trade-off Between Security and Privacy

Previous research on the trade-off between security and privacy emphasized that the will-ingness to trade one's own freedom, (mostly associated with privacy) for increased secu-rity depends on the level of increased security [43]. This study revealed that only a tangible crime reduction (at least 10%) is positively perceived as increase in security. Although, the analysis of relative importance initially revealed only a slightly higher importance of crime reduction (23.1%) in contrast to the handling of recorded footage (20.2%), for the scenario selection, secure scenarios were clearly preferred compared to scenarios that focused on privacy. Sensitivity analyses showed that security is much more preferred in a direct confrontation of security and privacy. Thus, security issues are more important criteria for the acceptance of crime surveillance than privacy issues, provided the technology is efficient, causes a noticeable decrease in crime rate, and consequently a gain in security.

4.4   Impact of User Profiles

In our study, we identified groups of participants with differing preferences in scenario decisions and differing needs for security and crime threat perception. Consistent with one of our preceding studies [30, 45], we found that especially different levels of perceived crime threat (PCT) affect evaluations of surveillance technologies. This finding is in line with recent studies, in which fear of crime influences the evaluation of the built environment [46]. Most noteworthy, people with a high PCT (the "worried" group, also characterized by a high security need) accept surveillance technologies at private locations while all other locations are rejected. In contrast, "unworried" or "undecided" respondents with a low or average PCT clearly prefer public locations for surveillance technology implementation and include security- and privacy considerations into their preference decision. The heterogeneity of preference patterns indicates that residents and their individual characteristics have to be known and considered in the design and integration of surveillance technologies in urban as well as private environments.

4.5   Limitations and Further Research

The applied conjoint analysis approach was useful for assessing preferences of different surveillance technology scenarios. However, it has some limitations to be considered in future studies.

First, estimated preference ratings do not mirror actual behavior, i.e., confirmation or rejection might be lower or higher in real situations (e.g., [47, 48]). A second limitation belongs to the limited number of attributes. A compromise had to be made between an economic research design with a limited number of attributes and the complexity of the research issue under study. Therefore, future studies will include further aspects, e.g., length of data storage, or use adaptive conjoint approaches (e.g., ACBC) allowing for bigger attribute numbers.

Further, some aspects have to be criticized in terms of content. The very similar evaluation of needs for privacy showed that the item content might have been too alike. In further studies, we will use more specific and tangible items concerning needs for privacy, which might lead to a more precise differentiation of respondents with different privacy needs.

Also, the study should be replicated in larger and more representative samples regarding, e.g., age, gender, and place of residence. Further, we assume that the evaluation of crime surveillance technologies is affected by current events such as terrorist attacks and offenses in the local environment. Thus, periodically longitudinal studies represent an interesting and necessary approach for further research. Certainly, the outcomes reflect a European perspective, only mirroring one cultural context which cannot be easily transferred to other countries in which the extent of crime might be completely different. Presumably, especially the trade-off between privacy and security might be different in societies with higher crime rates and a higher vulnerability of residents towards public assaults. Thus, we aim for a replication in other countries to compare crime surveillance needs and desires of city residents depending on their origins and cultures.

References

1. La Vigne, N.G., Lowry, S.S., Markman, J.S., Dwyer, A.M.: Evaluating the use of public surveillance cameras for crime control and prevention. Final Technical report, The Urban Institute, Justice Policy Centre, Washington, DC (2011)
2. Ziefle, M., Schneider, C., Valeé, D., Schnettler, A. Krempels, K.H., Jarke, M.: Urban Future outline (UFO): a roadmap on research for livable cities. ERCIM News 98 (2014). http://ercim-news.ercim.eu/en98/keynote-smart-cities
3. Dailey, K.: The rise of CCTV surveillance in the US. BBC News Magazine, 28 April 2013 (2013) http://www.bbc.com/news/magazine-22274770
4. La Vigne, N.G., Lowry, S.S.: Evaluation of camera use to prevent crime in commuter parking facilities: a randomized controlled trial. Technical report. Urban Institute, Washington, DC (2011)
5. Whitaker, R.: The End of Privacy: How Total Surveillance is Becoming a Reality. The New Press, New York (1999)
6. Cerezo, A.: CCTV and crime displacement: a quasi-experimental evaluation. Eur. J. Criminol. 10, 222–236 (2013)
7. Gerrard, G., Thompson, R.: Two million cameras in the UK. CCTV Image Mag. 42, 10–12 (2011)   http://www.securitynewsdesk.com/wp-content/uploads/2011/03/CCTV-Image-42-How-many-cameras-are-there-in-the-UK.pdf
8. Patton, J.W.: Protecting privacy in public? Surveillance technologies and the value of public places. Ethics Inf. Technol. 2, 181–187 (2000). https://doi.org/10.1023/a:1010057606781
9. Welsh, B.C., Farrington, D.P., Taheri, S.A.: Effectiveness and social costs of public area surveillance for crime prevention. Ann. Rev. Law Soc. Sci. 11(1), 111–130 (2015). https://doi.org/10.1146/annurev-lawsocsci-120814-121649
10. Jho, W.: Challenges for e-governance: protests from civil society on the protection of privacy in e-government in Korea. Int. Rev. Admin. Sci. 71(5), 151–166 (2005). https://doi.org/10.1177/0020852305051690
11. Joh, E.E.: Privacy protests: surveillance evasion and fourth amendment suspicion. Arizona Law Rev. 55, 997–1029 (2013). https://ssrn.com/abstract=2285095
12. Song, M., Dacheng T., Maybank, S.J.: Sparse Camera Network for Visual Surveillance—A Comprehensive Survey. Cornell University Online Library. https://arxiv.org/pdf/1302.0446.pdf (2013)
13. Hampapur, A., Brown, L., Connell, J., Ekin, A., Haas, N., Lu, M., Merkl, H., Pankati, S., Senioa, A., Shu, C.-F., Tian, Y.L.: Smart Video Surveillance – Exploring the concept of multiscale spatitemporal tracking. IEEE Signal Process. Mag. 22(2), 38–51 (2005). https://doi.org/10.1109/MSP.2005.1406476
14. Isnard, A.: Can surveillance cameras be successful in preventing crime and controlling anti-social behaviours? In: Character, Impact and Prevention of Crime in Regional Australia Conference. Townsville, Australia, 2nd–3rd August 2001 (2001) http://www.aic.gov.au/media_library/conferences/regional/isnard1.pdf
15. Cameron, A., Kolodinski E., May, H., Williams, N.: Measuring the effects of video surveillance on crime in Los angeles. report for california research Bureau. School of Policy Planning and Development, CA, USA. https://www.library.ca.gov/crb/08/08-007.pdf (2008)

16. Firmino, R.J., Kanashiro, M., Bruno, F., Evangelista, R., da Costa Nascimento, L.: Fear, security, and the spread of CCTV in Brazilian cities: legislation, debate, and the market. Journal of Urban Technology **20**(3), 65–84 (2013). https://doi.org/10.1080/10630732.2013. 809221
17. Sousa, W.H., Madensen, T.D.: Citizen acceptance of police interventions: an example of CCTV surveillance in Las Vegas, Nevada. Crim. Justice Stud. **29**(1), 40–56 (2016)
18. Wiecek, C., Saetnan, A.R.: Restrictive? Permissive? The Contradictory Framing of Video Surveillance in Norway and Denmark. Technical report, Department of Sociology and Political Science, Norwegian University of Science and Technology, Trondheim (2002). http://www.urbaneye.net/results/ue_wp4.pdf
19. Rogers, E.M.: Diffusion of Innovations. NY Free Press, New York (2003)
20. Davis, F.D., Bagozzi, R.P., Warshaw, P.R.: User acceptance of computer technology: a comparison of two theoretical models. Manage. Sci. **35**(8), 982–1003 (1989). https://doi.org/ 10.1287/mnsc.35.8.982
21. Venkatesh, V., Morris, M.G., Davis, G.B., Davis, F.D.: User acceptance of information technology. MIS Q. **27**(3), 425–478 (2003). https://doi.org/10.2307/30036540
22. Arning, K., Gaul, S., Ziefle, M.: "Same same but different" how service contexts of mobile technologies shape usage motives and barriers. In: Leitner, G., Hitz, M., Holzinger, A. (eds.) USAB 2010. LNCS, vol. 6389, pp. 34–54. Springer, Heidelberg (2010). https://doi.org/10. 1007/978-3-642-16607-5_3
23. van Heek, J., Arning, K., Ziefle, M.: Where, wherefore, and how? Contrasting two surveillance contexts according to acceptance. In: 6th International Conference on Smart Cities and Green ICT Systems (Smartgreens 2017). SCITEPRESS – Science and Technology Publications, pp. 87–98 (2017). https://doi.org/10.5220/0006325400780090
24. van Heek, J., Arning, K., Ziefle, M.: Safety and privacy perceptions in public spaces: an empirical study on user requirements for city mobility. In: Giaffreda, R., Cagáňová, D., Li, Y. (eds.) Internet of Things 2014 LNICST, vol. 151, pp. 97–103. Springer, Berlin (2015). https://doi.org/10.1007/978-3-319-19743-2_15
25. Luce, R.D., Tukey, J.W.: Simultaneous conjoint measurement: a new type of fundamental measurement. J. Math. Psychol. **1**, 1–27 (1964). https://doi.org/10.1016/0022-2496(64) 90015-x
26. Orme, B.: Interpreting the results of conjoint analysis, getting started with conjoint analysis: strategies for product design and pricing research, pp. 77–89. Research Publications, LLC Madison (2009)
27. Sawtooth Software. CBC/HB System for Hierarchical Bayes Estimation. Version 5.0. Sawtooth Software Research Paper Series (2009)
28. Rao, V.R.: Applied Conjoint Analysis. Springer, New York (2014). https://doi.org/10.1007/ 978-3-540-87753-0
29. Park, H.H., Oh, G.S., Paek, S.Y.: Measuring the crime displacement and diffusion of benefit effects of open-street CCTV in South Korea. Int. J. Law Crime Justice **40**(3), 179–191 (2012). https://doi.org/10.1016/j.ijlcj.2012.03.003
30. Saetnan, A.R., Lomell, H.R., Wiecek, C.: Controlling CCTV in public spaces: is privacy the (only) issue? Reflections on Norwegian and Danish observations. Surveill. Soc. **2**(2–3), 396–414 (2004)
31. van Heek, J., Arning, K., Ziefle, M.: How fear of crime affects needs for privacy and safety —acceptance of surveillance technologies in smart cities. In: Proceedings of the 5th International Conference on Smart Cities and Green ICT Systems, 23–25 April 2016, Rome, Italy, pp. 32–43 (2016)

32. Denning, P.J.: The Invisible Future: the Seamless Integration of Technology into Everyday Life. McGraw-Hill Companies, Inc., New York (2001) https://dl.acm.org/citation.cfm?id=504949
33. Welsh, B.C., Farrington, D.P.: Public area CCTV and crime prevention: an updated systematic review and meta-analysis. Justice Q. 26(4), 716–745 (2009). https://doi.org/10.1080/07418820802506206
34. Sawtooth Software: Testing the CBC Design (2015, 2017). https://www.sawtoothsoftware.com/help/lighthouse-studio/manual/hid_web_cbc_designs_6.html
35. Sawtooth Software: Market Simulators for Conjoint Analysis. Sawtooth Software Research Paper Series (2009)
36. Sawtooth Software: Survey Software & Conjoint Analysis - CBC Latent Class Technical Paper. Sawtooth Software Research Paper Series (2004)
37. Welsh, B.C., Farrington, D.P.: Making Public Places Safer: Surveillance and Crime Prevention. Oxford University Press, Oxford (2009). https://doi.org/10.1093/acprof:oso/9780195326215.001.0001
38. Himmel, S., Ziefle, M., Arning, K.: From living space to urban quarter: acceptance of ICT monitoring solutions in an ageing society. In: Kurosu, M. (ed.) HCI 2013. LNCS, vol. 8006, pp. 49–58. Springer, Heidelberg (2013). https://doi.org/10.1007/978-3-642-39265-8_6
39. Arning, K., Ziefle, M.: "Get that camera out of my house!" Conjoint measurement of preferences for video-based healthcare monitoring systems in private and public places. In: Geissbühler, A., Demongeot, J., Mokhtari, M., Abdulrazak, B., Aloulou, H. (eds.) ICOST 2015. LNCS, vol. 9102, pp. 152–164. Springer, Cham (2015). https://doi.org/10.1007/978-3-319-19312-0_13
40. Pavone, V., Esposti, S.D.: Public assessment of new surveillance-oriented security technologies: beyond the trade-off between privacy and security. Public Underst. Sci. 21 (5), 556–572 (2012). https://doi.org/10.1177/0963662510376886
41. Slobogin, C.: Public privacy: camera surveillance of public places and the right to anonymity. Mississippi Law J. 72, 213–299 (2002). https://doi.org/10.2139/ssrn.364600
42. Sheldon, B.: Camera surveillance within the UK: enhancing public safety or a social threat? Int. Rev. Law Comput. Technol. 25(3), 193–203 (2011). https://doi.org/10.1080/13600869.2011.617494
43. Bowyer, K.W.: Face recognition technology: security versus privacy. Technol. Soc. Mag. 23 (1), 9–19 (2004). https://doi.org/10.1109/MTAS.2004.1273467
44. Kim, J.E., Boulos, G., Yackovich, J., Barth, T., Beckel, C., Mosse, D.: Seamless integration of heterogeneous devices and access control in smart homes. In: 8th International Conference on Intelligent Environments. Guanajuato, México, 26th–29th June 2012 (2012). https://doi.org/10.1109/IE.2012.57
45. Visser, M., Scholte, M., Scheepers, P.: Fear of crime and feelings of unsafety in European countries: macro and micro explanations in cross-national perspective. Sociol. Q. 54(2), 278–301 (2013). https://doi.org/10.1111/tsq.12020
46. Foster, S., Wood, L., Christian, H., Knuiman, M., Giles-Corti, B.: Planning safer suburbs: do changes in the built environment influence residents' perceptions of crime risk? Soc. Sci. Med. 97, 87–94 (2013). https://doi.org/10.1016/j.socscimed.2013.08.010
47. Ajzen, I., Fishbein, M.: Attitude-behavior relations: a theoretical analysis and re-view of empirical research. Psychol. Bull. 88, 888–918 (1977). https://doi.org/10.1037/0033-2909.84.5.888
48. Ajzen, I., Fishbein, M.: Understanding Attitudes and Predicting Social Behavior. Prentice-Hall, Englewood Cliffs (1980)

A Strategic Urban Grid Planning Tool to Improve the Resilience of Smart Grid Networks

Eng Tseng Lau[1]($^{\boxtimes}$), Kok Keong Chai[1], Yue Chen[1], and Alexandr Vasenev[2]

[1] School of Electronic Engineering and Computer Science,
Queen Mary University of London, Mile End Road, London E1 4NS, UK
{e.t.lau,michael.chai,yue.chen}@qmul.ac.uk
[2] Faculty of Engineering, Mathematics and Computer Science, University of Twente,
Drienerlolaan 5, 7522 NB Enschede, The Netherlands
a.vasenev@utwente.nl

Abstract. The unresponsive and poor resilience of the traditional city architecture may cause instability and failure. Therefore, strategical positioning of new urban electricity or city components do not only make the city more resilient to electricity outages, but also a step towards a greener and a smarter city. Money and resilience are two conflicting goals in this case. In case of blackouts, distributed energy resources can serve critical demand to essential city components such as hospitals, water purification facilities, fire and police stations. In addition, the city level stakeholders may need to envision monetary saving and the overall urban planning resilience related to city component changes. In order to provide decision makers with resilience and monetary information, it is needed to analyze the impact of modifying the city components. This paper introduces a novel tool suitable for this purpose and reports on the validation efforts through a stakeholder workshop. The outcomes indicate that predicted outcomes of two alternative solutions can be analyzed and compared with the assistance of the tool.

Keywords: Stakeholder workshop · System design · Monetary cost Grid resilience · Smart grid

1 Introduction

The increased interconnectivity and deployment of smarter grids, distributed energy resources (DER), as well as increased consumer demands and critical facilities but with limited amount of storage technology available to store excessive amount of generated energy make energy such a limited resource. The robustness and resilience of the grid can be formulated to evaluate the way to share a limited resource between multiple stakeholders. To find the optimal arrangements, stakeholders need to collaboratively plan an overall grid system. Additionally, robustness and resilience management is important for stakeholders to evaluate

© Springer Nature Switzerland AG 2019
B. Donnellan et al. (Eds.): SMARTGREENS 2017/VEHITS 2017, CCIS 921, pp. 150–167, 2019.
https://doi.org/10.1007/978-3-030-02907-4_8

the improved grid system on possible undesirable events. This is because enhancing the robustness and resilience may (or may not) incur additional monetary costs.

Furthermore, the integration of a renewable into the grid is an another dimension of challenge that concerns multiple domains. One might consider the renewable energy-related landscape [3] aims to reduce greenhouse gas emission [21]. In addition, to find a suitable location for a biogas plant, the distances from the site to the biomass sources must be accounted for [7]. In the case of solar investments, an important concern is the interplay between the urban form and solar energy inputs [1]. Importantly, utility planners should consider how the grid can operate during contingency events (see e.g. [4,12]).

Additionally, there is a need to account for grid resilience – the ability of the grid to withstand a failure in an efficient manner [5]. Specifically, it concerns supplying electricity to critical infrastructures (e.g., hospitals) during blackouts, as well as the ability to quickly restore normal operation state [5]. Threat analysis related to non-adversarial and intentional threats (e.g., [18,19]) can highlight which components may deserve particular attention. DER can also be used to compensate for the discontinuity of electricity produced by intermittent renewables. However, optimizing the cost of dispatches of DG units is needed to ensure that this task performed efficiently. Stakeholders also may need to consider both monetary and resilience aspects to account how a city benefits from installation of new components such as DER. The considerations include the mitigation of fault and attack, threat ranking, the monetary cost and resilience analysis, and the impact on different critical infrastructures. In this regard, a decision making tool – Overall Grid Modelling (OGM) was built to demonstrate effect of grid component changes based on the perspectives of strategic planning within stakeholders.

This paper reports initial features and functionalities of the OGM tool related to several state-of-the-art tools for modelling and controlling smart grids in terms of its practicability and efficiency. The OGM tool validation efforts are conducted through a stakeholder workshop. These aspects are relevant to evaluate limits and possible overlaps in functionality, and the reasonably expected scalability of the OGM tool. It is an approach supported by the OGM tool, where the tool simulation provides a perception towards decision makers of the grid elements that they wish to optimize.

The overall organisation structure of the paper is as follows: Sect. 2 reviews the state-of-the-art modelling tools to ensure that the OGM tool is aligned with standard core functionalities of the existing tools. Section 3 presents the methodology of OGM tool usages. Section 3.1 reports the methodology of the stakeholder workshop organized that validates the functionalities of the OGM tool. Section 4 presents the findings through the stakeholder workshop. Finally, Sects. 5 and 6 discuss and conclude the findings.

2   State-of-the-Art Modelling Tools

In this section, a state-of-the-art modelling and controlling smart grid tools are reviewed. The review aims to identify functionalities that can be represented to users. The improvements in the interface can be studied in terms of the readability, the way how the output results presented, whether the tool provides a clear implication (positive or negative aspects) to users, and how the tool suits the users' needs and requirements with respect to the output results. The functionalities of the smart grid tools are cross-related in order to ensure that the OGM tool is well-aligned with standard core functionalities of existing smart grid tools.

DNV GL has developed a microgrid mathematical optimization tool [6] to evaluate the full integration of distributed generations, electrical, thermal storages, technological updates, building automation and customers' behavioural usages. The software module also includes the detailed policy drives, climate, technology cost projections and tariffs at which referring specifically to a particular geographical location. The holistic-based simulation aims at maximizing the economic value and reliability of electrical system and energy. The model simulates the day-ahead energy prices, demand forecasts, weather forecasts, dynamic performance of the buildings, storage, CHP, distributed generation, and demand management mechanism that optimizes the energy economics during the day. The optimization problem is formulated through the Mixed Integer Linear Programming (MILP) approach. The overall reliability of the grid is assessed by perturbing the grid with outages or contingencies through the relevant utility statistics (SAIDI, SAIFI). The optimization tool is also capable of maximizing the uninterruptable and critical load that can be served from available resources during the outage period.

The Massachusetts Institute of Technology (MIT) has built a laboratory-scale microgrid based on the earlier model developed from computer simulation studies [17]. The institute aims to evaluate the transition of voltage that may lead to voltage instability (the disconnection and reconnection to the central power grid). The project focuses on a laboratory-scale power system that combines the energy generation and storage devices to serve local customers at low level grid. The Masdar Institute corporates with MIT by concentrating on analytical-based weighted multi-objective optimization methods. The analytical methods analyze the system configuration and operation planning simultaneously in order to determine the costs and emissions. The method generates a set of optimal planning/designs and operating strategies that minimizes costs and emissions simultaneously.

Siemens PTI provides a consultant service, software and training program to optimize system networks for generation, transmission and distribution and power plants for smart grids [16]. The consulting services offer expertise in system dynamics and threat analysis, energy markets and regulation, control systems, power quality, and steady-state and dynamic system evaluations. The software solutions with completed power system analysis tools include PSS®E, PSS®SINCAL PSS®NETOMAC, PSS®ODMS, PSS®MUST, and MOD®. Value propositions considered are: Reliability, fuel savings, and environmental benefits.

Microgrid Master Controller software developed by Etap Grid performs the detailed modelling, simulation and optimization of electrical systems [8]. The software controller predicts and forecasts energy generations and loads. The controller also integrates and automatically controls the microgrid elements, such as PVs, energy storages, back-up generations, wind, gas turbines, CHP, fuel cells, and demand management. The software automatically optimizes the grid during grid-connected or islanded grid operations. The economic cost calculation is the main value proposition in Etap Grid, as the software aims to lower the total cost of ownership by reducing the average cost of electricity from the national electricity price.

Argonne National Laboratory (ANL) offers a range of resilient-based tools, techniques, and engineering methods to optimize the interdependencies of energy and global security needs [2]. These include the power infrastructure modelling tool that inspects the impact of power outages in the large grid, and power system restoration optimization tool combined with AC power flow-based cascading failure/outage. The tool models the tendency of islanding operations, either synthetic based or natural threats. Example of applications include: identification of system vulnerabilities and implementation of preventative measures; critical power infrastructure, resiliency analysis; and system dependency/interdependency analysis with non-power infrastructure systems. The integrated system restoration optimization module supports restoration planning and operational decision-making in the transmission and distribution systems. The cascading failure module considers system monitoring, protection, control and further simulates the most important cascading techniques. The module further provides cascading risk analysis and generates credible cascading scenarios for restoration purposes.

2.1   Summary of Smart Grid Modelling Tool Functionalities

Table 1 summarizes important features of the mentioned modelling tools. Even though numbers of tools exist to model grids, they lack important features to enable an interactive resilience analysis, such as user interfaces and resilience calculations modules. Moreover, most of the tools do not provide user interfaces. This can hamper the interactive tool navigation and analysis as required by users. More importantly, having user interface enables interactions with users from various backgrounds. At the moment, only the DNV GL tool accounts for critical loads. Loads are important especially in times of blackouts urban-level loads can be more critical than others. Therefore, prioritization of critical loads are particularly relevant for a resilience analysis tool.

In contrast, the OGM tool introduced below, particularly focus on addressing the interaction needs and the aforementioned shortcomings. Through the implemented mathematical optimization module, the important features such as the simulation of outage, islanding operation, cost and resilience analysis are performed. The users are able to manipulate/control the tool and changes in the resilience coefficient are demonstrated that reflect grid structural changes implemented by users whenever a new case/scenario is applied (i.e., adding or remove a local generator). The methodology, policy and the development governing the OGM is available in the documentations [10,11]. The tool enables the decision makers to manipulate/control the grid component changes and varieties of resilience coefficient metric and cost analysis are illustrated through the alterations within the grid components. The tool provides decision makers the best option in terms of grid planning, and also the information on how the introduction of a city component increases grid resilience and also account for possible monetary savings. In line with International Electrotechnical Commision [9], the OGM tool supports simulation of electricity continuity planning and also ensuring the cost concerned through the interventions for benefits of business planning.

Table 1. Summary of modelling tools in comparison with the OGM tool. Adapted from [13].

Tool		DNV GL	MIT	Masdar Institute	Siemen PTI	Etap Grid	ANL	OGM tool
Functionality	Mathematical optimization	✓		✓	✓	✓	✓	✓
	User interface ready	✓			✓			✓
	Grid topology ready				✓			✓
	Prototype based		✓					
	Demand forecasts	✓				✓		✓
	Generation and storage modelling	✓	✓	✓	✓	✓	✓	✓
	Account for critical loads	✓					✓	✓
	Support of threat ranking						✓	✓
	Islanding operation	✓	.			✓	✓	✓
	Scenario/case studies	✓	✓		✓	✓		✓
	Outage/contingency simulations	✓			✓	✓	✓	✓
	Cost analysis	✓		✓	✓	✓		✓
	Emission analysis	✓		✓				
	Resilience analysis						✓	✓
	Reliability analysis	✓	✓	✓	✓		✓	
	Power flow analysis		✓		✓	✓	✓	

3   The OGM Tool

The OGM tool development was based on the agile process, where the processes of specification, design, implementation and evaluation strategies are concurrent, and as an iterative approach. The tool is developed in a series of increments where the user will evaluate each increment and make proposals for later improvements.

The OGM tool incorporates a GUI (see Table 1). To facilitate continuous interactions in a user-friendly and easily controllable manner, the user is also able to simulate several use-case scenario in order to observe the output changes directly where the components can be introduced/removed/moved within the grid (i.e. the addition/removal of particular consumption profiles, critical loads, generators, storages, renewables, outage simulation and islanding analysis). The tool allows concurrency in updating new trends of input information provided by the user using the existing model. The OGM tool is aimed for decision makers (Municipal authority planner, DNO, Developers, Critical Infrastructure Operator, Business and Citizen Representative) with various technical/conceptual background that aims to be easily-interpretable, without incorporating complex power-flow model and analysis. The expertise of the decision makers is essential to account for sound strategic grid planning.

An example of a network topology tree (or the system architecture) is shown in Fig. 1, where the architecture included a number of city grid components. The distribution of grid components as in Fig. 1 is presented in Table 2.

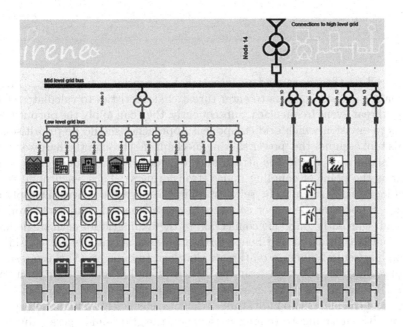

Fig. 1. The baseline system architecture of the OGM tool.

The tool simulates outage consequences using the input of known outage scenario through the known critical loads and specifics of generation profiles. Then, computations modules in the OGM will process the outage scenario. Output results will demonstrate the monetary savings and resilience indicator through the component changes. Through the output results, the decision makers will select most suitable alternative for grid outage mitigations and repeat the simulation if needed.

The threat analysis and ranking is another distinctive feature of the OGM tool, where the tool provides the analysis of non-adversarial threats (e.g., [18,

Table 2. Number of distributed generators, energy storages, types of consumer profiles and their populations included.

Node no.	Number of generators		Number of energy storage	Profiles included	Populations
	Non-renewable	Renewable			
1	2	0	0	Households	2500
2	3	0	1	Offices	2
3	3	0	1	Hospitals	2
4	2	0	0	Supermarkets	3
5	2	0	0	Warehouses	8
6	0	0	0	-	-
7	0	0	0	-	-
8	0	0	0	-	-
9	1	0	0	-	-
10	0	0	0	-	-
11	1	2	0	-	-
12	0	1	0	-	-
13	0	0	0	-	-
14	0	0	0	-	-

19]) as well as threats related to intentional disruptive actions (e.g., [14,20]). This enables decision makers to enter threat characteristics to calculate relative value of threat event frequencies. Subsequently, they can apply the output threat analysis to envision which grid component approaches should be prioritised.

This tool assumes the 'power sharing mechanism' through the hardware solution to island a microgrid (de-attach and re-attach it to the main grid) can be located at the point of coupling nodes (transformers). Thus, each node with a critical load might strive to be self-sustaining: balance the (critical) supply and demand. A node can be either connected or disconnected completely from the main grid. Currently, only one connection to the main grid for each single nodes in the tool is considered and hence forth a meshed network is not considered. The tool particularly focuses on threats that lead to outages: (1) those resulting in the disconnection of a node from the main grid; and (2) outage of a component (e.g., a DER as an electricity generation component).

The tool calculates two indicators – resilience coefficients and monetary costs (with or without savings) – to inform users how the grid would operate during an outage event. Resilience is the ability of a power system to remain or withstand a failure, and to restore quickly to the normal operating state [5]. In order to justify the grid resilience, a resilience performance metric is used. The resilience coefficient in this paper is computed based on the extents in which the amount of energy demand within consumers are met when there is an outage in the grid [5]. A grid is robust and resilient when the computed resilient coefficient is high, or is maintained throughout the outage period. The resilient coefficient is determined as the mean fraction of the demand served for the outage node divided by the overall demand to be served. Similar to [13], the resilience coefficient in this case is therefore the fraction of demand served for ith consumer ($P_{i,t}$) divided by the total demand $D\ (P_{I,t})$ in the contingency state at time t:

$$\alpha_R(t) = \frac{P_{i,t}}{P_{I,t}}. \tag{1}$$

The monetary cost C in this paper is calculated as the difference between the business-as-usual traditional grid operation cost (without capability of islanding, and also without implementation of DGs, energy system storages and renewables (C_{BAU})) and the optimized grid operation cost (when DERs are activated). The negative monetary cost value computed indicates that the monetary saving is not achieved - the particular improvement incurs additional costs.

$$C(t) = C_{BAU} - C_{optimized}. \tag{2}$$

Figure 2 shows the example of resilience coefficient and monetary costs calculated for the grid as described in Fig. 1. The top panel presents the plot of monetary savings in comparison to the business-as-usual and the optimized grid planning. The bottom panel illustrates the distribution of resilience coefficient metric. Negative monetary savings indicate additional costs, whereas positive savings indicate the cost saving of the improvement in the grid operation. The resilience coefficient would be between 0–1 (the resilience coefficient is computed as zero at a particular time interval when no outage occurs) because of the fraction of demand served over the overall demand during an outage event.

The GUI implementation of the OGM is developed using IntelliJ IDEA, the Java IDE software. The dual-simplex is used for the numerical grid optimization of the Linear Programming problem. The lp_solve 5.5.2.3 [15] is applied as the library file for Java that is called to perform the optimization algorithm for the OGM tool.

3.1   Methodology

The stakeholder workshop was conducted to validate the applicability and the scalability of the OGM tool, using the expertise of the stakeholders with years of experience in real-life domains. In the beginning of the workshop, mini-lectures on the logic and assumptions behind the development of the OGM tool, as well as the smart grid technologies were delivered to introduce stakeholders to major ideas of smart grids, as well as the current issues and challenges. The OGM tool was demonstrated to stakeholders to clarify the idea how modelling tools can be used to improve the resilience of the overall grid.

The configuration as defined in Fig. 1 was simulated which further enabled fellow stakeholders to modify the grid components with the intention of improving the resilience of the grid. During the workshop session, exercise handouts were given to three stakeholders who represented different stakeholder roles (City planner, Distribution Network Operator, Citizen & Business representatives). The stakeholders need to collaborative decide how to introduce new components or modifying the existing components to improve resilience of the grid. The system architecture in Fig. 1 and Table 2 was used as the baseline configuration,

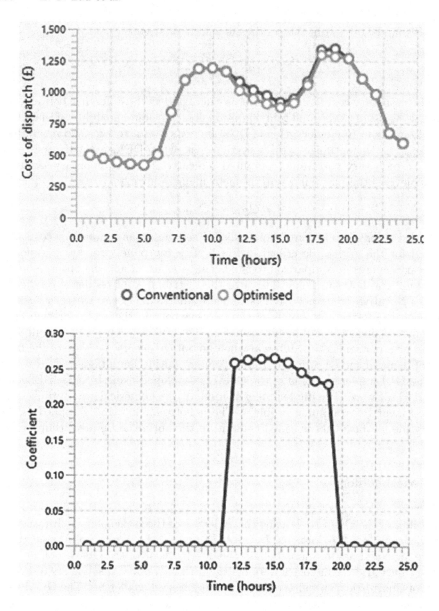

Fig. 2. Resilience coefficient and monetary costs calculated for the grid: top panel – plot of monetary savings in relation to the business-as-usual and the optimized solution; bottom panel – the distribution of resilience coefficient. Outage starts at 1200 with durations of eight hours.

where the amount of renewable sources are low. In addition to the description of the grid architecture, stakeholders were briefed on the changes that the grid context might undertake. It was suggested that the populations within the city are increased, and towards the decarbonization plan. Specifically, amount of city components would be as follows: Households = 4500; Offices = 3; Hospitals = 3; Supermarkets = 5; Warehouses = 12.

After providing the information, stakeholders were asked to discuss what grid updates might be introduced to ensure that a city can withstand a blackout with less negative impact as possible. The aim of this exercise is to investigate how the manipulation of the OGM tool can guide the fellow stakeholders to improve the resilience of a complex urban grid, in the context of collaborative decision making in the situation of uncertainty.

Two different outage scenarios (4 and 8 h) were chosen to examine the impact of grid component changes on the resilience of the city in sustaining both the shorter or longer outages. In addition, the outage in every single node is also examined in order to examine the outage effects on the changes of the supply and across individual consumer and the overall demand, as well as the changes in the monetary savings and resilient coefficient in the grid level city as shown in Table 3. The 'economic-islanding' capability during the normal grid operation is enabled that employs DERs to provide demand management capability at times of high electricity price, rather than drawing the electricity from the main grid [11]. Questionnaires were disseminated to fellow stakeholders at the end of the workshop.

Table 3. Type of grid operations and the indicators applied.

Grid operation			Indicators	
			Resilient coefficient	Monetary cost
Normal (Economic islanding enabled)				✓
Outage	**Duration** (hrs)	**Type**		
	4	single	✓	✓
	8	single	✓	✓
	4	complete	✓	✓
	8	complete	✓	✓

4   Validating the Tool

4.1   Results

In order to access the effectiveness of the collaborative decisions as made by fellow stakeholders, normal and failure of grid operations are simulated for each node, and also the complete grid outage. Outages occur due to the grid failures (e.g. a line-disconnection between the microgrid and main grid level, and also the line disconnection within the microgrid nodes). When there is a failure event, the islanding capability is triggered to ensure uninterrupted operation during a utility system outage through the N-1 compliance [11]. Decisions placed and the performance of the implemented decisions by stakeholders are compared with the baseline case in terms of resilience coefficients and monetary savings. The decisions were evaluated using the OGM tool and the timeline for the simulation is allowed for 24 h. The grid with various operating conditions were simulated for the baseline case and two solutions as placed by stakeholders in considering decarbonization strategy.

After some discussions, stakeholders proposed the first solution based on the modification of grid components in the baseline case (Fig. 1), as shown in Fig. 3. The updates were:

1. Remove a generator from Node 2;
2. Remove a generator from Node 3;
3. Add a PV generator in Node 2;
4. Add a wind generator in Node 2;
5. Add an energy storage system in Node 1.

The second solution as proposed by stakeholders, using the baseline case (Fig. 1) were presented in Fig. 4, which were:

1. Remove an energy storage system in Node 3;
2. Add a generator in Node 3.

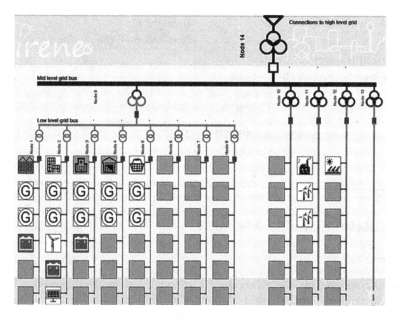

Fig. 3. The first solution of system architecture as proposed by stakeholders.

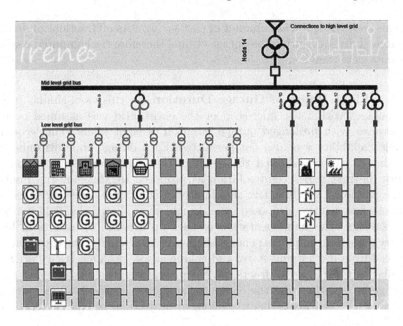

Fig. 4. The second solution of system architecture as proposed by stakeholders.

Case 1 – Normal Operation. In this case, assuming no failure occurred, the normal mode of operation was applied. The monetary cost and resilience coefficient achieved for baseline, first and second solutions as proposed by stakeholders were shown in Table 4. Based on Table 4, the first solution proposed by stakeholders achieved higher amount of monetary savings than the first solution, and also higher than the Baseline case. Higher amount of cost savings achieved during the 'economic-islanding' normal mode of grid operations. The resilience coefficients were all zeros as the grid resilience was not considered during the normal operation mode (without any outage events). The simulation, however excluded the addition of start-up investment, installation and maintenance costs of individual DERs.

Case 2 – Four Hours of Outage Duration. In Case 2, the outages in microgrid or the entire grid was assumed occur at 0900 for the duration of four hours. The capability of 'economic-islanding' was disabled in the case of outage events. Table 5 showed the result of the simulation using the baseline, followed by the first and the second solution. Negative sign indicated that additional costs were introduced (no monetary savings are achieved). Overall the baseline scenario promoted highest amount of cost savings than the decisions as imposed by stakeholders. This was due to the introduction of renewables that required higher amount of costs for generations compared with conventional generators. However, cost savings were reduced in the first solution, where energy storages were employed. As energy storages generated zero cost during the discharging state, this created significant amount of cost savings. As all fractions of demands were successfully met during the outage events, therefore the computed resilience coefficients were identical.

Case 3 – Eight Hours of Outage Duration. In this case, similar to Case 2, the outages within the microgrid or the entire grid was assumed occur at 0900 however with prolonged outage duration of eight hours. The 'economic-islanding' capability was also not permitted. Each outage node disconnections was evaluated. Table 6 showed the result of the simulation using the baseline scenario, the first and second solution as proposed by stakeholders. Similarly as in the previous case, Negative sign indicated additional costs are introduced. Overall the first solution proposed by stakeholders promoted the highest amount of cost savings. The implementation of back-up generations results in higher monetary costs than employing energy storages (where energy storages generated zero cost during discharging state, and charge at low peak electricity price).

Similar to Case 2, as all fractions of demands were successfully met during the outage events, therefore the computed resilience coefficients were identical, but with varied coefficients due to prolonged projections of outage durations.

Table 4. Case 1 – cost savings and resilience coefficient in normal mode.

	Baseline	First solution	Second solution
Cost savings ($£$)	885.72	**1023.26**	890.76
Resilience coefficient	0	0	0

Table 5. Case 2 – cost savings and resilience coefficient in outage mode.

Outage node	Cost savings ($£$)			Resilient coefficient		
	Baseline	First	Second	Baseline	First	Second
Node 1	−299.75	−489.9	−489.9	0.111	0.111	0.111
Node 2	−556.84	−546.98	−546.98	0.430	0.430	0.430
Node 3	−291.56	−286.57	−486.07	0.240	0.240	0.240
Node 4	−410.3	−400.01	−400.01	0.144	0.144	0.144
Node 5	−428.09	−437.74	−437.74	0.074	0.074	0.074
Grid outage	−296.76	−325.23	−400.23	1	1	1
Total savings $£$	**−2283.3**	−2486.43	−2760.93	–	–	–

Table 6. Case 3 – cost savings and resilience coefficient in outage mode.

Outage node	Cost savings (£)			Resilient coefficient		
	Baseline	First	Second	Baseline	First	Second
Node 1	−670.87	−659	−659	0.111	0.111	0.111
Node 2	−1571.89	−955.64	−955.64	0.430	0.430	0.430
Node 3	−867.42	−622.1	−622.1	0.240	0.240	0.240
Node 4	−710.95	−646.53	−646.53	0.144	0.144	0.144
Node 5	−690.21	−689.65	−689.65	0.074	0.074	0.074
Grid outage	1817.43	1850.51	2118.89	1	1	1
Total savings £	−5411.34	**−4307.06**	−4637.06	–	–	–

5 Discussion

Overall, the stakeholder workshop was successfully conducted and two different scenarios of grid configuration changes were proposed by stakeholders, in comparison with the baseline case. The stakeholder workshop indicated that the proposed tool can support extensive collaboration between stakeholders who actively engage in discussions with each others for increasing the robustness of the electricity network. During the workshop, stakeholders suggested several ideas for improving the OGM tool, such as to account for the capital costs of investments, integrate flexibility to allow for city configurations, improve the user-friendly interface, store output parameters for comparisons based on different component alterations, and also to breakdown cost savings to reflect where changes affect the whole grid system.

The questionnaire feedback was delegated to fellow stakeholders at the end of the workshop. The questionnaire feedback was shown in Table 7. The outcomes of the workshop showed that the tasks related to grid component changes in responding to decarbonization strategy could be effectively performed in an understandable manner. Results can be compared and a better alternative based on the comparisons could be selected.

One of the stakeholder with electricity market knowledge noted that no expert knowledge was required to use the OGM tool. Additionally, another stakeholder praised the calculations and the scope of the OGM tools in performing the necessary tasks. The stakeholders positively noted the practicability of the demand management capability in the OGM tool, assumptions on uninterruptible loads, the efficiency of OGM tool in running/re-running a simulation, the ease of understanding the performance metric 'resilience coefficient' in measuring the performance of different grid topologies/configurations, and being useful as a collaborative-decision making system. One of the stakeholder recognized an opportunity of the tool to assist with network congestion as a very important aspect that is of value to utility companies. However, the immediate benefit could be realized if the tool will be road tested with some utility companies so

Table 7. Questionnaire results.

Question	Rating scale (1 – Very negative, 7 – Very positive)						
	1	2	3	4	5	6	7
	Number of respondents						
Q1. Knowledge on smart grids	0	0	1	1	0	1	0
Q2a. Practicability of demand management capability	0	0	0	0	2	0	1
Q2b. Practicability of controlled generations	0	0	0	1	1	1	0
Q2c. Practicability of islanded operation during outage	0	0	0	1	2	0	0
Q2d. Practicability of disconnected load during outage	0	0	0	2	1	0	0
Q2e. Practicability of critical loads	0	0	0	0	1	1	1
Q2f. Practicability of uninterruptible loads	0	0	0	0	2	1	0
Q3a. Effectiveness of OGM tool in addressing outage	0	0	0	2	1	0	0
Q3b. Effectiveness of the demand forecast	0	0	1	0	0	2	0
Q4a. Speed of OGM tool to run/re-run a simulation	0	0	0	1	0	1	1
Q4b. Speed of OGM tool to construct/re-construct grid components	0	0	1	1	0	1	0
Q4c. Speed of OGM tool to run/re-run demand forecast	0	0	0	1	2	1	0
Q5a. Level of knowledge required in using the tool	0	0	1	0	1	1	0
Q5b. Level of easiness in using the tool	0	0	0	1	1	1	0
Q6. Reason for rating as 5 or above in Q5	–No expert knowledge required to use the tool –Tool is for special market, so industry knowledge is required						
Q7. Understandable of resilient coefficient metric	0	0	2	0	1	0	0
Q8. Practicability of resilient coefficient metric	0	0	2	1	0	0	0
Q9. Practicability of evaluating electricity network	0	0	0	1	1	1	0
Q10. Fast in providing simulation analysis	0	0	0	0	1	2	0
Q11a. Usefulness of the tool in addressing outage in urban electricity network	0	0	0	0	1	2	0
Q11b. Usefulness of the tool as a collaborative decision support system	0	0	0	0	0	3	0
Q11c. Usefulness of the tool in establishing collaborative frameworks among stakeholders	0	0	0	0	1	2	0

that it could be proven and validated. This would help define the next steps of activity and help make the tool appealing to a range of potential market sectors.

However, one of the stakeholder (business and citizen representative) argued that specialised industry knowledge was required in order to fully understandable in using the OGM tool. This was because business and citizen representative might have low level electricity and smart grid background knowledge. Still, the time needed to construct/re-construct the grid components was found inefficient. Also, the last stakeholder explicitly voted that high level of knowledge was required in using the tool. One of the stakeholder mentioned that even though the OGM tool was almost immediately applicable, there should be a need to recognition that cross connections would also exist in addition to the vertical hierarchy (single line electricity connection). As a tool to explore islanding, the tool would need to additionally consider the transition from grid to microgrid and back again. Additionally, at some instance stakeholders noted the unrealistic practicability of using the metric 'resilient-coefficient' in tool simulations.

6   Conclusion

This paper presents an approach of using a strategic urban grid planning tool to improve the resilience of smart urban electricity networks. The approach allows decision makers to envision and manipulate grid component changes and further examine the resilience coefficient metric and the potential monetary savings across the grid. New analysis results are demonstrated whenever a grid component modification is applied. The approach is supported by the OGM tool. The tool simulates and provides a feedback towards decision makers of the grid elements that they wish to improve.

The tool was validated during a stakeholder workshop. Different cases and solutions were proposed by stakeholders were calculated to show the trade-off in between the resilient coefficient and monetary savings. The OGM tool was found useful to point out those complex aspects as proposed that should be considered to minimize such trade-offs.

In summary, the idea and logic of using the tool for grid planning are well-received. The survey feedback gathered would not only further supports and complements the analysis, but also to improve the efficiency and practicability the OGM tool.

As the agility concept is strongly supported by the OGM tool development, the continuous improvement strategy based on feedbacks obtained are implemented into the tool in responding to the requirements from fellow decision makers. The practicality and efficiency of the tool are continuously enhanced to improve the overall experience in using the tool to support the grid planning through the information provided by decision makers. Future major improvements would be related to moving towards the meshed grid topology, as the current tree representation of the grid architecture implemented in the OGM tool is one of the limitations pointed out by stakeholders.

Acknowledgement. This work was partially supported by the Joint Program Initiative (JPI) Urban Europe via the project IRENE (Improving the Robustness of Urban Electricity Network). Grant Reference: ES/M008509/1. Further information about project IRENE is available in the weblink: http://ireneproject.eu.

References

1. Amado, M., Poggi, F.: Solar urban planning: a parametric approach. Energy Procedia **48**, 1539–1548 (2014)
2. Argonne National Laboratory (ANL): Resilient infrastructure capabilities (2016). http://www.anl.gov/egs/group/resilient-infrastructure/resilient-infrastructure-capabilities. Accessed 19 Jan 2017
3. Barjis, J.: Collaborative, participative and interactive enterprise modeling. In: Filipe, J., Cordeiro, J. (eds.) ICEIS 2009. LNBIP, vol. 24, pp. 651–662. Springer, Heidelberg (2009). https://doi.org/10.1007/978-3-642-01347-8_54
4. Bennett, B.: Understanding, Assessing, and Responding to Terrorism: Protecting Critical Infrastructure and Personnel. Wiley, Hoboken (2007)
5. Bollinger, L.A.: Fostering climate resilient electricity infrastructure (2015). http://repository.tudelft.nl/islandora/object/uuid:d45aea59-a449-46ad-ace1-3254529c05f4/datastream/OBJ/download. Accessed 06 Dec 2016
6. DNV GL: Microgrid optimizer - a holistic operational simulation tool to maximize economic value or electrical power reliability (2016).. http://production.presstogo.com/fileroot7/gallery/DNVGL/files/original/3a1dd794f6ff46b9a279175c15af0f11.pdf. Accessed 05 Dec 2016
7. Dugan, R., McGranaghan, M.: Sim city. IEEE Power Mag. **9**(5), 74–81 (2011)
8. ETAP Grid: Power technologies international (2015). http://etap.com/Documents/Download%20PDF/ETAP-Grid-2015-LQ.pdf (2015). Accessed 19 Jan 2017
9. IEC: White paper - microgrids for disaster preparedness and recovery with electricity continuity and systems. Technical report, IEC WP Microgrids, Switzerland (2014)
10. IRENE: D2.2 - Root causes identification and societal impact analysis. Technical report (2016)
11. IRENE: D3.1 - System architecture design, supply demand model and simulation. Technical report (2016)
12. Jung, O., et al.: Towards a collaborative framework to improve urban grid resilience. In: Proceedings of 2016 IEEE International Energy Conference (ENERGYCON), 4–8 April, pp. 1–6. IEEE (2016). https://doi.org/10.1109/ENERGYCON.2016.7513887
13. Lau, E.T., Chai, K.K., Chen, Y., Vasenev, A.: Towards improving resilience of smart urban electricity networks by interactively assessing potential microgrids. In: Proceedings of the 6th International Conference on Smart Cities and Green ICT Systems (SmartGreens 2017), Porto, Portugal, 22–24 April 2017, pp. 1–8 (2017). https://doi.org/10.5220/0006377803520359
14. Le, A., Chen, Y., Chai, K.K., Vasenev, A., Montoya, L.: Assessing loss event frequencies of smart grid cyber threats: encoding flexibility into FAIR using bayesian network approach. In: Hu, J., Leung, V.C.M., Yang, K., Zhang, Y., Gao, J., Yang, S. (eds.) Smart Grid Inspired Future Technologies. LNICST, vol. 175, pp. 43–51. Springer, Cham (2017). https://doi.org/10.1007/978-3-319-47729-9_5

15. lp_solve: Introduction to lp_solve 5.5.2.5 (2015). http://lpsolve.sourceforge.net/5. 5/. Accessed 19 Oct 2016
16. Siemens PTI: Power technologies international (2016). http://w3.siemens.com/ smartgrid/global/en/products-systems-solutions/software-solutions/planning-data-management-software/PTI/Pages/Power-Technologies-International-(PTI). aspx. Accessed 19 Jan 2017
17. Stauffer, N.: The microgrid - a small-scale flexible, reliable source of energy (2012). http://energy.mit.edu/news/the-microgrid/. Accessed 19 Jan 2017
18. Vasenev, A., Montoya Morales, A.L.: Analysing non-malicious threats to urban smart grids by interrelating threats and threat taxonomies. In: Proceedings of 2016 IEEE International Smart Cities Conference (ISC2), Trento, Italy, 12–15 September 2016, pp. 1–4. IEEE (2016)
19. Vasenev, A., Montoya Morales, A.L., Ceccarelli, A.: A Hazus-based method for assessing robustness of electricity supply to critical smart grid consumers during flood events. In: Proceedings of the 11th International Conference on Availability, Reliability and Security, ARES 2016, Salzburg, Austria, 31 August–02 September 2016, pp. 223–228. IEEE (2016)
20. Vasenev, A., Montoya, L., Ceccarelli, A., Le, A., Ionita, D.: Threat navigator: grouping and ranking malicious external threats to current and future urban smart grids. In: Hu, J., Leung, V.C.M., Yang, K., Zhang, Y., Gao, J., Yang, S. (eds.) Smart Grid Inspired Future Technologies. LNICST, vol. 175, pp. 184–192. Springer, Cham (2017). https://doi.org/10.1007/978-3-319-47729-9_19
21. Zubelzu, S., Alvarez, R., Hernandez, A.: Methodology to calculate the carbon footprint of household land use in the urban planning stage. Land Use Policy 48, 223–235 (2015)

Intelligent Vehicle Technologies

Scenario Interpretation for Automated Driving at Intersections

David Perdomo Lopez[1]([✉]), Rene Waldmann[1], Christian Joerdens[1],
and Raul Rojas[2]

[1] Automated Driving, Volkswagen Group Research, Wolfsburg, Germany
{david.perdomo.lopez,rene.waldmann,christian.joerdens}@volkswagen.de
[2] Department of Mathematics and Computer Science, Freie Universität Berlin,
Berlin, Germany
rojas@inf.fu-berlin.de

Abstract. Driving at urban intersections is a very tough issue due to the complexity of the scenario. The driver is required to understand the traffic rules, predict the motion of other vehicles and, accordingly, make the proper decision. In this sense, automated driving systems in such environments become an important objective from a research point of view. Particularly, understanding the surrounding of the ego vehicle represents a challenging task. In this paper we propose an approach that simplifies the interpretation of the scenario. This concept aims to break down the whole maneuver in a set of primary situations. Accordingly, this facilitates the decision making at intersection and the following planning along the desired driving corridor.

Keywords: Automated driving · Scenario interpretation
Selfs driving systems

1 Introduction

In the last decades the development of driver assistance and automated driving systems has become a very emerging field of research in the last years. The authors in [1] review the most promising approaches and techniques used in these kind of systems.

The concept of scenario interpretation is quite widespread in the autonomous driving research community. Nevertheless, it is possible to find some differences in the literature. Geyer et al. propose in [2] a definition for some relevant terms in the automated driving context (situation, scene, scenario, etc.). As can be seen in Fig. 1, they define the scenery as the combination of all possible single static elements (e.g. road network, number of lanes, crosswalks, position of traffic lights, speed limits, etc.). The scene consist of the scenery and the information of all dynamic objects. The situation contains the scene and optional ego vehicle information. The situation describes the current state, which could persist several

© Springer Nature Switzerland AG 2019
B. Donnellan et al. (Eds.): SMARTGREENS 2017/VEHITS 2017, CCIS 921, pp. 171–189, 2019.
https://doi.org/10.1007/978-3-030-02907-4_9

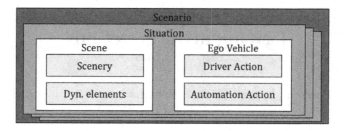

Fig. 1. Definition of the term scenario according to [2].

seconds until some conditions or criteria are filled. However, different states over time are described by the scenario, so that it contains at least one situation.

Another definition for terms like driver-situation, traffic situation, scenario, etc. is proposed in [3]. A driving situation is described with static and dynamic parameters. Some static parameters are the road network, traffic rules, priority, etc. On the other hand, dynamic parameters are for example objects, traffic lights phases, etc. Moreover, some other diverse parameter are used to describe the scenario (weather, road conditions, etc.).

Furthermore, a coherent review and comparison of these terms are presented in [4]. Nevertheless, Ulbrich et al. also propose their own definitions.

Due to the complexity of scenarios at urban intersections, it becomes obvious that a proper scenario interpretation is required. In recent years several methods have been proposed to tackle this problem. Vacek et al. [5] present an approach for a case- and rule-based situation interpretation using description logic. The raw data from the sensors is stored and transformed into a higher level representation. The different expected behavior of other vehicles generates the linkage of other cases over time with corresponding probabilities for every different situation. Since the number of different options becomes very large at intersections, the computational cost for the description logic reasoning constitutes the main drawback of this approach. Logic description is also used by Hüelsen et al. [6] to describe an ontology that represents the road networks, objects, their relations and the corresponding traffic rules. The goal is to reason relations, objects, traffic rules (e.g. *hasRightOfWay* or *hasToYield*) using inference services. Even keeping only necessary information for reasoning, the main drawback of this approach are the high computational costs. Therefore, this approach is insufficient for real-time computation..

Geyer et al. [7] present a method based on the cooperation between the driver and the system with the Conduct-By-Wire (CBW) concept. Depending on the current driving situation, and the required information, the so-called gates are identified. A driving situation is described with three types of parameters: Static (road network, traffic rules, priority, etc), dynamic (objects, traffic lights phases, etc.) and diverse (weather, road conditions, etc.). The system analyzes the required information at the gates. Consequently, different automation levels are set to make the cooperation between the system and the driver easier.

To determine which information is needed, an occupancy map and entry directions at the intersection are set. The CBW approach was also used by Schreiber and Negele [8] to develop of a maneuver catalog from the driver point of view. The focus is to analyze what the driver is expected to do. This information is combined with a set of maneuvers that should cover every possible traffic and driving maneuver. The authors in [9] present two methods for priority conflict resolution (priority charts and priority levels) using a vehicle-to-vehicle (V2V) communication system as a requirement. The first method uses vectors to describe the turning possibilities of all vehicles and their corresponding priority signs. Then, an auxiliary table containing all possible vectors associated with Boolean values is used to indicate if the ego vehicle has to move or stop. This table contains 111 different cases without considering the traffic signs combinations (3 for one vehicle, 27 for two vehicles, and 81 for three vehicles). On the other hand, the second proposed method aims to determine whether the ego vehicle can continue or must wait by interpreting the different priority levels (using an auxiliary truth table to detect potential conflicts with other vehicles). The authors propose a flowchart to handle the right of way problem. These two proposed methods depend on a specific topology (in this case a two road intersection). Moreover, V2V communication is required. Although the focus of [10] is not to turn automatically at urban intersections, the authors propose a maneuver-based planning for automated vehicles. Based on the desired maneuver (or set of maneuvers over the time) the proposed system plans the proper lane change by approaching the intersection. The approach was tested in a multi-lane road network without other road users.

In this paper, the introduction describes the general concept of *scenario interpretation* for automated driving and gives an overview of related work at intersections. Then, the problem is described in Sect. 2. Hereafter, the objective of Sect. 3 is to explain the proposed approach in detail. And finally we conclude the paper in Sect. 4.

2   Problem Description

The main problem is focused on understanding the perceived information around the ego vehicle. This interpretation should enable to plan the proper vehicle motion at urban intersections. In other words the surrounding of the ego vehicle has to be described, and then, the scenario interpretation gives a proper meaning to this description.

In fact, the scenario interpretation at intersection involves, inter alia, the following tasks:

- Filtering relevant information
- Using the information of the road network with corresponding logical correspondences
- Predicting the intention of other vehicles
- Handling occlusions
- Achieving risk assessment

- Considering logical traffic rules
- Handling the right of way
- Handling localization uncertainty
- Etc.

For example, in the first example of Fig. 2(A) the ego vehicle (in blue) is turning to the left and another vehicle (in red) is approaching the intersection. It becomes obvious that it is crucial to know on which lane the other car is driving to determine a possible collision with the ego vehicle: if the red car is driving on its most left lane, it is just allowed to turn to the left, so that a collision with the ego vehicle is not expected. Alternatively, if the other car is not driving on its most left lane, its path has a conflict with the ego's driving corridor. Thus, if the position of other vehicles (or ego vehicle) is not accurate enough (e.g. due to location uncertainty), the scenario interpretation module has to manage the uncertainty of the information in order to understand how critical the situation is.

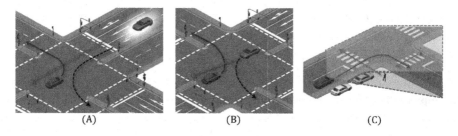

(A)   (B)   (C)

Fig. 2. Three examples of scenarios at intersections based on [15]. **(A) Intention prediction of an oncoming vehicle with inaccurate position:** the ego vehicle (in blue) is turning to the left (blue path) while an oncoming vehicle (in red) is approaching the intersection. The uncertainty of its measured position is represented by a yellow blob surrounding the vehicle. If it is driving forward (red path), both paths intersect. Otherwise (black dotted path), there is no collision between both driving corridors. **(B) Intention prediction of an oncoming vehicle with accurate position:** the ego vehicle is turning left and the other vehicle (which is already in the intersection and its position is accurate enough) could drive forward or turn left. **(C) Handling occlusion while approaching an intersection:** the ego vehicle aims to turn to the right. Due to an obstacle (e.g. another vehicle), the occlusion hiders to detect a crossing pedestrian at the right side of a zebra crossing. The green and red colored regions indicate the perceptible and non perceptible areas, respectively. (Color figure online)

A proper intention prediction is crucial (depending on the road network and its turning possibilities) even considering a perfect accuracy of the position of both vehicles. As shown in Fig. 2(B), the ego vehicle is turning left and the other car could perform two maneuvers: driving forward or turning left. In this case, the accuracy of the state of the other car (e.g. yaw, velocity, etc.) is relevant to achieve a proper intention prediction.

Furthermore, handling occlusions is an important task of the scenario interpretation module. It is not only crucial to understand the provided information, but also to take into account which information is missing. For example, in Fig. 2(C) the ego vehicle (blue) is approaching the intersection and an obstacle (a parked car colored in white) hinders to detect a possible pedestrian. For this given scenario, the first pedestrian (behind the obstacle) is not detected due to the occlusion, but a proper scenario interpretation should be able to interpret the occlusion as a critical missing information. Consequently, it is not clear if more pedestrians are approaching the crosswalk.

The complexity of understanding the preprocessed data depends on its quality. Namely, the more inaccurate the perception is, the more difficult is the interpretation of the provided data. But even if the perception provides accurate information about the surrounding of the ego vehicle, the problem is not simple. The large number of possible collisions with other road users at urban intersections makes the problem a very complex challenge. For this reason, the proposed concept aims to enable the decision making for automated driving at urban intersections in a simple manner. In this sense, the key of the problem is to define which information is important to describe the scenario, or preferably, how to classify all the possible situations depending on the available information. For this reason, we first try to make a conceptual description of the scenario. This elementary analysis is made very easily by answering three questions as a starting point for addressing the problem:

- Which maneuver is the ego vehicle making?
- How is the traffic flow controlled?
- How is the topology of the intersection?

Since the proposed approach is developed for automated driving systems, we can take for granted that the route of the ego vehicle is well known, and consequently, the driving corridor and its maneuver too.

However, this is not the only issue. Understanding the current regulation when approaching an intersection is a very important information to take into account. But unfortunately there is no standard regulation that controls the traffic flow at intersections in a unique manner for all the possible scenarios all over the world. Therefore, the presented work is this paper considers the regulation described in the *Vienna Convention on Road Signs and Signals* [11] and the german regulation [12] in particular. But in order to simplify the problem, we assume that the traffic flow at intersections can be controlled in three different ways: by the *right of way* rule, with traffic signs or traffic lights. In this context, the presented approach does not consider other inputs such as special vehicles, police officer indications, constructions, etc.

Another way to describe an intersection is to consider its topology. Other authors [13] have analyzed in detail the most common topologies to determine the relations between different topologies and traffic accidents. Considering the large number or possible different topologies, we deduce that a scenario interpretation based on specific topologies is not appropriate. Therefore, the proposed solution aims to achieve the maneuver independently on the intersection topology.

These considerations facilitate a simple classification, in which analyzing the possible conflicts with other road users is feasible. In other words, the ego vehicle intention and the control of the traffic flow yield different scenarios and potential conflicts with other vehicles or Vulnerable Road Users (VRUs). Figure 3 illustrates the different possible scenarios considering a simple intersection topology. This classification is an improved version of the method proposed by Fastenmeier [14]:

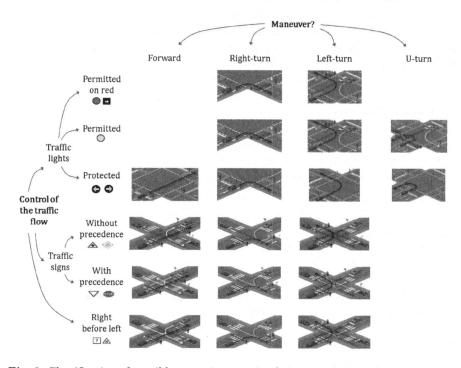

Fig. 3. Classification of possible scenarios at a simple intersection topology considering the desired maneuver and how the traffic flow is controlled. Every row represents a different maneuver of ego vehicle with its path (blue). Every column corresponds to a different manner to control the traffic flow. All possible paths of other vehicles with a potential collision with ego vehicle are colored depending on its priority. Other vehicles (or VRU) with a red path have priority with respect to ego vehicle. Other vehicles with yellow paths are required to give way to ego. The dotted arrows indicate paths of vehicles without an intersection with ego's path. See [15]. (Color figure online)

According to this classification, one can easily recognize the potential conflicts with other vehicles and its corresponding pass permission from an ego perspective. In other words, the different pass permission states indicate how the ego vehicle should handle the right of way:

Denied: a common circular red traffic light has been detected. The ego vehicle has to stop as long as the traffic light color is red.

Permitted: a common circular green light has been detected, so that the ego vehicle is allowed to turn. However, the ego vehicle has to give way to oncoming vehicles while turning left and VRU have priority in parallel conflicts.

Protected: a green arrow traffic light has been detected. According to [12], the path of the ego vehicle to complete the turning maneuver has no conflicts with other road users.

Permitted on Red: a static sign with a green arrow has been detected beside a red traffic light. Even if the traffic light indicates red, the ego vehicle is allowed to turn if there is no potential collision with other crossing/oncoming vehicles.

Right before Left: a traffic sign indicates that the rule *right before left* has to be applied or no traffic sign controls the traffic flow, and consequently, this rule is applied by default.

With Precedence: a priority road sign has been detected, so that other crossing vehicles are required to give way to the ego vehicle.

Without Precedence: a give-way or stop sign has been detected. The ego vehicle has to give way to other crossing vehicles. In case of a stop sign, the ego vehicle is required to stop even if there is no crossing vehicle.

3   Proposed Approach

The proposed approach aims to make the interpretation of the scenario (and further planning) easier by breaking it down into primary situations (see [15]). In order to give an overview of the concept, we first introduce the relevant submodules in the main system flowchart. Then we describe in detail how we define a scenario based on primary situations and finally we explain how the ego vehicle is guided to complete the desired maneuver using this concept.

3.1   Overview

From a general point of view, the basic conceptual flowchart of a self driving system can be simplified in four submodules: perception, scenario interpretation, planning and control. This is illustrated in Fig. 4.

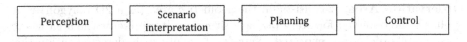

Fig. 4. Simplified conceptual flowchart of an automated driving system.

The perception module represents the low level processing of sensors and a priori data (e.g. image processing, object recognition and tracking, localization and mapping, etc.). The scenario interpretation, which is the focus of our approach, corresponds with the understanding of the processed data. Then, the planning calculates the proper trajectory and delivers it to the control module, which finally provides the adequate signals in terms of steering and acceleration.

The basic idea, which is based on [15], consist on achieving a scenario representation using the relevant information from the perception module. This interpretation should contain the essential information in order to make the proper decisions to guide the ego vehicle along the desired driving corridor. In this context, a scenario consists on mainly three important components: the current pass permission, which indicates how the ego vehicle should pass the intersection and under which conditions; the intention of the ego vehicle, and accordingly its maneuver; and a set of primary situations linked along the driving corridor. Furthermore it is assumed that some basic information such as the ego motion is well known.

The key issue is to use the classification show in Fig. 3 to define a set of primary situations based on the possible conflicts of the ego vehicle with other road users. For this reason we propose four different primary situations (and combinations of them), so that the main advantage is that the whole maneuver can be broken down into a set of expected primary situations (see Fig. 5):

A: there is a potential conflict with a perpendicular with VRU lane (e.g. a crosswalk, zebra crossing or bike lane) in front of the ego vehicle.
B: the driving corridor of the ego vehicle intersect a left-cross lane (e.g. at a T-form intersection without right-crossing lanes). $B1$ is not considered a primary situation on its own, but a mirrored version of B, in which the cross lane comes from the right side. In addition, $B2$ corresponds to a combination of B and $B1$ (e.g. at a X-form intersection).
C: the ego vehicle has a conflict with a parallel crosswalk, zebra crossing or bike lane. Perpendicular and parallel conflicts with VRUs by turning at intersections have to be handled in a different manner compared to situation A. For example, at an intersection controlled with traffic lights, when the state is permitted, the ego vehicle has precedence with respect to the VRUs crossing a perpendicular crosswalk. On the contrary, the ego vehicle has no precedence with respect to VRUs crossing a parallel crosswalk (this is explained graphically in Fig. 3).
D: the ego vehicle has a conflict with an oncoming vehicle.

Every situation should contain at least the following information (these terms are described in detail hereinafter):

Observation Area. It consists on the geometric area (as a 2D polygon) that has to be observed for every primary situation. It represents the area where relevant objects are expected. Namely, if an object is detected inside this area, it should be considered to predict a potential conflict with it.
Occupancy Probability. This is a discrete function indicating the probability over time, that the primary situation is occupied. There are two types of occupancies: real and virtual. These are calculated considering real detected objects or virtual expected objects, respectively. The concept of virtual objects is explained in detail in the following paragraphs.

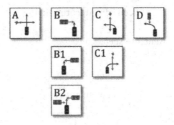

Fig. 5. Set of primary situations (A, B, B1, B2, C, C1, D).

Critical Area. This represents the area that is used for calculating the occupancy over time. In other words, the occupancy represents the probability that the critical area is occupied by other road users over time.

Distance to Situation. It corresponds to the distance along the desired driving corridor between the front bumper of the ego vehicle and the start of the primary situation.

Type. This indicates the type of primary situation (see Fig. 5).

Angle. It indicates the angle between the driving corridor of the ego vehicle and the intersecting lane at the point where both intersect.

The process of extracting all this information from the perception module and creating every primary situation corresponds to the first relevant step of our approach. An overview of this process is described graphically in Fig. 6 as a simplified flowchart.

In a nutshell, the outputs provided by the perception module are used to extract the relevant information and generate the scenario, which represents the input of the tactical decision making. This provides a target point as output indicating a desired position and velocity along the driving corridor of the ego vehicle. Namely, a target point represents *where* and *how fast* should the ego vehicle drive to complete the maneuver at the intersection, and correspondingly, is the result of the interpretation module. This is then used by the trajectory planner to calculate the optimal velocity profile to reach this target point.

3.2   Generating the Scenario Based on Primary Situations

The result of estimating the pass permission (see (a) in Fig. 6) indicates which pass permission is currently valid for the ego vehicle, i.e. under which conditions the ego vehicle is allowed to pass the intersection (denied, permitted, protected, etc.). Once this is estimated, which is not the focus of this paper, the first relevant step of our concept is to extract the set of primary situations along the ego driving corridor.

Extracting the primary situations consist on calculating the possible conflicts between the ego vehicle and other road users. This can be automatically calculated based on the road network information (i.e. considering the intersection points between the path of the ego vehicle and other intersecting paths). Then,

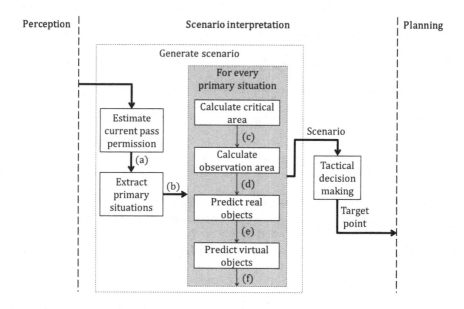

Fig. 6. Simplified flowchart representing the most relevant steps of the interpretation process.

every primary situation is linked representing the order in which consecutive single primary situations are expected. In this sense, a scenario (S) denotes the connections of primary situations (PS_i):

$$S = \{PS_1, PS_2, ..., PS_M\}, \tag{1}$$

where M is the number of different situations. In other words, M represents the total number of primary situations along the desired path. This is graphically described in Fig. 7 using a simple example.

In the given example, the ego vehicle is turning to the right. Using the road network information, it is very simple to calculate the conflicting point with the path of other road users. In fact, the only necessary information is the geometry of the paths and the type of intersecting lane (crossing vehicle lane or vulnerable road user such as crosswalk, bike lanes, etc.). In short, the distance from the bumper of the ego vehicle to the situation, the angle between the conflicting path and the type of situation is extracted from the a-priori road network. This information corresponds to (b) in Fig. 6.

Once the set of primary situations is extracted, the next step of the flowchart is done for every situation (see the blue rectangle in Fig. 6). First, the critical area is calculated based on static geometric information of the a priori road network. This area is in fact divided into three sub-areas (S_1, S_2 and S_3)) and every form depends on the type of the intersecting lane.

For vehicles, the first critical sub-area is a polygon of P points $S_1 = \{\overrightarrow{s_{11}}, ..., \overrightarrow{s_{1P}}\}$ that represents the overlapping area (i.e. where the lanes of ego

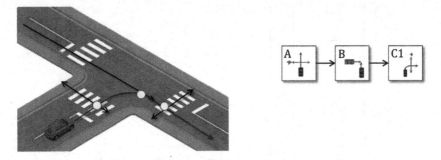

Fig. 7. Simple example of the extraction of three primary situations (A, B, C1). The path of the ego vehicle (colored in blue) intersect the other paths (colored in black). The conflicting point is marked with a yellow circle. (Color figure online)

and the other vehicles overlap) along the intersecting lane with the length d_1. This is illustrated in Fig. 8 with a red rectangle. The second critical sub-area S_2 indicates the area from the overlapping area to the start of the intersecting lane (see the intersecting lane colored in blue in Fig. 8 with the length d_2). The third critical sub-area S_3 is calculated with an empirical length d_3 and indicates the area before the other objects drive into the intersection.

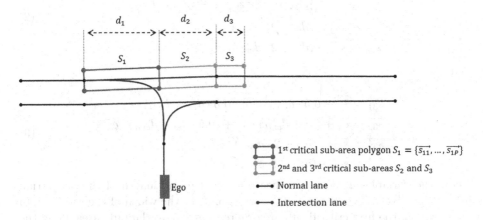

Fig. 8. Critical sub-areas for vehicles. (Color figure online)

In case of an intersecting lane for pedestrians (i.e. crosswalk or zebra crossing), the first critical sub-area polygon $S_1 = \{\overrightarrow{s_{11}}, ..., \overrightarrow{s_{1P}}\}$ represents the area that the ego vehicle would drive over the pedestrian lane (see red polygon in Fig. 9).

Since we take for granted that every pedestrian lane is straight, i.e. has no curvature, the area can be described as a rectangle ($P = 4$). The second and third critical sub-areas ($S_2 = \{\overrightarrow{s_{21}}, ..., \overrightarrow{s_{2P}}\}$ and $S_3 = \{\overrightarrow{s_{31}}, ..., \overrightarrow{s_{3P}}\}$) are

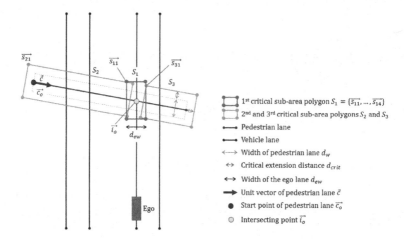

Fig. 9. Critical sub-areas for pedestrians. The first point of the polygons are marked at the upper-left corner of every sub-area.

calculated using the distance d_{crit}, which represents a constant extension of the area set empirically:

$$
\begin{aligned}
\overrightarrow{s_{21}} &= \overrightarrow{c_o} - d_{crit} \cdot \overrightarrow{c} + (0.5 \cdot d_w + d_{crit}) \cdot \overrightarrow{c_\perp} \\
\overrightarrow{s_{22}} &= \overrightarrow{i_o} - (0.5 \cdot d_{ew}) \cdot \overrightarrow{c} + (0.5 \cdot d_w + d_{crit}) \cdot \overrightarrow{c_\perp} \\
\overrightarrow{s_{23}} &= \overrightarrow{s_{22}} - (d_w + 2 \cdot d_{crit}) \cdot \overrightarrow{c_\perp} \\
\overrightarrow{s_{24}} &= \overrightarrow{s_{21}} - (d_w + 2 \cdot d_{crit}) \cdot \overrightarrow{c_\perp}
\end{aligned}
\tag{2}
$$

and

$$
\begin{aligned}
\overrightarrow{s_{31}} &= \overrightarrow{i_o} + 0.5 \cdot d_{ew} \cdot \overrightarrow{c} + (0.5 \cdot d_w + d_{crit}) \cdot \overrightarrow{c_\perp} \\
\overrightarrow{s_{32}} &= \overrightarrow{c_o} + (d_l + d_{crit}) \cdot \overrightarrow{c} + (0.5 \cdot d_w + d_{crit}) \cdot \overrightarrow{c_\perp} \\
\overrightarrow{s_{33}} &= \overrightarrow{s_{32}} - (d_w + 2 \cdot d_{crit}) \cdot \overrightarrow{c_\perp} \\
\overrightarrow{s_{34}} &= \overrightarrow{s_{31}} - (d_w + 2 \cdot d_{crit}) \cdot \overrightarrow{c_\perp}
\end{aligned}
\tag{3}
$$

where the variables d_w and d_l indicate the width and length of the pedestrian lane, respectively. The distance d_{ew} corresponds to the width of the ego lane. In other words, the first critical sub-area represents the overlapping area along the ego lane, while the second and third one represent the pedestrian lane (extended with the distance d_{crit}) at the left and right side of the ego vehicle, respectively.

On the other hand, another important concept of the primary situation is the observation area, which consist of a polygon with Q points describing the area where relevant objects could be ($O = \{\overrightarrow{o_1}, ..., \overrightarrow{o_Q}\}$). This is calculated depending on the type of the intersecting lane and the expected time that the ego vehicle needs to reach the situation t_{area}.

In the case of a primary situation for vehicles, the area is calculated along the path of other objects: between the end of the intersection lane and the calculated distance d_{obs} (see the green polygon in Fig. 10):

$$d_{obs} = \begin{cases} d_{min} & \text{if } v_{obj} \cdot t_{area} < d_{min} \\ d_{max} & \text{if } v_{obj} \cdot t_{area} > d_{max} \\ v_{obj} \cdot t_{area} & \text{else} \end{cases} \qquad (4)$$

where d_{min} and d_{max} are constrains set empirically that indicate the minimal and maximal distance of the observation area, respectively. v_{obj} represents the maximal expected velocity of a possible object and t_{area} corresponds to the time that the ego vehicle needs to reach the situation:

$$t_{area} = \frac{-v_{ego}\sqrt{v_{ego}^2 + a_{ego} \cdot d}}{a_{ego}}. \qquad (5)$$

To calculate the t_{area}, the distance from the front bumper of the ego vehicle d to the situation, the current ego velocity v_{ego} and acceleration a_{ego} are used. This results in a polygon of Q points describing the observation area ($O = \{\vec{o_1}, ..., \vec{o_Q}\}$), where the width corresponds with the width of the lane in the driving corridor of the other vehicles.

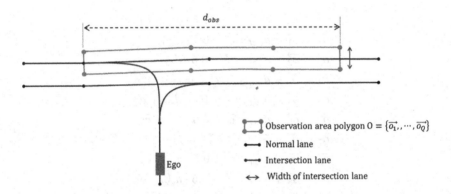

Fig. 10. Example of an observation area of a crossing vehicle primary situation.

On the contrary, the observation area for pedestrians is calculated based on a basic form described in Fig. 11.

This geometry aims to cover the area where pedestrians could be relevant for the situation. For this reason, the observation area is expressed as a polygon of twelve points $O = \{\vec{o_1}, \vec{o_2}, ..., \vec{o_{12}}\}$ calculated depending on the unit vector of the pedestrian lane $\vec{c}$, its origin $\vec{c_o}$, its length d_{length} and the extension distance d_{ext}:

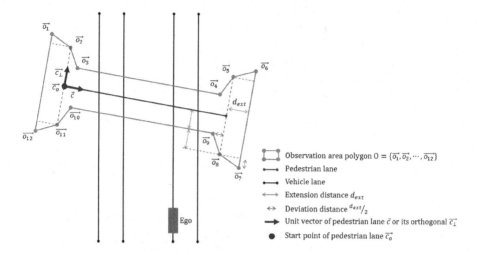

Fig. 11. Example of an observation area for a pedestrian primary situation.

$$\vec{o_1} = \vec{c_o} - d_{ext} \cdot \vec{c} + 2.5 \cdot d_{ext} \cdot \vec{c_\perp}$$
$$\vec{o_2} = \vec{o_1} + d_{ext} \cdot \vec{c} - 0.5 \cdot d_{ext} \cdot \vec{c_\perp}$$
$$\vec{o_3} = \vec{o_2} - d_{ext} \cdot \vec{c_\perp} + 0.5 \cdot d_{ext} \cdot \vec{c}$$
$$\vec{o_4} = \vec{o_3} + d_{length} \cdot \vec{c} - d_{ext} \cdot \vec{c}$$
$$\vec{o_5} = \vec{o_4} + d_{ext} \cdot \vec{c_\perp} + 0.5 \cdot d_{ext} \cdot \vec{c}$$
$$\vec{o_6} = \vec{o_5} + d_{ext} \cdot \vec{c} + 0.5 \cdot d_{ext} \cdot \vec{c_\perp}$$
$$\vec{o_7} = \vec{o_6} - 5 \cdot d_{ext} \cdot \vec{c_\perp} \tag{6}$$
$$\vec{o_8} = \vec{o_7} - d_{ext} \cdot \vec{c} + 0.5 \cdot d_{ext} \cdot \vec{c_\perp}$$
$$\vec{o_9} = \vec{o_8} + d_{ext} \cdot \vec{c_\perp} - 0.5 \cdot d_{ext} \cdot \vec{c}$$
$$\vec{o_{10}} = \vec{o_9} + (d_{ext} - d_{length}) \cdot \vec{c}$$
$$\vec{o_{11}} = \vec{o_{10}} - d_{ext} \cdot \vec{c_\perp} - 0.5 \cdot d_{ext} \cdot \vec{c}$$
$$\vec{o_{12}} = \vec{o_{11}} - d_{ext} \cdot \vec{c} - 0.5 \cdot d_{ext} \cdot \vec{c_\perp}$$

where $\vec{c}$ indicates the unit vector of the pedestrian lane and $\vec{c_\perp}$ its orthogonal vector. Obviously, the larger is the time that ego needs to reach the situation, the larger should be the extension of the observation area. In order to consider this dependency, the distance d_{ext} depends on t_{area}:

$$d_{ext} = \begin{cases} d_{maxExt} & \text{if } t > t_{area} \\ t_{area} \cdot \frac{d_{maxExt} - d_{minExt}}{t_{max}} + d_{minExt} & \text{if } 0 < t < t_{area} \\ d_{minExt} & \text{if } t < 0 \end{cases} \tag{7}$$

The variables d_{minExt} and d_{maxExt} indicate the minimal and maximal extension distance, respectively. t_{max} represents the maximal considered time. This dependency is illustrated in Fig. 12.

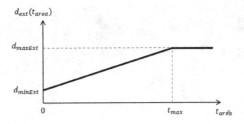

Fig. 12. The extension distance (d_{ext}) depending on the time that ego needs to reach the area (t_{area}).

The scenario with the extracted critical and observation area corresponds to (c) and (d) in the flowchart of Fig. 6, respectively. Using this information, the next step is the prediction of the objects detected inside the observation area. This prediction, which is not the focus of this paper, provides the probability that the situation is occupied over time. In other words, after the prediction, the situation contains the probability that it is not traversable over time (e).

A very important advantage of using the observation areas is that the system knows which regions should be considered. In this sense, if some region is not perceived by the sensors (e.g. due to occlusions), a virtual object can be created as a worst case situation, so that the system can react to this lack of information carefully by considering a virtual object. In this context, the next step consist on calculating those *non-perceptible areas* and generating a virtual object. Then, in the same way as for real detected objects, the occupancy probability function is calculated for virtual objects (see (f) in Fig. 6).

The idea is to imitate cognitive human reaction in a very simple way: setting a virtual object. This object is placed representing the worst case (i.e. a pedestrian is crossing so that a collision with the ego vehicle will occur). In order to explain this concept, Fig. 13 illustrates an example in which the ego vehicle is approaching a zebra-crossing before turning to the right (primary situation A). At the right side of the road, a parked vehicle causes an occlusion (i.e. a not perceptible area) at the observation area. Therefore, a virtual object is generated and the ego vehicle slows down its velocity (time n). The closer the vehicle gets, the smaller the *non-perceptible area* is (see the red polygon over time n, $n + 1$ and $n + 2$).

Once this information is extracted for every situation (see blue colored rectangle in Fig. 6) the scenario generated completely, so that the next step is to plan the maneuver by generating the corresponding target points.

3.3 Planning the Maneuver

Once the scenario is generated, the target points have to be set according to the available information, so that the ego vehicle is guided to achieve the whole maneuver. The set of generic target points are illustrated in Fig. 14.

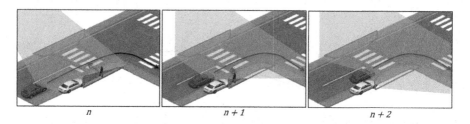

Fig. 13. Occlusion example illustrated over the time (n, $n+1$ and $n+2$). The ego vehicle (blue) is making a right turn maneuver (black path) and the other vehicle (white) hinder to perceive the observation area (green polygon) completely. The perceptible and not perceptible areas are colored in green and red respectively. The illustrated pedestrian corresponds to a generated virtual object (with its path marked as a dotted red arrow). (Color figure online)

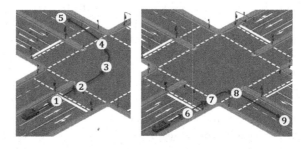

Fig. 14. Set of target points for a generic topology (see [15]).

The flowchart in Fig. 15 explains which information is needed to set the proper target points for every primary situation. For the sake of clarity, the diagram has been kept simple by considering only a very basic topology. In other words, the consideration of more complex topologies (e.g. handling T- or X-form intersections) is omitted to ease its representation and understanding. Furthermore, the difference between the pass permission *stop* or *give way* is not considered, and pass permission *permitted on red* is completely omitted.

As it can be seen in Fig. 15, an example is highlighted in red. In this example, the ego vehicle turns to the left without precedence. Once the ego vehicle approaches the intersection and has done the proper lane change(s), the first required information of the scenario is how the traffic flow is controlled, namely by traffic lights, traffic signs or the *right before left* rule. This determines the first main branching of the flowchart. Then, a primary situation A is expected depending on the existence of a perpendicular conflict with VRUs (Vulnerable Road Users). Since in the given example a yield sign was detected (see red path along the flow chart in Fig. 15), and the ego vehicle intends to turn left, both possible left and right crossing vehicles have the right of way (primary situation $B2$). Consequently, if a collision with crossing vehicles from both sides inside the corresponding observation areas is predicted, the target point 2 forces ego vehicle to stop in front of the critical sub-area as long as no collision is expected.

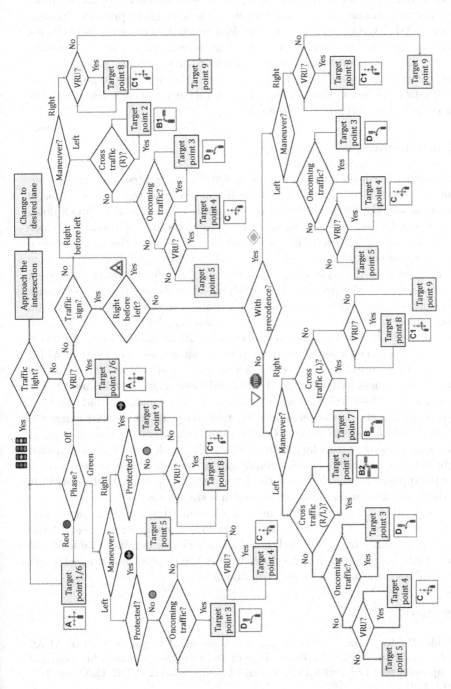

Fig. 15. Flowchart for automated turning at intersections based on primary situations. It explains the connections of the expected primary situations and their according target points step by step. The last target point guides the ego vehicle to the end of the turning maneuver (rectangles with green border). A given example is highlighted in red (left turn without precedence). See [15]. (Color figure online)

In other words, the velocity of the target point is set considering predicted probability that the primary situation is occupied. Then, the next primary situation D implies setting the target point 3 to avoid a collision with oncoming vehicles. But in case that no collision is predicted (e.g. because there are no oncoming vehicles in the corresponding observation area or the occupancy probability for the current time to area is zero $occ(t_{area} = 0.0)$), the next target point 4 (primary situation C) is set. Finally, the left turn maneuver is completed with the target point 5 if last situation is passable.

4   Conclusions

In this paper a scenario interpretation concept for automated driving at intersections has been introduced. This approach aims to make the interpretation of the scenario and the further decision making easier. The idea is to break down the problem into four primary situations (or combinations of them), in which every situation contains the required information to execute the whole maneuver at urban intersections in a simple manner. After explaining the meaning of the term *scenario interpretation*, an overview of the related work was given. Then, we addressed the problem by classifying the different possible scenarios at urban intersection. This classification was done considering the intention on the ego vehicle and how the traffic flow is controlled. This analysis identified the potential conflicts with other road users in a simple manner. Our approach generates a scenario as a set of expected primary situations over time, in which only the relevant information is needed. These primary situations are defined by the potential conflicts with other road users and some required information such as critical areas, observation areas, estimation of the probability that the situation is occupied, etc. Moreover, a flowchart to complete the desired maneuver at the intersection was presented. This diagram represents the combination of primary situations over time facilitating to plan the whole maneuver. Compared to state-of-the-art solutions, a very important advantage of our system is that it may be applied independently of the intersections topology. Furthermore, it offers the possibility of handling occlusions in a simple way.

Research work to optimize the process of estimating the pass permission at intersections has to be done. Furthermore, the computational cost of the proposed approach have to be analyzed. In this sense, a detailed evaluation of the proposed approach and its functionality over the time for real scenarios will be achieved and compared with other methods.

References

1. Okuda, R., Kajiwara, Y., Terashima, K.: A survey of technical trend of ADAS and autonomous driving. In: Proceedings of Technical Program-2014 International Symposium on VLSI Technology, Systems and Application (VLSI-TSA), pp. 1–4. IEEE (2014)
2. Geyer, S., et al.: Concept and development of a unified ontology for generating test and use-case catalogues for assisted and automated vehicle guidance. IET Intell. Transp. Syst. **8**(3), 183–189 (2014)

3. Domsch, C., Negele, H.: Einsatz von referenzfahrsituationen bei der entwicklung von fahrerassistenzsystemen; 3. Tagung Aktive Sicherheit durch Fahrerassistenz, pp. 07–08, April 2008
4. Ulbrich, S., Menzel, T., Reschka, A., Schuldt, F., Maurer, M.: Defining and substantiating the terms scene, situation, and scenario for automated driving. In: 2015 IEEE 18th International Conference on Intelligent Transportation Systems (ITSC), pp. 982–988. IEEE (2015)
5. Vacek, S., Gindele, T., Zöllner, J.M., Dillmann, R.: Using case-based reasoning for autonomous vehicle guidance. In: IEEE/RSJ International Conference on Intelligent Robots and Systems, IROS 2007, pp. 4271–4276. IEEE (2007)
6. Hülsen, M., Zöllner, J.M., Weiss, C.: Traffic intersection situation description ontology for advanced driver assistance. In: 2011 IEEE Intelligent Vehicles Symposium (IV), pp. 993–999. IEEE (2011)
7. Geyer, S., Hakuli, S., Winner, H., Franz, B., Kauer, M.: Development of a cooperative system behavior for a highly automated vehicle guidance concept based on the conduct-by-wire principle. In: 2011 IEEE Intelligent Vehicles Symposium (IV), pp. 411–416. IEEE (2011)
8. Schreiber, M., Kauer, M., Schlesinger, D., Hakuli, S., and Bruder, R.: Verification of a maneuver catalog for a maneuver-based vehicle guidance system. In: 2010 IEEE International Conference on Systems Man and Cybernetics (SMC), pp. 3683–3689. IEEE (2010)
9. Alonso, J., Milanés, V., Pérez, J., Onieva, E., González, C., De Pedro, T.: Autonomous vehicle control systems for safe crossroads. Transp. Re. Part C: Emerg. Technol. **19**(6), 1095–1110 (2011)
10. Lotz, F., Winner, H.: Maneuver delegation and planning for automated vehicles at multi-lane road intersections. In: 2014 IEEE 17th International Conference on Intelligent Transportation Systems (ITSC), pp. 1423–1429. IEEE (2014)
11. United Nations Economic Commission for Europe, U. N. E. C: Convention on Road Signs and Signals of 1968, Switzerland, Geneva (2006)
12. Forschungsgesellschaft für Straßen und Verkehrswesen (FGSV): Richtlinien für Lichtsignalanlagen (RiLSA) (2010)
13. Gerstenberger, M. (2015). Unfallgeschehen an Knotenpunkten: Grundlagenuntersuchung zu Ursachen und Ansätzen zur Verbesserung durch Assistenz. Ph.D. thesis, München, Technische Universität München, Dissertation (2015)
14. Fastenmeier, W., et al.: Autofahrer und Verkehrssituation. Neue Wege zur Bewertung von Sicherheit und Zuverlässigkeit moderner Strassenverkehrssysteme, Number 33 (1995)
15. Lopez, D.P., Waldmann, R., Joerdens, C., Rojas, R.: Scenario interpretation based on primary situations for automatic turning at urban intersections. In: Proceedings of the 3rd International Conference on Vehicle Technology and Intelligent Transport Systems, VEHITS, vol. 1, pp. 15–23. INSTICC, ScitePress (2017)

Fuel Optimal Control of an Articulated Hauler Utilising a Human Machine Interface

Jörgen Albrektsson[1,2(✉)] and Jan Åslund[1]

[1] Department of Electrical Engineering, Linköping University,
581 83 Linköping, Sweden
jan.aslund@liu.se
[2] Volvo Construction Equipment, 631 85 Eskilstuna, Sweden
jorgen.albrektsson@volvo.com

Abstract. Utilising optimal control presents an opportunity to increase the fuel efficiency in an off-road transport mission conducted by an articulated hauler. A human machine interface (HMI) instructing the hauler operator to follow the fuel optimal vehicle speed trajectory has been developed and tested in real working conditions. The HMI implementation includes a Dynamic Programming based method to calculate the optimal vehicle speed and gear shift trajectories. Input to the optimisation algorithm is road related data such as distance, road inclination and rolling resistance. The road related data is estimated in a map module utilising an Extended Kalman Filter (EKF), a Rauch-Tung-Striebel smoother and a data fusion algorithm. Two test modes were compared: (1) The hauler operator tried to follow the optimal vehicle speed trajectory as presented in the HMI and (2) the operator was given a constant target speed to follow. The objective of the second test mode is to achieve an approximately equal cycle time as for the optimally controlled transport mission, hence, with similar productivity. A small fuel efficiency improvement was found when the human machine interface was used.

Keywords: Off-road · Construction equipment
Human machine interface · Optimal control · Dynamic programming
Kalman filters

1 Introduction

Articulated haulers are today used in numerous applications in the construction industry when there is a need for efficient transportation of material even at tough off-road conditions. Utilising optimal control on machine level have a potential to increase fuel efficiency, reduce the environmental impact of CO_2 emissions and save money for the customer.

Commonly there is a set production target [ton/h] for the transport mission, i.e. the hauler should transport a certain amount of material in a set time. In Albrektsson et al. [1] a method to derive a Pareto front of minimum fuel

B. Donnellan et al. (Eds.): SMARTGREENS 2017/VEHITS 2017, CCIS 921, pp. 190–208, 2019.
https://doi.org/10.1007/978-3-030-02907-4_10

consumption vs. cycle time for an articulated hauler transport mission is proposed. With a set production rate a target cycle time for the hauler transport mission can be calculated. The Pareto front provides the means to determine a time penalty parameter and use it in a Dynamic Programming (DP) algorithm to calculate the optimal vehicle speed and gear trajectory for the transport mission.

The optimisation algorithm analyses the road elevation to enable an efficient use of the kinetic energy in the vehicle and to create a gear shift strategy that continuously selects the optimal gear, enabling the internal combustion engine (ICE) to work as efficient as possible, further reducing the fuel consumption. The use of dynamic programming to create a look-ahead cruise control have been examined in e.g. the work of Hellström et al. [2] and a gear shift strategy for an off-road vehicle was developed in the work of Fu et al. [3], both reporting promising results.

The work at hand is a continuation of [1] where the proposed methods are realised in a human machine interface (HMI) instructing the machine operator to follow the optimal vehicle speed trajectory. The performance of the HMI implementation is tested in real working conditions and the potential fuel efficiency gain is evaluated.

The article outline is as follows: Sect. 2 displays the theoretical basis consisting of three main parts: A model of an articulated hauler, a map module and an optimisation module. In Sect. 3 the design of the human machine interface is shown. The set-up of the performed tests to evaluate the performance of the method is presented in Sect. 4 and in Sect. 5 the result of the tests are displayed.

2   Theoretical Basis

The road related parameters needed to perform the optimisation of the transport mission is distance, road inclination, rolling resistance and a max vehicle speed limit. This data is created by the map module and then used as input to the optimisation module. Both the map module and the optimisation module rely on a vehicle model which is presented first in the section.

The presented work is a continuation of previous work and Sect. 2 is an abbreviated version of the theory presented in [1], where the main part of the equations also can be found.

2.1   Vehicle Model

The vehicle model is an important part of both the map module and the optimisation module. In the vehicle model the main longitudinal forces acting on the articulated hauler are considered, see Fig. 1. The notation used in Fig. 1 is: v = vehicle speed, α = road inclination, F_t = tractive force, F_a = aerodynamic force, F_r = rolling resistance force and F_g = force generated by road inclination. A drivetrain model connects the traction force to the torque generated by the internal combustion engine (ICE) of the hauler. In the map module the ICE torque is known by means of a CAN signal from the ICE's electronic control

Fig. 1. Articulated hauler with corresponding longitudinal forces. Picture from [1].

unit, enabling the calculation of the traction force. In the optimisation module the traction force is known by means of the vehicle model and road data and the drivetrain model is used to calculate the ICE torque which is translated into a fuel cost. Figure 2 displays the model of the drivetrain. The notation used is: T = torque [Nm], ω = angular velocity [rad/s], gr = gear [-], J = mass moment of inertia [kgm^2], i = gear ratio [-], η = efficiency [%].

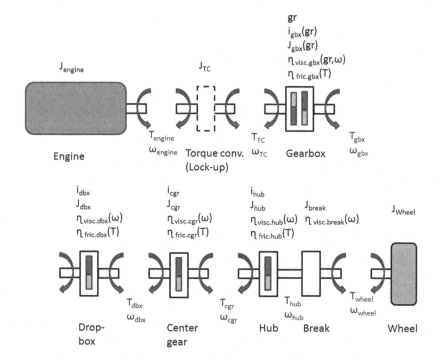

Fig. 2. Model of the drivetrain. Picture from [1].

An expression for the longitudinal dynamics of the hauler is derived through combining the drivetrain model, the external forces and Newton's second law of motion. In continuous time the expression is:

$$m_{tot}(gr) \cdot \frac{d}{dt} v(t) = F_t(t) - F_a(t) - F_r(t) - F_g(t) \qquad (1)$$

where gr = gear and t = time.

2.2   Map Module

The map module estimates and stores road related data with latitude and longitude coordinates as identification points. The main parameters estimated in the algorithm are: latitude (φ), longitude (λ), altitude (z), mean vehicle speed (v), road inclination (α), vehicle heading (β), rolling resistance coefficient (c_r), speed limit (v_{max}) and travelling direction (Dir.). Out of the estimated parameters, only $\varphi, \lambda, \alpha, c_r$ and v_{max} are used in the optimisation module. In Salholm [4] a method for estimation of road inclination for on-road commercial vehicles is exhaustively described. The method have been adopted for off-road conditions and enhanced to include rolling resistance in the work of [5] and [6]. The proposed method utilizes an extended Kalman filter (EKF) to work as an observer for the unmeasured parameter rolling resistance. The EKF also helps to limit the potential bias error developing when only using an inclination sensor to measure the road inclination [4], pp. 80–88.

Map Building Process. On a high level, the map building process can be described according to the steps below:

1. Operator drives the track between the loading and unloading site forth and back as fast (but safe) as possible while necessary sensor data is recorded.
2. The direction of the travel is detected and the data updated accordingly.
3. The collected data is processed in the map-building algorithm according to:
   (a) Calculation of applied brake force.
   (b) Translation of ICE torque into force at wheels.
   (c) Geographic and vehicle dependent data (measured and calculated) are merged in an Extended Kalman Filter (EKF).
   (d) Smoothing of estimates with Rauch-Tung-Striebel algorithm to remove potential lag.
   (e) Merge the estimates into a map utilising a fusion algorithm.
4. The highest recorded speed at each coordinate is used to set the max speed limit.

Sensor and Data Fusion. The road estimation algorithm utilises data recorded from the vehicle CAN: v, α, vehicle articulation (Φ), engine torque (T_{engine}) and from a GNSS sensor: $\varphi, \lambda, z, \beta$. When congregated, the data of the road is presented in a format similar to a map using coordinates.

Sensor and Data Fusion Methods. Utilising an extended Kalman filter (EKF) is the proposed method for sensor fusion. With a regular Kalman filter it is only possible to estimate the states of a linear process while the extended Kalman filter gives the possibility to estimate the states in a non-linear process [7]. Through the use of the EKF it is possible to estimate the states of a process enabling the estimation of rolling resistance, which is not directly measurable with the standard mounted sensors on an articulated hauler. The use of the extended Kalman filter in the proposed map building process follows to a large extent the guidance given in [7].

The Rauch-Tung-Striebel (RTS) smoother [8] is an efficient two-pass algorithm for fixed interval smoothing and is used to compensate for filtering delay and to include later measurements in the road estimation.

To merge data from different runs along the road a general data fusion method, as described in [9] p. 30, is used. The congregated road data is stored in a map for each coordinate pair along with the covariance matrix.

Estimation Model. This section presents a road model and the method used to estimate the road related parameters.

Time vs Spatial Sampling. Distance, rather than time, is used as the independent variable facilitating the fusion of data from several different runs along the road. To shift to distance as the independent variable the following conversion is used in the vehicle's longitudinal model.

$$\frac{dv}{dt} = \frac{dv}{ds}\frac{ds}{dt} = v\frac{dv}{ds} \Rightarrow \frac{dv}{ds} = \frac{1}{v}\frac{dv}{dt}, v \neq 0 \tag{2}$$

Road Model. Equation (3) describes the correlation between road altitude and road inclination angle.

$$\frac{dz}{ds} = sin(\alpha(s)) \tag{3}$$

Extended Kalman Filter (EKF) and Smoothing The guidance given in [7] is followed at the implementation of the EKF in the proposed map building process. The states to be estimated presented in continuous time are displayed in (4).

$$\hat{x}(t) = [\varphi(t) \ \lambda(t) \ z(t) \ v(t) \ \alpha(t) \ \beta(t) \ c_r(t)]^T \tag{4}$$

The explanation of the parameters can be found in the beginning of this section. Spatial samples are used instead of continuous time in the model. To shift to distance as the independent variable Eq. (2) is used. Equation (4) is translated into discrete notation, see Eq. (5), where k represents the index of the location.

$$\hat{x}_k = [\varphi_k \ \lambda_k \ z_k \ v_k \ \alpha_k \ \beta_k \ c_{r.k}]^T \tag{5}$$

Time update (a priori estimate).

1. Define two distances, one in meters and one in degrees:

$$\Delta s_{m.k} = \hat{v}_k \cdot T_s$$
$$\Delta s_{deg.k} = \frac{\Delta s_{m.k}}{r_{earth}} \cdot \frac{180}{\pi} \tag{6}$$

2. Project the state ahead (state equations).

$$\hat{x}_k^- = \begin{bmatrix} \varphi_{k-1} + \Delta s_{deg.k-1}cos(\alpha_{k-1})cos(\beta_{k-1}) \\ \lambda_{k-1} + \Delta s_{deg.k-1}cos(\alpha_{k-1})sin(\beta_{k-1}) \\ z_{k-1} + \Delta s_{m.k-1}sin(\alpha_{k-1}) \\ v_{k-1} + \frac{\Delta s_{m.k-1}}{v_{k-1}} \frac{F_{t.k-1} - F_{a.k-1} - F_{g.k-1} - F_{r.k-1}}{m_{tot}} \\ \alpha_{k-1} \\ \beta_{k-1} + \Delta s_{m.k-1} \frac{cos(\alpha.k-1)}{r_{turn.k-1}} \\ c_{r.k-1} \end{bmatrix} \tag{7}$$

With:

$$r_{turn.k-1} = l_1 cot(\Phi_{k-1}) + \frac{l_2}{sin(\Phi_{k-1})} \tag{8}$$

where r_{turn} is the turning radius of the vehicle, Φ is the articulation angle, l_1 and l_2 distances between axles and articulation point (front/rear).

3. Define the Jacobian: A[i,j] = df[i]/dx[j] and project the error covariance ahead:

$$P_k^- = A \cdot P_{k-1} \cdot A^T + Q \tag{9}$$

4. Define the measurement vector (a priori estimate):

$$y_k = [\varphi_{k.gps} \ \lambda_{k.gps} \ z_{k.gps} \ v_{k.CAN} \ \alpha_{k.CAN} \ \beta_{k.gps}]^T \tag{10}$$

Measurement equation:

$$y_k = H \cdot x_k + e_k \tag{11}$$

where the H matrix is:

$$H = \begin{bmatrix} I_6 \ h_{*7} = 0 \end{bmatrix} \tag{12}$$

Calculate the Kalman gain:

$$K_k = P_k^- H^T (H P_k^- H^T + R)^{-1} \tag{13}$$

5. Update estimates with measurement

$$\hat{x}_k = \hat{x}_k^- + K_k(y_k - H\hat{x}_k^-) \tag{14}$$

6. Update error covariance

$$P_k = (I - K_k H) P_k^-$$ (15)

7. Save $\hat{x}_k$, P_k, $\hat{x}_k^-$ and P_k^- at each coordinate [k] to be used in smoothing process.
8. Initiate smoothing with the last predicted values $(\hat{x}_{N+1|N}^-)$ and last predicted covariance matrix $(P_{N+1|N}^-)$, where N is the total number of measured data points. Run smoothing backwards along the track. Kalman smoothing gain:

$$K_k^s = P_{k|k} + A^T P_{k+1|k}^{--1}$$ (16)

Smoothed estimates:

$$\hat{x}_{k|N}^s = \hat{x}_{k|k} + K_k^s(\hat{x}_{k+1|N}^s - \hat{x}_{k+1|k}^-)$$ (17)

Smoothed error covariance matrix

$$P_{k|N}^s = P_{k|k}(P_{k+1|N}^s - P_{k+1|k}^-)K_k^{sT}$$ (18)

Fusion of Map Data. The first recorded track is used as reference and split into 5 m long sections. The φ and λ coordinates serve as identification points to which the corresponding parameter estimates are appended. The search for measurement points in the next recording of the track is limited to a rectangular area that is ± 1.5 m in the heading direction, ± 8 m orthogonal to the heading and to measurement points which have the same heading, β, ± 15 deg. The sign of α and the heading is switched (180 deg) before the search for matching measurements if the track was driven in opposite direction compared to the reference recording. If more than one measurement is within the search area the measurement closest to the reference point in the horizontal plane is chosen. A method for fusion of independent estimates, as described in [9] p. 30, is used when the new recording of the track is merged into the stored map. Equation (19) reveals how the states in the map are calculated.

$$\begin{aligned} P_k^f &= ((P_k^1)^{-1} + (P_k^2)^{-1})^{-1} \\ \hat{x}_k^f &= P_k^f \cdot ((P_k^1)^{-1}\hat{x}_k^1 + (P_k^2)^{-1}\hat{x}_k^2) \end{aligned}$$ (19)

2.3 Optimisation Module

This section presents a method to calculate the fuel optimal vehicle velocity and gear shift trajectories of an articulated hauler as it travels along a road with varying inclination and surface conditions. Machine data and the road dependent data developed in Sect. 2.2 serve as input to the optimisation algorithm.

Method and Objective. Developed in the 1950's by Richard Bellman, Dynamic Programming (DP) is a well known algorithm to solve optimal control problems. The Dynamic programming method fits the optimal control problem well since dimension is small and since road inclination and rolling resistance can be considered as a priori known disturbances by means of the earlier described map module. The DP algorithm is not described in-depth here, instead the reader is referred to [10] and e.g. [11].

While in [1] the objective was to derive a Pareto front for the complete mission it is in this case rather to derive optimal trajectories for a specific cycle time target. To avoid the *Curse of dimensionality*, see [12], the approach of [2] and [13] is applied, i.e. the trip time is added to the objective which becomes:

$$\text{minimise } M + \beta t \qquad (P1)$$

where $M =$ fuel consumption, $t =$ cycle time and the trade-off between fuel consumption and cycle time is represented by the scalar coefficient β.

State Space. The target is to control vehicle speed and gear thus speed and gear would be a natural choice as state variables. In [2] energy is proposed as state variable instead of speed since this damps the oscillatory behaviour of the control, if the preferred Euler forward method is used for discretisation. Thus energy and gear is chosen as state variables rendering in the state vector: $x_k = [\, e_k \ gr_k \,]^T$, where e = energy and gr = gear number. Denominating the control variables u, the control vector is $u_k = [\, u_{e.k} \ u_{gr.k} \,]^T = [\, e_{k+1} - e_k \ gr_{k+1} - gr_k \,]^T$.

Control Constraints. The velocity of the vehicle has a lower limit since the drivetrain model only is developed for driving with the torque converter in lock-up mode. The maximum speed limit is set by the map module. Equation (20) displays the limitations imposed on the vehicle speed.

$$v_{min} \leq v \leq v_{max} \qquad (20)$$

Limitations in the gearbox entails the need of a constraint on the maximum number of gear shifts which is set to 2 (both up and down shift).

$$gr_k - 2 \leq gr_{k+1} \leq gr_k + 2 \qquad (21)$$

Dynamic Model. The model in Eq. (1) from Sect. 2.1 is utilised as dynamic model in the optimisation algorithm. Reformulated into terms of energy and converted into spatial coordinates Eq. (1) becomes:

$$\frac{de}{ds} = F_t - F_a - F_r - F_g \qquad (22)$$

Discretisation. The data from the map module is discrete and represented by N steps with length h of which the sum equals the total distance of the transport mission, S. Additionally, the optimisation problem is solved numerically, thus discretisation is needed. The Euler forward method is used to discretise Eq. (22) and the discretised complete vehicle model is written

$$\frac{e_{k+1} - e_k}{h_k} = F_{t.k} - F_{a.k} - F_{r.k} - F_{g.k} \tag{23}$$

Similarly the fuel mass flow $\dot{m}_f$ is transformed into spatial representation using Eq. (2) and then discretised with the Euler method.

$$m_{f.k+1} = m_{f.k} + \frac{h_k}{v_k}\dot{m}_{f.k} \tag{24}$$

Cost Function. A central part of the DP algorithm is the cost function. In the proposed optimisation algorithm the cost function is based on calculating the equivalent fuel cost, m_f, for bringing the vehicle from one position on the road to the next position. During the transition both states, i.e. the kinetic energy (speed) and the gear, may change. A time penalty, β, as introduced in the beginning of the section and the cost for changing gear, $m_{f.gs}$, which is modelled approximately equal to the work that is lost speeding up or slowing down the engine to meet the next gear, are added to the cost function.

$$\zeta_k = m_{f,k} + \beta t_k + m_{f.gs,k} \tag{25}$$

3   Human Machine Interface

To be able to test the performance of the map module and the optimisation module, a human machine interface (HMI) was designed. At this stage the target of the optimal control implementation was solely to control the speed of the articulated hauler, omitting the control of gear shifts.

3.1   Hardware

As the Volvo Co-Pilot system [14] offers a robust platform with pre-defined interfaces to the vehicle CAN (computer area network) and includes a GNSS smart receiver, it was chosen to serve as hardware platform for the HMI used in the machine. The centre of the Co-Pilot system is the 10" tablet like interactive display which employs an Android operative system.

3.2   Implementation

Most of the algorithms in Sect. 2 were first developed and tested off-line on recorded data in the Matlab environment. The algorithms were translated into

C++ and successively improved enabling a substantial reduction in computational time. In the Volvo Co-Pilot system the signal handling and user interfaces are implemented in Java while the computationally heavier logic like the extended Kalman filter, smoother and fusion algorithms in the map module and the Dynamic Programming algorithm in the optimisation module are implemented in C++ using the Java Native Interface (JNI).

The data presented in the HMI is generated by the map module and the optimisation module. The data storage structure in the finished application consists of three files, see Fig. 3. From the collection of sensor data, the data is successively refined to a format which is ready to be displayed by the HMI.

Data (map) storage in application

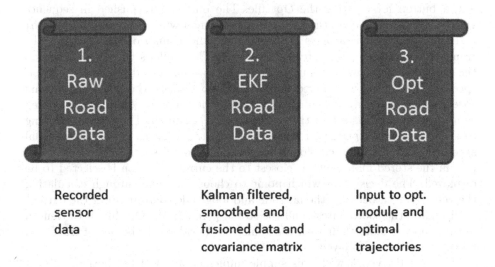

| 1. Raw Road Data | 2. EKF Road Data | 3. Opt Road Data |

Recorded sensor data

Kalman filtered, smoothed and fusioned data and covariance matrix

Input to opt. module and optimal trajectories

Fig. 3. Application data storage structure.

Map Module Implementation. The map is created/updated in a two stage process. In the first step the sensor signals are recorded, converted to an adequate formate and stored in a data file. This "raw" road data file is overwritten if a new recording of the road is made. At the finish of the recording of the road the stored raw data is processed in the EKF and the RTS smoother. If it is the first recording of the road, the data is split into 5 m long section and the data is stored along with the covariance matrix in a second file (EKF road data). If the road already has been recorded the points in the EKF road data file is matched with the points in the newly recorded raw road data file, see Sect. 2.2, and then merged utilising the earlier described fusion algorithm into a new EKF road data file. From the EKF road data file the coordinates, distance, road inclination, rolling resistance and max allowed speed are extracted into a third file (Opt road data) which is used as input to the optimisation.

Optimisation Module Implementation. The DP algorithm is implemented in C++ and is built both as a stand alone application for use on a PC but also implemented as a part of the HMI application in the Volvo Co-Pilot. The Opt road data file is used as input and when the optimisation is finished the optimal vehicle speed and optimal gear is appended to the respective coordinate pair in the same file.

User Interface (UI). The user interface is made in two layers (views). From the main view the map building is controlled, i.e. there is a button to start the recording of sensor data, a button that creates or updates the EKF file and a button for creating the Opt file. The buttons are pushed in sequence when creating the map/optimisation. From the main view the operator can start a second view which displays a speedometer and a map of the track for the transport (marked in light green), see Fig. 4. The hauler's current position on the track is marked with a purple dot. The speedometer displays the current speed of the vehicle with a red needle and also the (for the position) optimal speed, which is displayed as a green triangle. The target speed range, presented as a green triangle, is set to the optimal speed ± 2 km/h. Thus through trying to keep the vehicle speed within the displayed target speed range, the optimal speed trajectory is followed along the track. The target speed of the coordinate pair in the stored map which is closest to the current position is selected to be displayed. The decision of which point to choose is made through calculating the vector length between the latitude and longitude coordinates of the current position and the coordinates of all the stored points in the Opt file. If the length of the vector is more than 10 m the vehicle is considered to be out of track and no target speed is displayed.

A major drawback with this simple implementation of the optimal trajectories is that only the vehicle speed and not the gear is controlled. With this implementation the control of the gear is left to be handled by the normal gear shift strategy of the hauler, creating a discrepancy between the targeted optimal gear and the actual gear selected.

4   Machine Tests

The machine tests were performed on a Volvo A35G hauler with a nominal payload of 33.5 ton. The hauler was equipped with an interactive display and a GNSS smart receiver from the Co-Pilot system. A measurement system, based on the DeWeSoft data-logger system, was mounted in the hauler. To the DeWeSoft logger, a Vector CanCase capable of logging CAN communication, a separate GPS and an external high accuracy fuel flow meter were connected. The mass of the load was measured using the Volvo on-board weighing system and communicated via CAN.

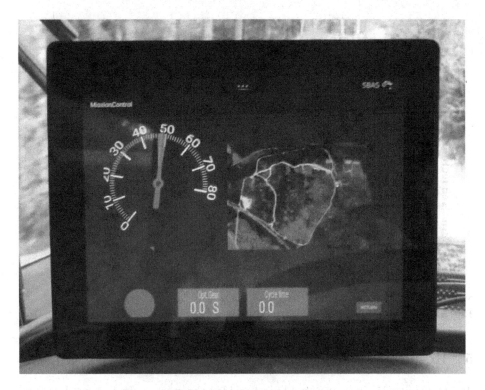

Fig. 4. Articulated hauler HMI.

The tests of the optimisation module were performed on a 1150 m long gravel road as a part of a larger test scheme. Initially, the road was travelled forth and back loaded and unloaded 3 times in each direction while the road characteristics were estimated using the map module. The vehicle trajectories were optimised using the code implemented on the Co-Pilot.

Evaluating the performance of an articulated hauler can be done in different ways. Two common measurements are productivity which measures transported mass per hour [ton/h] and fuel efficiency measuring transported mass per litre consumed fuel [ton/l]. While it is the potential fuel efficiency improvement of using optimal control that is to be proven, the productivity ought to be close to equal in the comparison measurements, hence, equal cycle time and equal load are required.

To achieve similar transport mission cycle times as in the tests where the optimal control strategy was applied, an estimation of a constant target vehicle speed to be kept during the transport mission was made. Two target test speeds were chosen: 30 km/h and 35 km/h. As a reference a full speed test were also carried out. As the tests were made as a part of a larger test schedule, the hauler was loaded with gravel at the start of each transport test cycle and then the load was dumped at the end. Hence, the load varied between the different tests rendering in undesired variations in productivity due to difference in load. The return tests were made with empty haulers and thus equal load.

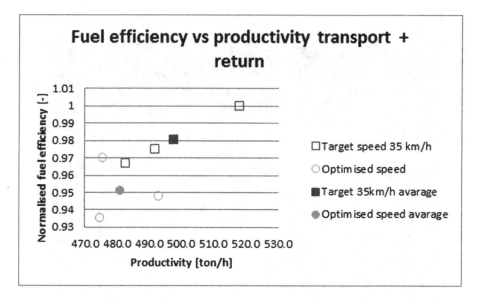

Fig. 5. Normalised fuel efficiency vs. productivity for transport and return with hauler.

5   Result

The tests with 35 km/h as speed target yielded the most equivalent productivity results when compared to the optimal speed trajectory tests, consequently the succeeding comparison is based on these results. At a first glance, the test results show an average decrease of fuel efficiency of approx. 3% at the same time as the productivity dropped with approx. 3% when the optimal speed trajectory was applied, see Fig. 5. It is worth noticing that in the result with the highest fuel efficiency and productivity the hauler was carrying significantly more load, 34.8 ton compared to between 32.1 and 33.4 in the other measurements ($\approx$6%). A dissemination of the results in Fig. 5 is shown in Figs. 6, 7, 8 and 9 in which vehicle speed and accumulated fuel consumption for 3 individual tests in each direction and test mode are presented.

5.1   Discussion

The comparison between the two test modes is off-set by several factors and there are preconditions in the test set-up leading to that the optimal speed mode does not reach it's full potential. The main factors are discussed below:

1. During the test a common comment from the operators was that the HMI ought to have an indication about the derivative of the target vehicle speed, i.e. an indication of whether the target speed is increasing or decreasing in the next step should be made or the target speed of the next coordinate point could be displayed. With the current set-up the information comes to late and

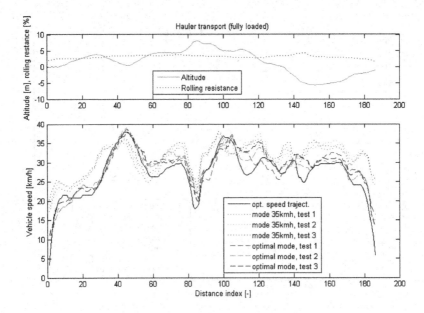

Fig. 6. Hauler transport vehicle speed.

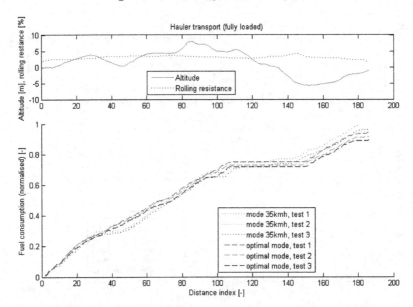

Fig. 7. Hauler transport accumulated fuel consumption (normalised).

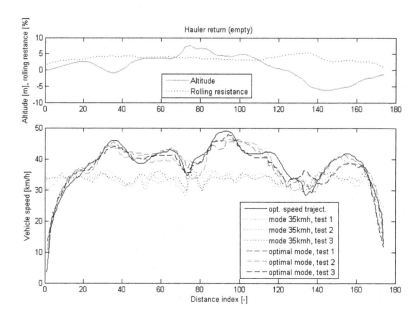

Fig. 8. Hauler return vehicle speed.

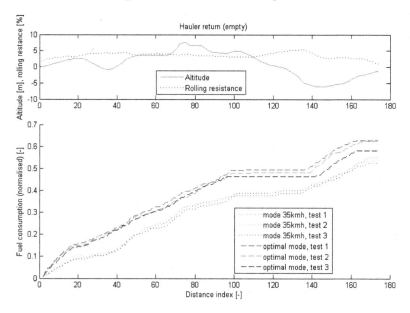

Fig. 9. Hauler return accumulated fuel consumption (normalised).

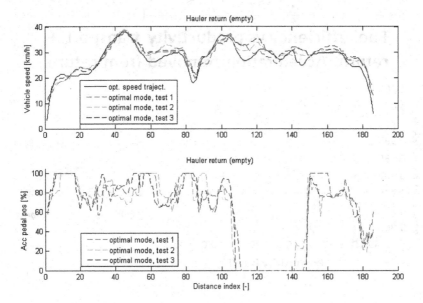

Fig. 10. Hauler transport vehicle speed and accelerator pedal position.

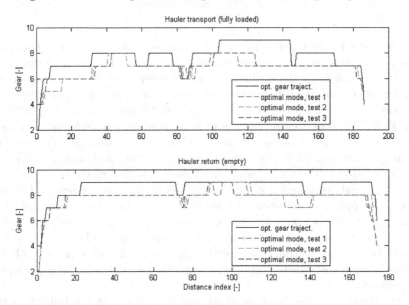

Fig. 11. Hauler transport and return selected gear compared with optimal gear trajectory.

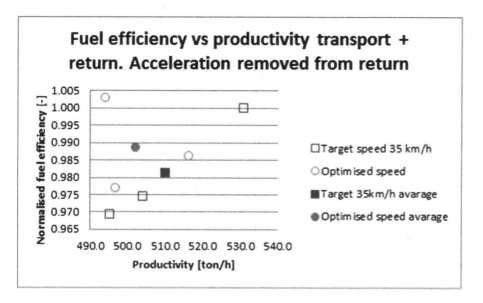

Fig. 12. Normalised fuel efficiency vs. productivity for transport and return with hauler after removal of acceleration in the return measurement.

the operator has to react to information that is already outdated. This limits the possibilities to plan the operation of the hauler for the operator and if there is a drift from the optimal speed trajectory the slow dynamics of the hauler makes it difficult to get back to an optimal speed. The phenomena can be seen in a graph showing the accelerator pedal position. As seen in Fig. 10 the signal is fluctuating more than desired, rendering in fuel efficiency losses from excessive transient operation of the engine. Additionally, the gear shift strategy is influenced by the operator input and thus a change in accelerator pedal position in the wrong position of the track may result in an undesired gear shift and the vehicle can be stuck in the wrong gear for a period of time.

2. While the optimisation of the hauler trajectories includes the control of the gear shift this was not in the scope of the test. Instead the haulers built in gear shift strategy was used. This lead to the use of much lower gears than what the desired (optimal) gear shift trajectory stipulates, see Fig. 11. The use of lower gears leads to higher engine speeds, affecting the fuel consumption negatively.

3. As the changes in road inclination are quick and frequent it is important that the resolution of the track representation is high enough to capture peaks and dips and at the same time give relevant speed information to the driver. With the current set up the nominal distance between each identification point is 5 m but with the additional allowance for the search area it can be up to 8 m, see Sect. 2.2. A higher resolution of the road or a method pinpointing the exact position of the local minima and maxima of the road altitude may be needed to avoid giving the new speed target too early or late

to the operator, resulting in a loss of kinetic energy and possibly incorrect gear shift. However, higher resolution will have a negative impact on the computational speed, especially in the optimisation module. Additionally, the earlier proposed notification of speed derivative may be another way to reduce the impact of low resolution.

4. In the analysis after the test it was discovered that the definition of where the loading/unloading zone ends and the track starts has a major influence on the test results. In the test, the loading and unloading zone was defined to be quite large, allowing a large area where the speed and movement of the hauler is uncontrolled. This means that when the track is entered and the control to a target speed begins, the speed already varies significantly and the accumulated fuel consumption is affected proportionally. This can be seen in e.g. Figure 8 between distance index 1 and 20. In the tests where the target speed was set to 35 km/h the speed of the hauler is close to 35 km/h already at the start of the measurement, omitting the very costly acceleration of the hauler, see the accumulated fuel signal in Fig. 9. If e.g. the first part of the measurements, up to the point where the speed of the optimised trajectory controlled haulers is essentially equal to the target speed of the second test mode (approx. 33 km/h at distance index 12), is removed and consequently the acceleration part is omitted, a small but noticeable fuel efficiency improvement is found, see Fig. 12. Hence, it is important to minimise the size of the loading/unloading zone to be able to control the speed of the haulers as much as possible and, in this case, to attain comparable measurements.

Some soft results are worth mentioning. The software implementation and the hardware worked well during the test. Even though the dynamics of the hauler is quite slow it is in most cases possible to follow the optimised speed trajectory fairly well as displayed in Figs. 6 and 8. This would be further improved with the inclusion of a notification of the speed derivative in the HMI. The test shows that relatively constant cycle times are achieved while following the optimised trajectory. The deviation is ±1 s from the average cycle time of 141 s for the hauler transport and ±1.5 s from the average 91.5 s for the hauler return.

6 Conclusion

A method to estimate road characteristics and to calculate fuel optimal speed and gear trajectories for an articulated hauler travelling on an off-road track has been developed and implemented in a human machine interface. The HMI instructs the hauler operator to follow the optimal velocity trajectory while the gear shift is left to the gear shift strategy built into the machine. The HMI was implemented on a tablet like interface with capability to read vehicle CAN messages and with a GNSS sensor attached.

The system was tested in real working conditions in a Volvo A35G hauler. The performance of the system was evaluated through assessing the difference in fuel efficiency of a transport mission where the vehicle speed had been optimised compared to a transport mission with equal productivity achieved through setting a constant speed target. While a to wide definition of the loading/unloading zone makes a peer to peer comparison of the two modes difficult, there are signs of fuel efficiency gains if using the optimal control strategy.

During the tests the implementation of both the road estimation method and the optimisation method in the Co-Pilot system worked well. A learning from the test is that it would be beneficial for the hauler operator to get information on the derivative of the speed signal to be able to better plan the operation of the hauler. Another soft result is that the variation in cycle time between transport cycles is limited when using the optimal control strategy.

Next level of complexity would be to design a cruise control enabling control over the gear shifts, this would better make use of the full potential of the optimisation algorithm and would be an interesting continuation of the presented work.

Acknowledgements. The authors acknowledge Volvo CE and FFI - Strategic Vehicle Research and Innovation, for sponsorship of this work.

References

1. Albrektsson, J., Åslund, J.: Road estimation and fuel optimal control of an off-road vehicle. In: Proceedings of the 3rd International Conference on Vehicle Technology and Intelligent Transport Systems, vol. 1, pp. 58–67 (2017)
2. Hellström, E., Åslund, J., Nielsen, L.: Design of an efficient algorithm for fuel-optimal look-ahead control. Control Eng. Practice **18**(11), 1318–1327 (2010)
3. Fu, J., Bortolin, G.: Gear shift optimization for off-road construction vehicles. Elsevier Procedia - Soc. Behav. Sci. **54**, 989–998 (2012)
4. Sahlholm, P.: Distributed road grade estimation for heavy duty vehicles. Doctoral thesis, Royal Institute of Technology, Stockholm (2011)
5. Almesåker, B.: Iterative map building for gear shift decision. Master thesis, Uppsala University (2010)
6. Saaf, M., Hana, A.: Map building and gear shift optimization for articulated haulers Master thesis, Mälardalen University (2011)
7. Welch, G., Bishop, G.: An introduction to the Kalman filter. University of North Carolina at Chapel Hill (2006)
8. Rauch, H.E., Striebel, C.T., Tung, F.: Maximum likelihood estimates of linear dynamic systems. AIAA J. **3**(8), 1445–1450 (1965)
9. Gustafsson, F.: Statistical sensor fusion. Studentlitteratur AB, edit. 2:1 (2012)
10. Bellman, R., Dreyfus, S.: Applied Dynamic Programming. Princeton University Press, Princeton (1962)
11. Guzzella, L., Sciaretta, A.: Vehicle Propulsion Systems, 3rd edn. Springer, Heidelberg (2013). https://doi.org/10.1007/978-3-642-35913-2
12. Bellman, R.: Adaptive Control Process: A Guided Tour. Princeton University Press, Princeton (1961)
13. Monastyrsky, V.V., Golownykh, I.M.: Rapid computations of optimal control for vehicles. Transp. Res. **27B**(3), 219–227 (1993)
14. AB Volvo Press information: Volvo Co-Pilot ensures a precise and profitable performance (2016). https://www.volvoce.com/global/en/news-and-events/news-and-press-releases/2016/volvo-co-pilot-ensures-a-precise-and-profitable-performance/

Intelligent Offloading Distribution of High Definition Street Maps for Highly Automated Vehicles

Florian Jomrich[1,3]([✉]), Aakash Sharma[2], Tobias Rückelt[3], Doreen Böhnstedt[3], and Ralf Steinmetz[3]

[1] Adam Opel AG, Bahnhofsplatz, 65423 Rüsselsheim, Hessen, Germany
florian.jomrich@opel.com
[2] Department of Computer Science, UiT: The Arctic University of Norway,
Hansine Hansens veg 54 Breivika, 9019 Tromsø, Norway
aakash.sharma@uit.no
[3] KOM Multimedia Communications Lab, Technical University of Darmstadt,
Rundeturmstr. 10, 64283 Darmstadt, Hessen, Germany
{tobias.ruckelt,doreen.bohnstedt,ralf.steinmetz}@kom.tu-darmstadt.de

Abstract. Highly automated vehicles will change our personal mobility in the future. To ensure the safety and the comfort of their passengers, the cars have to rely on as many information regarding their current surrounding traffic situation, as they can obtain. In addition to classical sensors like cameras or radar sensors, automated vehicles use data from a so called High Definition Street Map. Through such maps, the vehicles are provided with continuous updates regarding their future driving environment on a centimeter accurate level. The required amount of data, which is necessary therefore, motivates the development of more efficient data transmission concepts. In this paper we present HD-Wmapan extension of our previous work the Dynamic Map Update Protocol. Based on each vehicle's current context the Dynamic Map Update Protocol achieves a highly data efficient transmission of map updates compared to existing distribution approaches. HD-Wmapfurther reduces the costs of such transmissions by enabling map data to be shared via ad hoc communication between the vehicles. To evaluate the capabilities of HD-Wmapwe perform a first simulation of the morning commuting traffic within the area of Cologne, Germany. In this scenario HD-Wmapachieved an ad hoc map data off loading quota from cellular networks of up to 25.5%. These results demonstrate the gains of our approach to realize efficient map distribution via ad hoc communication, releasing load from wireless Internet access networks.

1 Introduction

To ensure the safety and the comfort of the passengers of future highly automated vehicles, researchers and engineers let the cars rely on a multitude of different sensors. Besides onboard systems like cameras, radar sensors or lidar

© Springer Nature Switzerland AG 2019
B. Donnellan et al. (Eds.): SMARTGREENS 2017/VEHITS 2017, CCIS 921, pp. 209–228, 2019.
https://doi.org/10.1007/978-3-030-02907-4_11

scanners [46], the vehicles further rely on an additional "virtual" sensor, the so called High Definition Street Map (HD-Map) [22]. This map is a centimetre accurate [40] virtual representation of the vehicle's surrounding world. With the aid of this detailed information, the autonomous cars can plan their future driving manoeuvres in advance. The capabilities of the map thereby extend the sensing range of the car's on-board systems [2,4] and provide an additional independent view on the current traffic situation. This is especially helpful for modern localisation and perception algorithms [20], as they are able to verify their personal sensing information with existing map material [7]. Nearly all companies and research groups, which are involved into the topic of highly automated driving, rely upon the HD-Map to further enhance the driving capabilities of their own autonomous vehicles. Examples therefore are Google [28], HERE [42], TomTom [44], Continental [13] and car manufacturers like BMW [2,6] and Tesla [34]. Due to its high precision and its high dynamic information content about the current surrounding traffic situations, the HD-Map gets outdated very quickly. Even standard navigation map material is outdated after a short period of time and profits from regular updates [35]. With the HD-Map, this situation is becoming even more critical. Update processes, which could wait for months now have to be finished within minutes. These kind of updates are only possible through a continuous datastream, which lets Hammerschmidt [13] speak of a "living map". Thus, currently existing update procedures for standard navigation maps are not feasible anymore in the context of HD-Maps. New concepts have to be investigated to let autonomous vehicles fully rely on HD-Maps. Within our previous work we addressed this problem by the presentation of the Dynamic Map Update Protocol [21]. The protocol enables the autonomous vehicles to receive map updates specifically for their personal requirements based on their own current and future driving context. Within the evaluation of this work, we show that the protocol outperforms existing state of the art map update approaches in terms of data efficiency and processing load.

This current work now represents an extension of the protocol from [21] to further reduce the costs for update transmissions and the general network load. The protocol itself has been initially designed to synchronize with a centralized map server. All vehicles communicate individually with this server, when requesting their updates via a cellular communication interface. This, however, introduces high transmission costs for the map updates and requires permanent access to a map server, representing a (potential) single point of failure. Within the scope of the present paper, we extend the protocol to enable vehicles to dynamically share already received map updates between each other via ad hoc network communication. This extension, which we call HD-Wmap, reduces cellular network transmission costs and also improves the load at the centralized map server, as it only has to provide map updates where no direct ad hoc data exchange could happen in advance. The general technique for offloading data transmissions on different communication channels is a well established paradigm, that has been evaluated in many different publications for the vehicular context (see Sect. 2.2). To the best of our knowledge, the presented work is the first, that adapts this concept for the distribution of navigation map material required for autonomous driving via ad hoc communication between the vehicles.

The outline of the remaining paper is described in the following. In Sect. 2, current state of the art map update concepts are discussed. Furthermore we give an introduction to related data-offloading approaches. Afterwards, the general working principle of the Dynamic Map Update Protocol is summarized in Sect. 3. This helps to follow the further extensions of the protocol, introduced in Chap. 4. In Chap. 5 we give an overview about our new and adapted simulation scenario, which reflects the morning rush hour traffic of the city of Cologne with up to 40.000 vehicles. Based on this scenario, the obtained results of the extended Dynamic Map Update Protocol are presented in comparison to a solely centralized approach of the protocol. We conclude the paper with a summary of the presented adaptation and the achieved improvements, as well as an outlook regarding future work.

2   Related Work

First, we give an overview about related work in the area of map updates, which provides the basis for the development of our Dynamic Map Update Protocol. In the second section we introduce related work in the domain of ad hoc communication and data offloading, which is related to our new extensions of the protocol.

2.1   Research on Map Updates

Electronic navigation devices for vehicles have been introduced to the market more than 25 years ago [19] and are now a frequently used feature in series. Most navigation systems can be grouped into one of two different categories: offline and online navigation. Most of the built-in car navigation devices are offline systems. They can operate completely independent of any kind of data connection, as they solely rely on an internal storage space when calculating the route to a given destination. This storage contains the complete map data in an efficient binary format. It enables fast read access and, therefore, improves significantly the performance of routing algorithms. This advantage, however, comes at the cost of a major disadvantage for the rapidly outdated HD-Maps. The binary map cannot be updated via partial replacements of the data, as Min [31] expressed in their work. Map updates must be provided as a unit of the whole map material in a single file. This might be one of the reasons why mapping companies only provide an updated version of their own map material after several months (e.g. Tom-Tom[1]). This is an inacceptable circumstance in the context of HD-Map material, because it has to be updated within minutes to ensure the function and safety of the autonomous vehicle.

---

[1] http://uk.support.tomtom.com/app/content/id/9/locale/en_gb/page/4.

In contrast to offline systems, online navigation systems do not have to rely on an own large internal storage space. Such systems are, for example, represented by current smartphone applications like Google Maps[2]. Each time a new route is calculated, they request the newest map material from a dedicated map server by wireless data transmission. Thus, this approach ensures that the map material is always up to date. This, however, requires a lot of redundant data to be transmitted each time a route is calculated. This is especially true if certain routes are requested frequently (e.g. the owner's daily commute to work). In the context of HD-Maps with their high degree of detail and thus increased size, redundant data transfer even becomes a more severe problem. Unlike a human driver the highly automated vehicle has to always rely on the information provided by the HD-Map and request its route guidance for every trip. Furthermore, it might not always be the case that a data connection is available, which will render the online navigation unusable.

In conclusion, both approaches, offline and online navigation, contain certain disadvantages.

To address these problems there has been a strong interest in research. Several different approaches enabling so called partial and incremental map updates have been published [3, 5, 9, 24, 27, 30, 31].

The general idea behind a partial map update is to divide the whole map material into smaller chunks, so called map tiles. These map tiles are then further addressed as individual maps that can be updated independently from each other.

Incremental map updates realize a sequential distinction of the map materials construction steps over time. Through special data structures (e.g. databases like PostgreSql[3]); the history of changes within the map can be reproduced. This enables the map server to provide highly data efficient map updates, which only contain the map changes required by the vehicles. Both approaches are usually combined to partial, incremental map updates. Those updates are then provided through a wireless connection (e.g. cellular) to the vehicles. A reference example of such a system is the work by Min [31]. However partial and incremental map updates introduce further challenges. A main challenge is to ensure the consistency of the map material after an update has been conducted, as stated by Asahara et al. [3]. Roads that traverse different map tiles might become unroutable if a map tile update causes inconsistency. To solve this problem the authors propose a procedure that checks the neighbouring map tiles regarding consistency. Engineers of Hitachi Automotive Systems, Ltd. [14] improve this approach further as they specifically generate connected map objects that ensure the consistency of the updated map. This prevents situations in which the update of one single map tile after another, as suggested by Asahara et al., would lead to a cascade of updates of the surrounding map tiles.

---

[2] https://maps.google.de/.

[3] https://www.postgresql.org/.

In contrast to the other presented publications, the ActMap Project [5] conducted by Bastiaensen et al. suggests to update only the map tiles on the car's current route.

As it minimizes the amount of necessary map updates the most, we considered the approach by Bastiensen et al. as the reference algorithm for comparison with our own Dynamic Map Update Protocol.

The design of the Dynamic Map Update Protocol has been influenced by all the aforementioned concepts. Our protocol enhances them by enabling specific map updates regarding the context of each individual vehicle as presented in Sect. 3 and in [21].

2.2 Data Offloading via Ad Hoc Communication

HD-Wmapimproves the protocol from [21] further by leveraging the capabilities of vehicles to create so called Vehicular Ad hoc Networks (VANETs).

Our approach, as introduced in detail in Sect. 4, relies on the capability of the cars to communicate directly with each other. Technologies introduced in the 802.11p WiFi standard [32] allow them to exchange data, when they are in the transmission range of each other.

In the context of mobile vehicular networks a strong research effort has been conducted to leverage this functionality. Many publications propose different approaches to offload data streams from costly cellular networks to free wireless communication.

Several of the existing approaches [11, 25, 29] therefore rely on the deployment of dedicated road side units in the roaming area of the vehicles. These units are able to communicate with the cars via WiFi technology. They are directly connected to the Internet via a cable connection to offload the network traffic. In our opinion, the wide deployment of such units (as it would be required for the deployment of map updates) is highly questionable, because they introduce additional installation and maintenance costs. Thus, the advantage of these road side units, compared to the already existing towers of the cellular network, is questionable. The work by Lee et al. [23] tried to improve this situation by only relying on the deployment of so called relay nodes for their offloading approach. In contrast to the road side units, these nodes do not possess a direct connection to the Internet, but are used as static nodes to hold and forward data between the vehicles. We argue that this still involves additional costs, which should be avoided. In contrast to the mentioned approaches, our extension of the Dynamic Map Update Protocol therefore relies solely on the direct ad hoc communication between the cars, as it does not involve further costs regarding any kind of additional hardware infrastructure.

Lee et al. stated that the installation of relay nodes is a necessary requirement, as the probability of two vehicles meeting each other with the same data requirements, which they can share, might otherwise be too low. We argue that this is especially not the case for map data updates. People tend to roam in their local neighborhood more than traveling far distances [33]. Thus, the chance to meet a vehicle that has already obtained required map data, is getting more and more probable when the car advances to this area.

To the best of our knowledge, we are the first who present an approach that disseminates map data between vehicles via ad hoc communication. Related papers propose ad hoc communication for map data sharing as a potential application [12,38,39] without getting into further detail or presenting a specific solution approach, as we do in our paper.

3   General Working Principle of the Dynamic Map Update Protocol

For better understanding of the performed enhancements of our previous work, we briefly summarize the general working principle of the Dynamic Map Update Protocol. For further details we refer to our previous work [21].

The Dynamic Map Update Protocol bases on the existing map update concepts presented in the Related Work section. It improves these concepts by introducing contextual relevance into the update process. In contrast to a human, a highly automated driving vehicle always has to rely on data detailing a specific given road when driving. The start and the destination of a trip are always known before the trip or have to be assumed, for example by a most probable path calculation as explained by Ress et al. [37] or Burgstahler et al. [10]. The Dynamic Map Update Protocol leverages this knowledge about the travelling path to decrease the amount of data to be transmitted when requesting a map update. To achieve a low transmission overhead, as offline navigation systems and the high up-to-dateness of online navigation systems, the protocol establishes a hybrid navigation approach as illustrated in Fig. 1. We assume that the highly automated vehicle has limited persistent storage to save a certain amount of map material for its navigation purposes. Thus, it is not dependent on an always available data connection. However, to ensure that the car can rely upon the most up to date map material at every time, it is equipped with a cellular communication module. Through this module, the car checks initially at a dedicated map server if its map tiles for its current route are up-to-date. Therefore, it calculates the desired route based on its stored map material (that might be outdated). In a second step, the vehicle transmits the start and destination points of its route, as well as the used map tile IDs and their version to the server. Then the server compares its personal map database with the one of the car. In the now following update step, in contrast to Bastiaensens [5] map update approach, our protocol does not directly update all the map tiles, which are identified as outdated along the path of the vehicle. As a main contribution of the protocol and illustrated in Fig. 1 the server is able to distinguish between map updates which are mandatory or optional for the vehicle's route. Mandatory map updates directly influence the current route on which the car should reach its destination. For optional updates, this is not the case. These kind of updates might be changes within the map tile that concern streets, which the car does not use on its planned route. Successively, the server provides the mandatory updates to the car and informs it about the optional map updates. Thus, the car can "decide" if it wants to request those map updates as well or delay their transmission to a later point in time. That way, the Dynamic Map Update Protocol is able to reduce the amount of transmitted map data significantly.

To distinguish between individual map tiles, the Dynamic Map Update Protocol uses the indexing structure of Geohashes [43]. A Geoshash is a string specifically generated to identify a certain geographic area in the world. Its length denotes the size of its addressed area. We leverage this property of the Geohash to further optimize the updating procedure of our protocol. Inspired by the general working principle of modern routing algorithms [31], we distinguish the map material into different layers. Depending on their personal type, streets are then added to one of those layers. Highway streets for example span a longer distance to interconnect cities, compared to smaller urban streets (see Fig. 2). Thus, for the exemplary evaluation of the Dynamic Map Update Protocol in [21] highway streets are grouped together in map tiles of larger size (assuming a Geohash size of 4 resembling a covered area of 40 km × 20 km), as illustrated in Fig. 2. In contrast urban streets are composed in smaller map tiles (assuming a Geohash size of 5, which covers an area of about 5 km × 5 km). The concept of different map layers allows the Dynamic Map Update Protocol to provide specific updates regarding the current streets on which the vehicle is travelling, neglecting unnecessary information. Furthermore the protocol overhead to exchange map material is reduced in this way.

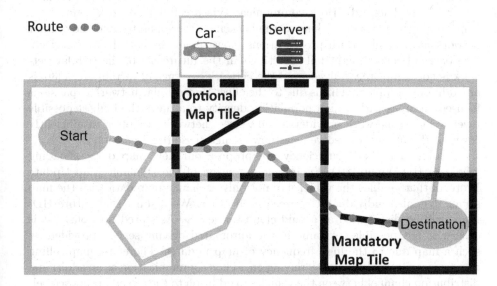

Fig. 1. Example for the general principle of the Dynamic Map Update Protocol [21].

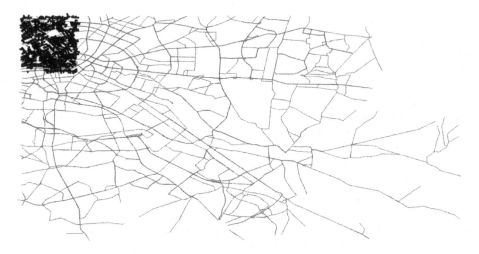

Fig. 2. Size of a city street layer map tile (bold) in comparison to a highway layer map tile [21]). ©OpenStreetMap contributors.

4 Improving Map Data Distribution Through Ad Hoc Transmission Offloading

The concept of the Dynamic Map Update Protocol [21] is based on the assumption of a bidirectional exchange of map data between a vehicle and a central map server (in the backend). Both communicate via a cellular network connection. This approach is feasible as long as the backend server operates and the cellular network is available. Both preconditions, however might not always be given. The backend server might fail sometimes in the future. Also, the cellular network is not completely fail-safe and the network coverage of certain areas [36] is an additional problem. This is due to the effect that cellular network expansions focused on areas with a high population density to achieve the highest possible revenue for the network providers. Thus, street networks are often insufficiently covered. Highly automated vehicles now change these requirements.

To overcome this insufficiency, we propose our new map data offloading schemata HD-Wmap. HD-Wmapis an extension of the Dynamic Map Update Protocol that enables the vehicles to not only exchange map data with the map server, but also individually between each other. We assume that future HD-Maps will be distributed as a paid cloud service [8]. As stated previously, it is otherwise not possible to maintain the automated driving service provided by such a map due to the high frequency of map updates. Therefore, map selling companies will highly profit by relying on ad hoc communication as an additional distribution channel between the vehicles in addition to the already proposed cellular transmission from a backend server [21], which is hosting the map. Ad hoc communication will significantly improve the service quality and reliability of such map data providers. This is especially true in scenarios of high load like in a traffic jam or the daily rush hour. Direct communication will then help to reduce the load on the cellular networks by lowering their costly data traffic. Also the processing load on the map server will be reduced as the cars can share already provided map material instead of requesting it several times. Furthermore, ad hoc communication enables the distribution of map updates even if there is no cellular infrastructure available. To realize the sharing of map data HD-Wmaprelies on technologies assumed to be available in highly automated vehicles as presented in the following.

4.1 Compatibility to Standards

Different technologies exist to enable ad hoc communication between vehicles, e.g. the WiFi-based 802.11p [32] standard or an upcoming extension of the LTE cellular communication, the so called LTE V2X [41]. In the following, we explain the general principle of HD-Wmapbased on the 802.11p standard, as it is the more advanced standard. LTE V2X provides similar functionalities, that have been inspired by the discussions related to 802.11p.

To be able to realize a certain set of applications for advanced driver assistance systems [15] as specified by the European Telecommunications Standards Institute (ETSI)[4] the vehicles continuously exchange standardized messages with each other. This includes Cooperative Awareness Messages (CAM) [16], Service Awareness Messages (SAM) [17] and Decentralized Environmental Notification Basic Services (DENM) [18]. CAMs are transmitted to inform the surrounding vehicles about the transmitting vehicles current position, speed and further direction. They are continuously broadcasted with a rate between 1 and 10 Hz depending on the currently present overall load on the data channel. SAMs are transmitted to inform vehicles about certain application services availability. Furthermore, DENMs supply vehicles with the capability to inform others about certain events currently experienced in the traffic. This could be for example an information about an ongoing construction side. All these messages include optional containers that facilitate the extensions required by HD-Wmap. Thus, we propose to extend one of these messages to include the necessary additional information required to initiate the map data sharing procedure between the vehicles, as explained in the following sections.

4.2 Home Zone Concept

Highly automated vehicles most frequently need the map material of the routes on which they have to drive regularly. These are for example the commuting routes of their owners, the local neighbourhood and nearby towns. These areas are therefore considered as home zone in the following. The home zone is assumed to be updated frequently at the beginning of each trip, including all HD-Map layers, e.g. by downloading updates for it at home through a WiFi connection. This complete information of the area differs from the partial and layer-specific

---

[4] http://www.etsi.org/.

updates of vehicles from apart. As home-zoned vehicles roam in this area the most, they are predestined to share their data with others. Thus, home zones provide the backbone for efficient map distribution via ad hoc communication, releasing load from wireless Internet access networks.

4.3   Procedure of HD-Wmap to Offload Map Data

Algorithm 1. Actions performed in a map exchange by HD-Wmap.

HD-Wmapprocedure actions:

1. Check which updates are required for the current trip.
Gather necessary Geohashes and tile versions from the backend server.

2. (a) If (distance to map tile > x) request it via ad hoc.
   (b) Else download it via cellular.

3. Answer requests, if data is available in the own internal storage,
e.g. recently downloaded or as part of the home zone.

The working principle of HD-Wmapis summarized by the procedure actions in Algorithm 1. We explain them in this section based on an example as illustrated by the Figs. 3 and 4. In the example two different cars a sedan and a cabriolet each drive an own trip (see Fig. 4). The home zones of the vehicles are assumed to be the areas, which include the two most left, respectively the two most right map tiles of the example map database as shown in Fig. 3. As first step (1 in Algorithm 1) in the procedure of HD-Wmapeach vehicle requests the mandatory and optional updates of map material for its current route from the dedicated update server. In our example these are two mandatory map updates, one for each car, indicated by black color in Fig. 4. The vehicles first try to gather those updates via an ad hoc transmission (2(a)). A certain time is necessary for a car to be able to obtain an outdated map tile via ad hoc communication. Thus, the car has to start the request process for it at a certain distance (e.g. several kilometres) in advance, before it reaches this location. Therefore the vehicle is sending a request message to its neighbours with a certain frequency. We consider such a request as an extension of one of the already standardized messages as explained in Sect. 4.1. The additional container to be added in the message just has to include two additional parameters, the Geohash ID of the requested map tile and its required version status.

If one of the cars within the proximity of the requester has obtained this map tile, it starts to provide its data to him (3). This is for example the case for the sedan, as the cabriolet can share map data of its home zone when both cars meet in the upper middle map tile, as illustrated by Fig. 4. The vehicles themselves are not capable to provide smaller delta updates like the map data server, because they cannot store the whole version history of a map tile within their limited internal storage space. Thus they have to provide the whole map tile of a layer of the HD-Map to their neighbours. Depending on the covered area, such a map tile can be a small number of megabytes in size [1]. The ad hoc communication realized by the 802.11p standard [32], however, has been specifically designed for small amounts of data (several hundreds of bytes) to be transmitted at once. Thus we propose to split up a single map tile into smaller chunks of data, which are then transmitted individually, to ensure a reliable data reception. Depending on the time, in which the cars are in range to each other, they can share several of those data chunks up to several complete map tiles.

Only if the requesting car does not receive all the required data chunks of a map tile via ad hoc transmission until it reaches a certain minimum distance to the requested map tile, it will then request the remaining data parts directly from the map server through the cellular network (2(b)). This update will than be added as well to the internal map storage of the car as it is relevant for its current trip and thus can be provided further to other vehicles in the surrounding. For example, this is the case for the cabriolet, as its outdated map tile could not be provided by the sedan. We consider the cellular communication as a fall back option in HD-Wmap, that is available, but comes at the price of high transmission costs, respectively lower efficiency.

To investigate the performance of the HD-Wmapextension we present the first obtained results from a simulation of the city area of Cologne in the following Evaluation Sect. 5.

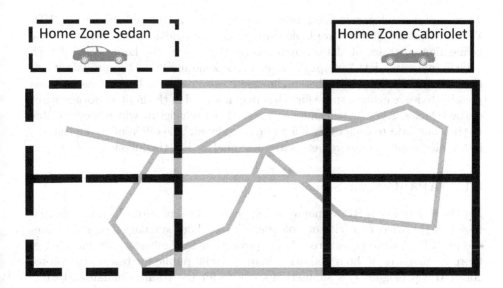

Fig. 3. Example for two different home zones of two cars.

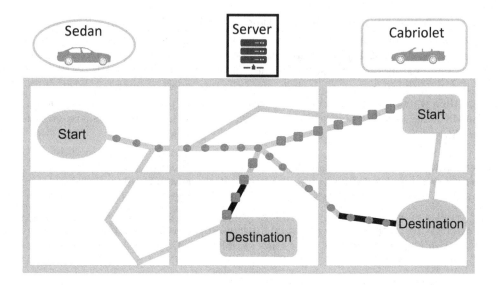

Fig. 4. Example scenario for HD-Wmapto illustrate the ad hoc sharing of map data.

5   Evaluation

There are several parameters, which are influencing the performance of HD-Wmap. This includes the available throughput bandwidth of the current ad hoc connection to transmit data. Furthermore, the size of the home zone of each vehicle influences HD-Wmap. A larger home zone increases the probability to share map material with others. The size however should be kept as small as possible to leave enough space for other map material in the limited storage space of the vehicles. Also the transmission range of the vehicles by which they are able to transmit data to each other has to be considered. The different configurations of the aforementioned parameters are now analyzed in the following.

5.1   TAPASCologne Scenario

For the evaluation of HD-Wmapwe relied on a traffic scenario, which was created based on the area of the city of Cologne. We based our simulation on map material available by the OpenStreet-Map[5] project, as to the best of our knowledge, there is currently no high definition map material public for testing the protocol. TAPASCologne[6] is a simulation scenario for the traffic simulator SUMO[7] resembling the daily traffic in the area of the city of Cologne in Germany. A dataset for the morning rush hour between six and eight o'clock is public available. We extracted the corresponding street map states for each day of August

---

[5] https://www.openstreetmap.org/.

[6] http://sumo.dlr.de/wiki/Data/Scenarios/TAPASCologne.

[7] http://sumo.dlr.de/.

2016 as a lower bound estimation for changes in future HD-Maps. Each of the three scenarios presented in the following has been conducted 30 times, based upon 30 different states of outdated map data. These difference files have been created by comparing the map data of the first day of August to the data of the remaining 30 days. With the help of SUMO we extracted the trip of each of the roaming cars with a timely resolution of 10 s. As in our previous work [21] we assumed a grid size of 5 km × 5 km (resembling a Geohash of size 5) for all map tiles present within the simulated area of Cologne.

5.2 Scenario Configuration

As the provided city scenario of Cologne is still rather small with a coverage area of about 40 km × 40 km with respect to a map tile size of 5 km × 5 km, we configured the vehicles in our simulation to immediatly request their required map tiles. Therefore, we assumed a circular transmission range of messages of 300 m [26,45] for the 802.11p technology [32] to be commonly achievable. The home zone area of each vehicle was simulated for two different sizes. The home zone size zero only includes the map tile from which each vehicle starts. A home zone of size one adds the eight surrounding map tiles to the list of stored data. We use a simplified transmission model, which is independent from technology and allows the evaluation of different transfer rates and map sizes. In this model, a partial transmission of one data chunk is finished each time step. Thus, the transfer rate and the map size can be modeled with a varying number of chunks to be transfered. A car can share map tile layers as soon as their download is finished. Chunks that could not be received via ad hoc communication in time are downloaded instantly via a cellular connection. This is a simplification regarding the simulation accuracy that will be extended in future work. With an area of about 5 km × 5 km covered by a map tile of Geohash size five, we assumed a data size of the tile of about 10 megabytes [1]. In our opinion this is expected to be an average reasonable size of a map tile covering city streets. Due to the different kinds of street networks, this value, however, might change depending on the exact represented location in a real implementation of HD-Wmap. To transmit such a map tile for example with a slow transmission speed of only 0.5 Mbit/s would require 160 s. In our simulation this time is mapped onto the transmission of 16 consecutive data chunks due to the time progress of 10 s per simulation step. To ensure a transmission at even worse network connectivity conditions we investigated values of data chunks required for a full map tile as illustrated by Fig. 5. To cover different scenarios, we vary the number of chunks from 1 to 70 in our evaluation.

5.3 Percentage of Offloaded Map Data

As the first evaluation metric we analyze the percentage of map tiles, which can be received via ad hoc communication in comparison to the remaining amount of data that has to be transmitted via the cellular network. The achieved savings are presented in Fig. 5. The investigated range reaches from only one necessary data chunk representing a data connection with a high bandwidth of 8 Mbit/s by assuming a map tile size of 10 megabytes, up to 70 resembling a very poor data connection of only 114 Kbit/s. In this the further parameters were fixed as a home zone size of zero with a transmission range of 300 m.

HD-Wmapthereby achieved a sharing quota of up to 25.5% in average under the best transmission conditions. It clearly shows the effectiveness of the approach to off load the transmission of map data from the cellular network via ad hoc communication. This quota only decreases by 1/3 to around 17.8% when the ad hoc channel capacity is reduced to 1/70 of its initial value. This indicates, that the offloading is limited by other factors besides throughput that require further investigation in the future.

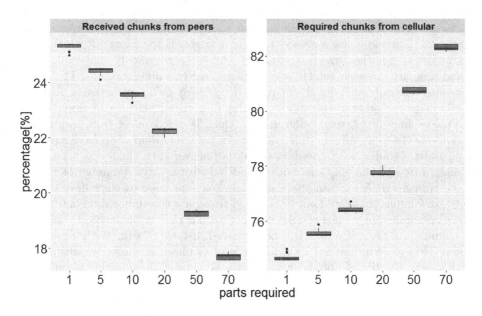

Fig. 5. Investigation of different amounts of data chunks to be required per map tile.

5.4   Variation of the Transmission Range

The second parameter of the simulation, that we investigated is the transmission range of the vehicles as shown by Fig. 6. Initially set to a distance of 300 m we further reduced this value to 200, 100 and 50 m. The amount of required data chunks was set to 20 and the home zone size to 0.

In comparison the reduction of the transmission range to 1/3 or 1/7 of the initial value only led to a reduction of the sharing functionality by 7.2% respectively 18.4%. This reflects the simulated scenario of Cologne in the morning rush hour with a probably high portion of commuting trips in the amount of all performed trips. Commuting trips mostly follow common main roads and lead to congestion due to a high density of vehicles, which allows efficient sharing of map data even with a largely reduced transmission range.

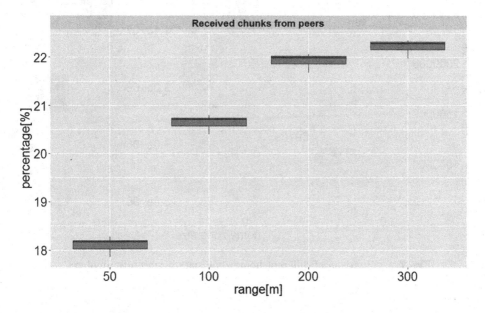

Fig. 6. Comparison of different transmission ranges.

5.5   Variation of the Home Zone Size

The third and final parameter, which we varied in our evaluation is the size of the home zone. Furthermore a transmission range of 300 m and a required amount of 20 data chunks was fixed. As stated previously a home zone of size zero only includes the map tile that covers the start position of the vehicle. A home zone size of one adds the eight surrounding map tiles to it. As expected, the amount of total required chunks to perform a car's trip can be reduced significantly by increasing its home zone size as indicated by the rightmost graph in Fig. 7. Interestingly, the amount of chunks which could be shared via ad hoc communication also decreases with an increased home zone size (see leftmost graph in Fig. 7 and the graphs of Fig. 8). This is presumably due to a high percentage of short trips to be simulated in the morning commuting scenario. A larger home zone of size one leads to the situation that cars, which only perform trips, with a range between 5 and 10 km do not have to request any more map tiles, as it would have been the case for a home zone size of zero. The smaller amount of vehicles, which perform larger trips in the small Cologne scenario to the outskirts of the map, however, might not find a suitable exchange

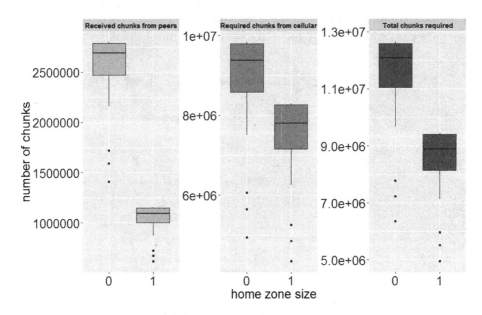

Fig. 7. Comparison of different home zone sizes in absolute numbers.

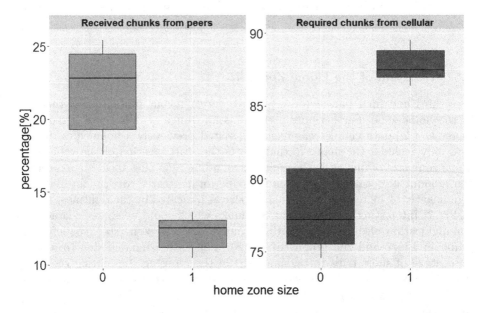

Fig. 8. Comparison of different home zone sizes in percentage.

partner due to a decreased density of cars in these areas. Finally this leads to an increased percentage of necessary cellular transmission for all considered trips, which require updates to their database. We expect a larger scenario with longer travel distances or a smaller map tile size and a more homogeneous distribution of vehicles to show higher offloading potential.

6 Conclusions and Future Work

This paper presents HD-Wmap, an extension of the Dynamic Map Update Protocol [21]. The Dynamic Map Update Protocol has been designed to enable efficient context based map data updates, which are required for Advanced Driver Assistance Systems and highly automated vehicles. Our proposed extension HD-Wmapimproves the achieved results further by introducing the capability to the cars to share personal map data with the vehicles in their proximity. This is achieved by relying on ad hoc communication technology in the vehicular context, like the 802.11p standard [32]. To the best of our knowledge, we are the first work to propose a concrete concept to share map data between vehicles via this technology. To ensure the fair and efficient data sharing between the vehicles we propose the concept of a so called home zone in which the vehicles roam the most. This area is assumed to be updated and stored prioritised in the car's internal storage space. Vehicles preferably request map updates via ad hoc from surrounding ones and use the cellular updates from our previous work as backup procedure.

Within the simulated scenario of Cologne, HD-Wmapachieved an ad hoc map data off loading quota of up to 25.5%. Especially the design and the selection of a proper home zone area for each vehicle denote very important factors to improve the ad hoc data sharing-efficiency of the map updating process.

We plan to extend our initial home zone concept. For example, daily commuting trips of the vehicles should be considered to be included in the home zone as well to improve their sharing capabilities through HD-Wmap. Future simulations will span larger areas to reveal the complete efficiency improvement of the offloading strategy of HD-Wmap, which we expect to be even higher.

References

1. Here HD Live Map: So sehen autonome Fahrzeuge die Welt. http://winfuture. de/videos/Software/Here-HD-Live-Map-So-sehen-autonome-Fahrzeuge-die-Welt-15605.html. Accessed 16 Aug 2017
2. Aeberhard, M., et al.: Experience, results and lessons learned from automated driving on Germany's highways. In: IEEE Intelligent Transportation Systems Magazine, vol. 7, pp. 42–57 (2015). http://ieeexplore.ieee.org/lpdocs/epic03/wrapper. htm?arnumber=7014396
3. Asahara, A., Tanizaki, M., Morioka, M., Shimada, S.: Locally differential map update method with maintained road connections for telematics services. In: 2008 Ninth International Conference on Mobile Data Management Workshops, MDMW 2008, pp. 11–18. IEEE (2008). http://ieeexplore.ieee.org/xpls/abs_all.jsp? arnumber=4839079

4. Barker, P.: Industry expert explains why autonomous cars need a map - HERE 360 (2015). http://360.here.com/2015/05/07/brad-templeton-autonomous-cars-maps/. Accessed 16 Aug 2017
5. Bastiaensen, E., et al.: ActMAP: real-time map updates for advanced in-vehicle applications. In: Proceedings of 10th World Congress on ITS, Madrid (2003)
6. Bender, P., Ziegler, J., Stiller, C.: Lanelets: efficient map representation for autonomous driving. In: 2014 IEEE Intelligent Vehicles Symposium Proceedings, pp. 420–425. IEEE (2014). http://ieeexplore.ieee.org/xpls/abs_all.jsp?arnumber=6856487
7. Boensch, R.: With high resolution into the autonomous world. VDI Nachrichten Ausgabe 03. http://www.vdi-nachrichten.com/Technik-Wirtschaft/Hochaufloesend-in-autonome-Welt. Accessed 16 Aug 2017 (2016)
8. Bonetti, P.: HERE introduces HD live map to show the path to highly automated driving - HERE 360. http://360.here.com/2016/01/05/here-introduces-hd-live-map-to-show-the-path-to-highly-automated-driving/. Accessed 16 Aug 2017
9. Cooper, A., Peled, A.: Incremental updating and versioning. In: Proceedings of 20thInternational Cartographic Conference, pp. 2804–2809 (2001). http://icaci.org/files/documents/ICC_proceedings/ICC2001/icc2001/file/f19007.pdf
10. Burgstahler, D., Peusens, C., Böhnstedt, D., Steinmetz, R.: Horizon.KOM: a first step towards an open vehicular horizon provider. In: Proceedings of the 2nd International Conference on Vehicle Technology and Intelligent Transportation Systems (VEHITS 2016), p. 6 (2016)
11. Dimatteo, S., Hui, P., Han, B., Li, V.O.K.: Cellular traffic offloading through WiFi networks. In: 2011 IEEE Eighth International Conference on Mobile Ad-Hoc and Sensor Systems, pp. 192–201, October 2011
12. Dressler, F., Hartenstein, H., Altintas, O., Tonguz, O.K.: Inter-vehicle communication: Quo vadis. IEEE Commun. Mag. **52**(6), 170–177 (2014)
13. Hammerschmidt, C.: With data from the cloud to the living map. VDI Nachrichten Ausgabe 03 (2016). http://www.vdi-nachrichten.com/Technik-Wirtschaft/Mit-Daten-Cloud-lebenden-Karte. Accessed 16 Aug 2017
14. Hitachi, A.L.: Map update service/solution (2016). http://www.hitachi-automotive.co.jp/en/products/cis/02.html. Accessed 16 Aug 2017
15. ETS Global Institute: Intelligent transport systems (ITS); vehicular communications; basic set of applications; definitions ETSI TR 102 638, V1.1.1 (2009)
16. ETS Global Institute: Intelligent transport systems (ITS); vehicular communications; basic set of applications; part 2: specification of cooperative awareness basic service ETSI TS 102 637-2, V1.1.1 (2010)
17. ETS Global Institute: Intelligent transport systems (ITS); users and applications requirements; part 1: facility layer structure, functional requirements and specifications ETSI TS 102 894-1, V1.1.1 (2013)
18. ETS Global Institute: Intelligent transport systems (ITS); vehicular communications; basic set of applications; part 3: specifications of decentralized environmental notification basic service ETSI EN 302 637-3, V1.2.1 (2014)
19. Ishikawa, K., Ogawa, M., Azuma, S., Ito, T.: Map navigation software of the electro-multivision of the 1991 toyoto soarer. In: 1991 Vehicle Navigation and Information Systems Conference, vol. 2, pp. 463–473. IEEE (1991). http://ieeexplore.ieee.org/xpls/abs_all.jsp?arnumber=1623657
20. Jo, K., Sunwoo, M.: Generation of a precise roadway map for autonomous cars. IEEE Trans. Intell. Transp. Syst. **15**, 925–937 (2014). http://ieeexplore.ieee.org/lpdocs/epic03/wrapper.htm?arnumber=6680774

21. Jomrich, F., Sharma, A., Rückelt, T., Burgstahler, D., Böhnstedt, D.: Dynamic map update protocol for highly automated driving vehicles. In: Proceedings of the 3rd International Conference on Vehicle Technology and Intelligent Transport Systems, pp. 68–78 (2017)
22. Lawton, C.: Why an HD map is an essential ingredient for self-driving cars - HERE 360 (2015). http://360.here.com/2015/05/15/hd-map-will-essential-ingredient-self-driving-cars/. Accessed 16 Aug 2017
23. Lee, M., Song, J., Jeong, J., Kwon, T.: Dove: data offloading through spatio-temporal rendezvous in vehicular networks. In: 2015 24th International Conference on Computer Communication and Networks (ICCCN), pp. 1–8, August 2015
24. Lee, S., Lee, S.: Map generation and updating technologies based on network and cloud computing: a survey. Int. J. Multimed. Ubiquitous Eng. 8, 107–114 (2013). http://www.sersc.org/journals/IJMUE/vol8 no4 2013/11.pdf
25. Li, Y., Jin, D., Wang, Z., Zeng, L., Chen, S.: Coding or not: optimal mobile data offloading in opportunistic vehicular networks. IEEE Trans. Intell. Transp. Syst. 15(1), 318–333 (2014)
26. Lin, L.: ETSI G5 technology: the European approach (2013). http://www.drive-c2x.eu/tl-files/publications/3rd. Accessed 16 Aug 2017
27. Liu, Y., Yang, Z., Han, X.: Research for incremental update data model applied in navigation electronic map. In: 2010 2nd International Conference on Advanced Computer Control (ICACC), vol. 1, pp. 261–265. IEEE (2010). http://ieeexplore.ieee.org/xpls/abs_all.jsp?arnumber=5487031
28. Madrigal, A.C.: The Trick That Makes Google's Self-Driving Cars Work - The Atlantic - GOOGLE (2014). http://www.theatlantic.com/technology/archive/2014/05/all-the-world-a-track-the-trick-that-makes-googles-self-driving-cars-work/370871/. Accessed 16 Aug 2017
29. Malandrino, F., Casetti, C., Chiasserini, C.F., Fiore, M.: Offloading cellular networks through its content download. In: 2012 9th Annual IEEE Communications Society Conference on Sensor, Mesh and Ad Hoc Communications and Networks (SECON), pp. 263–271, June 2012
30. Min, K. W., An, K.H., Kim, J.W., Jin, S.I.: The mobile spatial DBMS for the partial map air update in the navigation. In: 2008 11th International IEEE Conference on Intelligent Transportation Systems, ITSC 2008, pp. 476–481. IEEE (2008). http://ieeexplore.ieee.org/xpls/abs_all.jsp?arnumber=4732538
31. Min, K.: A system framework for map air update navigation service. ETRI J. 33, 476–486 (2011)
32. Morgan, Y.L.: Notes on DSRC amp; WAVE standards suite: its architecture, design, and characteristics. IEEE Commun. Surv. Tutor. 12(4), 504–518 (2010)
33. Pasaoglu, G., et al., European Commission, Joint Research Centre, Institute for Energy and Transport: Driving and parking patterns of European car drivers: a mobility survey. Publications Office (2012). OCLC: 847460656
34. Perkins, C.: Tesla is mapping out every lane on Earth to guide self-driving cars (2015). http://mashable.com/2015/10/14/tesla-high-precision-digital-maps/#v3xHW6i4QSqU. Accessed 16 Aug 2017
35. Plack, J.: The unmatched quality of HERE Maps content - HERE 360 (2013). http://360.here.com/2013/03/27/the-unmatched-quality-of-here-maps-content/. Accessed 16 Aug 2017
36. Pögel, T., Wolf, L.: Analysis of operational 3G network characteristics for adaptive vehicular connectivity maps. In: 2012 IEEE Wireless Communications and Networking Conference Workshops (WCNCW), pp. 355–359. IEEE (2012). http://ieeexplore.ieee.org/xpls/abs_all.jsp?arnumber=6215521

37. Ress, C., Balzer, D., Bracht, A., Durekovic, S., Löwenau, J.: Adasis protocol for advanced in-vehicle applications. In: 15th World Congress on Intelligent Transport Systems, p. 7 (2008). http://www.cvt-project.ir/en/Admin/Files/eventAttachments/ADASISv2-ITS-NY-Paper-Finalv4_149.pdf
38. Saad, A., Abdalrazak, T.R., Hussein, A.J., Abdullah, A.M.: Vehicular ad hoc networks: growth and survey for three layers. Int. J. Elect. Comput. Eng. **7**(1), 271 (2017)
39. Schoch, E., Kargl, F., Weber, M.: Communication patterns in vanets. IEEE Commun. Mag. **46**(11), 119–125 (2008)
40. Schumann, S.: Why we're mapping down to 20cm accuracy on roads - HERE 360 (2014). http://360.here.com/2014/02/12/why-were-mapping-down-to-20cm-accuracy-on-roads/. Accessed 16 Aug 2017
41. Shi, Y.: LTE-V: a cellular-assisted V2X communication technology. In: ITU Workshop (2015)
42. Stevenson, J.: Making the invisible visible with the HD Live Map (2016). http://360.here.com/2016/09/23/making-the-invisible-visible-with-the-hd-live-map-video-demo/. Accessed 16 Aug 2017
43. Suwardi, I.S., Dharma, D., Satya, D.P., Lestari, D.P.: Geohash index based spatial data model for corporate. In: 2015 International Conference on Electrical Engineering and Informatics (ICEEI), pp. 478–483. IEEE (2015). http://ieeexplore.ieee.org/xpls/abs_all.jsp?arnumber=7352548
44. TomTom: TomTom Enables Autonomous Driving (2016). http://automotive.tomtom.com/uploads/assets/993/1475151547-autonomous-driving-product-info-sheet.pdf. Accessesd 15 May 2016
45. Weigele, M.: Standards: WAVE/DSRC/802.11p (2008). http://www.cvt-project.ir/Admin/Files/eventAttachments/109.pdf. Accessed 16 Aug 2017
46. Ziegler, J., Bender, P., Schreiber, M., Lategahn, H., Strauss, T., et al.: Making bertha drive - an autonomous journey on a historic route. IEEE Intell. Transp. Syst. Mag. **6**, 8–20 (2014). http://ieeexplore.ieee.org/lpdocs/epic03/wrapper.htm?arnumber=6803933

Classification of Automotive Electric/Electronic Features and the Consequent Hierarchization of the Logical System Architecture
From Functional Chains to Functional Networks

Johannes Bach[✉], Stefan Otten, and Eric Sax

FZI Research Center for Information Technology, Haid-und-Neu-Str. 10-14,
76131 Karlsruhe, Germany
bach@fzi.de

Abstract. In the established Automotive Systems Engineering (ASE) practice, an important factor in handling the complexity of product development is the partitioning of the vehicle into different domains. The current technological advances enable increasingly complex features for assisted and automated driving that reach across these different domains and are difficult to handle by the existing approaches. To cope with these challenges, new innovative methods, procedures and techniques are required that integrate well with the established practice. In this contribution, we analyze existing and future automotive features and classify them in a comprehensive taxonomy. Based on this characterization, established industrial and new research approaches for logical system architectures are consolidated. The introduction of new levels of hierarchy in the logical system architecture facilitates the attribution of specific design schemes and engineering approaches to the related functional elements. This approach facilitates the management of features with high internal variety and wide distribution over several subsystems. The systematic approach provides a novel rationale for the evolution from functional chains to functional networks in the automotive industry.

1 Introduction

Many current and future innovations in the automotive industry emerge from the field of Advanced Driver Assistant Systems (ADAS) and automated driving. Examples for this trend depict Predictive Cruise Control (PCC) [10], Lane Keeping Assist (LKA) [71] and automated valet parking [59] features. These features leading the way, automated driving becomes a reality [13]. A main driver of this trend are steadily increasing high-performance computing capabilities and innovative sensor and communication technologies. These innovations revolutionize the established Electrics/Electronics (E/E) system architectures to handle the substantial raise in functional complexity regarding utilized algorithms, the

B. Donnellan et al. (Eds.): SMARTGREENS 2017/VEHITS 2017, CCIS 921, pp. 229–255, 2019.
https://doi.org/10.1007/978-3-030-02907-4_12

amount of processed information and the high distribution of functions over several Electronic Control Units (ECUs) all over the vehicle. This evolution necessitates improvements and advances in development methods and approaches to cope with the new complexity.

Several roles, teams and organizations participate in the development of an automotive system. Scattering over different development locations leads to collaborative development [89]. The development in automotive vehicles is historically structured into different domains [67], such as powertrain, safety and chassis. For Original Equipment Manufacturer (OEM), well-established and long-existing systems such as Electronic Stability Control (ESC) are iteratively optimized achieving a high-level of maturity. Supplier structures and adjacent business units, such as purchase or after sales, are shaped to the originated needs. As new highly-integrated features partly collude with the existing systems, the question of how to use legacy systems during development poses a challenge. To foster the reuse of specific functionalities of the established systems is a key issue for efficient development.

Nonetheless, legacy systems are often not comprehensively covered by novel research and development approaches for systems engineering that focus assisted and automated driving [48,80]. The presented functional architectures mainly focus on the automated driving or ADAS domain. Therefore, only focusing on assisting and automating functional aspects and applying hierarchization without consideration of the relevant conditions of adjacent domains. In our opinion, a comprehensive approach for the abstraction and description of the functional architecture with respect to different levels of integration and complexity of features is required.

To overcome these impediments, we present a taxonomy for existing automotive customer features across all domains, structuring them into different levels of complexity. The taxonomy forms a basis to provide a systematic approach for systems engineering with a focus on functional aspects. This systematic approach can be further elaborated to consider the impact on development processes.

This contribution is an extended version of our already published paper [7] and is structured as followed: Sect. 2 presents the state-of-the-art of ASE and architectures. Our cross-domain taxonomy and assessment of current and upcoming E/E features is elaborated in Sect. 3. Our proposed approach for logical architectures and hierarchization is given in Sect. 4. Sect. 5 demonstrates the applicability of the approach on exemplary automotive features. A conclusion and outlook on further activities is presented in Sect. 6.

2   State of the Art

Systems engineering is a discipline to "guide the engineering of complex systems" [40]. The term "System" is widely spread across different fields and application domains and several approaches for development are established. Within the automotive area, the system "Vehicle" is partitioned into different domains structuring the mechanical key components of the vehicle [88]. Figure 1 depicts the partitioning of the vehicle into specific automotive domains.

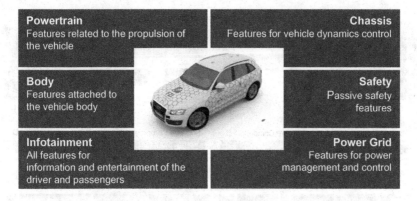

Powertrain
Features related to the propulsion of the vehicle

Chassis
Features for vehicle dynamics control

Body
Features attached to the vehicle body

Safety
Passive safety features

Infotainment
All features for information and entertainment of the driver and passengers

Power Grid
Features for power management and control

Fig. 1. Partitioning of the vehicle into specific automotive domains.

This modularization evolved from the product perspective and lead to corresponding organization structures to facilitate product engineering [88]. The domains originate from mechanical engineering and were expanded with electrical and information processing aspects. Within the different domains, several approaches for development processes, methods and tools are established and integrated into the overall product development process. These methods serve different needs and foci of the engineers, which differ from domain to domain. As upcoming customer features lead to fuzzy system borders, the different domains' development is moving closer together [33]. The integration and collaboration of domains is necessary without abandoning methodological flexibility and individuality.

This contribution considers the approaches and procedures of automotive E/E systems engineering, which consists of several different fields such as architectures, management, modeling and operation research [40]. In the automotive domain, the management of development processes is commonly based on the V-Model. The AutomotiveSPICE [5] specifies an established process reference that integrates the V-Model approach. In this contribution, we focus on the architecture and structuring of automotive embedded systems to facilitate the process of systems engineering.

2.1 System Architecture

Several approaches and methods for the structural description of system architectures [4,61,83] follow a model-based approach. The principle of abstraction contributes to reduced complexity [40] and facilitates system understanding [15]. It enables the structured analysis of specific topics, such as functional safety [1]. Common abstraction layers of automotive embedded systems are Logical Architecture, Software Architecture and Technical Architecture. A basic overview is given in Fig. 2.

The abstraction layers provide a partial description of the system based on different perspectives [96], using the principle of modularization of blocks and

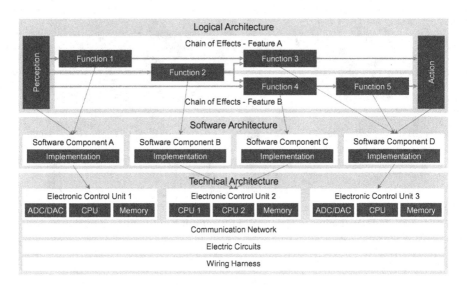

Fig. 2. Three abstraction levels of the automotive system architecture and mapping of functional behavior to software components and electronic control units [7].

connections. Also hierarchization and encapsulation of artifacts to describe different levels of detail is intended. Between the artifacts of different abstraction layers, interconnections and relations with a distinct semantic are present [61]. The example in Fig. 2 depicts relations, which describe the partitioning of functional entities into software components for integration on distinct ECUs.

Logical Architecture. The logical architecture is a breakdown of a feature "into interacting functional components" [63]. It represents the functional decomposition of a system into functional elements, which provide the functionality described in the corresponding requirements. The logical architecture focuses on the functional aspects, the logical interfaces and the coherence between the functional elements. It is completely independent from technical considerations or software specific issues. A common approach in the automotive area for the structuring of logical architectures is the usage of chains of effect to describe an overall approach from sensing to acting [70]. Demands for more elaborated concepts to improve the structuring of increasing complex features are initially addressed in [34] and [63]. Description of the functional element's internal behavior is highly dependent on the associated domain and not in scope of this contribution.

Software Architecture. The software architecture describes the different software components and the partitioning of the functional elements, including basic software (operating system and middleware) and communication [70]. A standardized middleware for software components allows reuse of the basic elements, for automotive embedded systems this is given by the AUTomotive Open System ARchitecture (AUTOSAR) [6]. It specifies a software framework and architecture consisting of basic software elements, a run-time environment (RTE) and application software components to enable reuse and scalability.

Improvements and extensions for AUTOSAR introduce adaptive deployment, service-oriented communication and dynamic scheduling and application execution as well as integration in new high-performance processor architectures. The related specification under the term "AUTOSAR adaptive" is currently under development within the AUTOSAR partnership [29].

Technical Architecture. The technical architecture specifies the integration level, which contains the hardware units to execute the defined software components [63]. This comprises the ECU, actuators and sensors and their interconnections. In ASE, the technical architecture is commonly further refined to represent specific E/E aspects, such as electric circuits and the wiring harness. The technical system architecture is based on a comprehensive E/E topology containing a segmentation into previously introduced domains, such as body, chassis and comfort. The current E/E architectures often reflect the organizational structure introduced by segmentation of the car's mechanical structure. Historically, single ECUs were introduced to perform independent functionality [43], connected with a single centralized gateway [77].

With increasing complexity and an increasing number of ECUs, domain-controlled E/E architectures with centralized domain-controllers were introduced [67,76]. This trend was an initial reflection to expanding system boundaries, more complex functional chains and higher integration of features. For each domain, master controllers were introduced to facilitate domain-comprehensive features. The evolution of technical system architectures is thus tightly coupled with the increasing interaction and networking of the logical architecture. The current development leads to centralized cross-domain E/E architectures based on high-performance computing units [33,55].

2.2   Architecture Concepts for Automated Driving

Research in the field of automated driving provides various approaches to describe the system architectures of current research concepts. Stiller [75] provides a cognitive oriented approach of perception, planning and action tasks. Different layers classify the abstract representation of functional elements. The architecture concept provided by Bauer [9] is categorized into a mission layer, a coordination layer and a behavior layer. Each layer consists of elements of the world model class, the planning class and the HMI class. The utilized sensors, actuators and the driver form the system environment. The influence of human-machine interactions on system architecture is discussed by Flemisch [28]. Based on the psychological categorization of the Dynamic Driving Task (DDT) into navigation, guidance and control, the automation system provides an interface on each level. Matthaei [48] proposes a "functional system architecture for an autonomous on-road motor vehicle". It applies a similar categorization into a strategic level, a tactical level and an operational level and a further distinction between localization, perception and mission accomplishment. An implemented system architecture for automated driving, using production vehicle sensors and additional prototyping sensors, is presented by Aeberhard [2]. Buechel [19] presents the prototype of an automated electric vehicle. The proposed software architecture consists of the three components data fusion, trajectory planning and trajectory controller, which is mapped to a centralized E/E architecture.

3   Taxonomy for Current and Upcoming Electrics/Electronics Features

Today's technical compendiums of carmakers are crammed with a high variety of customizable features. A significant proportion of those features is based on E/E functionality. With rising complexity of and dependencies between features, the established ASE methods and abstraction concepts are reaching the limits of their capabilities. To identify boundaries and necessary extensions of current systems engineering methods we start with establishing a comprehensive overview of current and upcoming E/E features. Our goal is to integrate well-established series features of the automotive industry and current feature concepts of research groups within one consistent taxonomy.

For this contribution, 52 features with significant E/E share were evaluated. The evaluated features available in series cars represent an abstracted set of the offered features of major car companies. We analyzed the online presence of BMW[1], Daimler[2], Ford[3], Peugeot[4], Toyota[5] and VW[6] to select the most common features. Research features were selected to cover a range as wide as possible.

The classification of the selected features applies three main criteria. The first criteria evaluates the hardware dependency of the examined feature. It evaluates the importance of additional and feature specific mechanical or electrical hardware to fulfill the desired functionality. The second and third criteria assess the

---

[1] BMW Technology Guide, Bayerische Motoren Werke Aktiengesellschaft, http://www.bmw.com/com/en/insights/technology/technology_guide/index.html.

[2] Welcome to the Mercedes-Benz TechCenter, Daimler AG, https://techcenter.mercedes-benz.com/en/index.html.

[3] Advanced technology at your fingertips, Ford Motor Company, http://www.ford.com/cars/focus/features/#page=FeatureCategory4.

[4] Technologies & Innovations, Automobiles Peugeot, http://www.peugeot.com/en/technology.

[5] Toyota Technology, Toyota Motor Sales, U.S.A., Inc., http://www.toyota.com/technology/.

[6] Technik auf den Punkt gebracht., Volkswagen AG, http://www.volkswagen.de/de/technologie/technik-lexikon.html.

product complexity. Aspects of product complexity are described by Renner [68]. For the taxonomy the aspects variety and connectivity are evaluated. Figure 3 outlines the applied criteria.

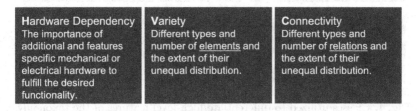

Fig. 3. Evaluation criteria evaluated for the presented classification of E/E features.

For each criteria, the selected features are evaluated and rated according to the five levels presented in Table 1.

Table 1. Used evaluation levels for the assessment of the criteria hardware dependency, variety and connectivity.

--	-	o	+	++
Very weak	Weak	Average	Strong	Very strong

To further differentiate the features we use the established vehicle domains introduced in Sect. 1 and the level of driving automation. The engineering association SAE International defines six levels of automation [69]. These levels define the extent and capabilities of E/E features necessary to take over the longitudinal and lateral vehicle motion control. For example assisting features are specified by SAE automation level 1 [69] as features that "perform either longitudinal or lateral vehicle motion control [...]". By SAE definition, features with automation level 3 and upwards perform the complete DDT with or without fallback and within or without a specific Operational Design Domain (ODD). This classification does not consider the variety of ADAS features, that do not perform any motion control. To allow distinction between passive advisory and active supporting features on level 0, we introduce the two additional classes advisory and supporting. Figure 4 presents the four sub-classes used to differentiate high-level features of the taxonomy.

An overview of our proposed taxonomy is presented in Fig. 5. It distinguishes the features by the three main categories Integrated Features, Distributed Features and Cross-Linked Features.

Integrated features are closely related to a specific mechanical domain of the vehicle. They represent the E/E content necessary to accomplish the targeted operation of physical components of the vehicle. They are strongly hardware

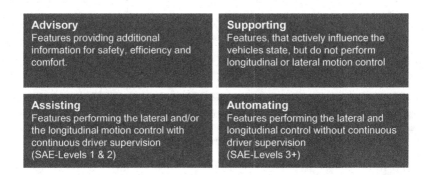

Fig. 4. Classes utilized to differentiate high-level features of the taxonomy.

dependent and characterized by a weak variety and connectivity. Distributed features combine individual components of different domains to enable additional capabilities. These features do not necessarily require additional mechanical hardware components. Their functional behavior can be expressed as the sequential combination of available information and usable actuators to provide added value. Features with an average rating in hardware dependency, variety and connectivity are classified as distributed features. Cross-linked features connect various functional elements and depend on the joined manipulation of the behavior of independent and domain separated components. They conflate various sources of information to achieve a comprehensive representation of the vehicle's state and surroundings. Cross-linked features are characterized by a weak hardware dependency, a strong variety and a strong connectivity. This representation forms the basis for cognitive and predictive features, including but not limited to high automation levels.

3.1 Integrated Features

As stated above, the integrated feature level subsumes the E/E content to operate the physical components of the vehicle. This entails a close proximity to specific mechanical units and commonly involves the usage of a dedicated ECU. Most sensors and actuators required for the assigned task of the feature are directly attached to the dedicated ECU. Integrated features are mainly based on proprioceptive sensors. Proprioceptive sensors obtain information about the internal state of the vehicle [14]. Out of the 52 evaluated features 24 are assigned to the integrated features class. The evaluation results of these features are presented in Table 2.

The powertrain domain contains "all functions controlling the generation of driving power and its conversion into propulsion". The taxonomy includes the features automatic transmission, engine control, traction control and hill assist as a representative feature set of the powertrain domain. The safety domain on integrated feature level includes the passive safety features airbag control and seat belt tensioner. More sophisticated active safety features are classified as

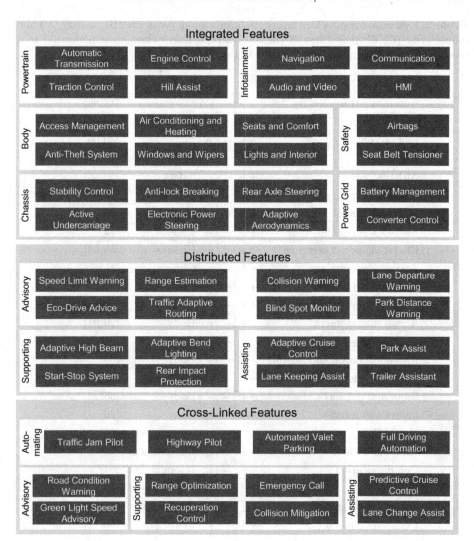

Fig. 5. Taxonomy of current and upcoming E/E features. Integrated features are grouped by vehicle domains, distributed and cross-linked features by level of interference.

distributed features. The chassis domain includes features to control the vehicle dynamics, providing a safe and attractive driving experience. Stability control, anti-lock breaking and Electronic Power Steering (EPS) describe features that mainly support safe and comfortable driving. Rear axle steering, damping and suspension control and adaptive aerodynamics particularly support agility. The body domain encompasses all features attached to the vehicle body, like lights, windows, wipers, seats and air conditioning as well as the car's access management and anti-theft system. The infotainment domain is the fifth vehicle domain

Table 2. Evaluation results for the criteria hardware dependency (H), variety (V) and connectivity (C) of 24 features of the integrated features class, their allocation to a specific vehicle domain and sources used in the evaluation.

Feature	H	V	C	Domain	Source
Automatic transmission	++	-	-	Powertrain	[41,86]
Traction control	+	-	o	Powertrain	[60,66]
Engine control	++	+	--	Powertrain	[30,66]
Hill assist	o	--	-	Powertrain	[17,95]
Airbags	++	o	-	Safety	[30,66]
Seat belt tensioner	++	-	-	Safety	[30,66]
Stability control	+	o	-	Chassis	[44,81,82]
Anti-lock breaking	++	--	--	Chassis	[82]
Electronic power steering	++	-	--	Chassis	[38,54]
Active undercarriage	++	-	o	Chassis	[3,46]
Rear axle steering	++	-	-	Chassis	[90]
Adaptive aerodynamics	++	--	-	Chassis	[73,79]
Access management	++	-	o	Body	[26,91]
Anti-theft system	++	--	--	Body	[25]
Air conditioning and heating	++	-	-	Body	[66]
Windows and wipers	++	-	--	Body	[62,66]
Seats and comfort	++	--	-	Body	[66]
Lights and interior	++	--	--	Body	[30]
Navigation	+	-	-	Infotainment	[66]
Audio and video	++	--	o	Infotainment	[30]
Communication	++	--	o	Infotainment	[64]
HMI	+	-	o	Infotainment	[18]
Battery management	++	-	--	Power grid	[65]
Converter control	++	-	-	Power grid	[66]

based on Weber's definition. It summarizes the features for navigation, communication, audio and video entertainment and Human Machine Interface (HMI). Battery management and converter control represent features that are part of 48 volt grids of hybrid electric vehicles as well as high voltage grids of fully electric vehicles.

3.2   Distributed Features

Most of the currently available ADAS are represented by the distributed features class. The functional behavior of distributed features resembles a chain of effects, the aforementioned sequential combination of available information and usable

Table 3. Evaluation results for the criteria hardware dependency (H), variety (V) and connectivity (C) of 16 features of the distributed features class, their allocation to a specific automation sub-class and sources used in the evaluation.

Feature	H	V	C	Sub-class	Source
Speed limit warning	-	+	o	Advisory	[24]
Eco-drive advice	-	+	o	Advisory	[53,58,84]
Range estimation	-	o	o	Advisory	[12,97]
Traffic adaptive routing	-	o	+	Advisory	[27,31,39]
Collision warning	o	+	-	Advisory	[23,74]
Blind spot monitor	o	o	-	Advisory	[49]
Lane departure warning	o	o	-	Advisory	[66,87]
Park distance warning	o	-	-	Advisory	[37]
Adaptive high beam	o	o	-	Supporting	[72]
Adaptive bend lighting	+	o	o	Supporting	[66]
Start-stop system	o	+	o	Supporting	[42,94]
Rear impact protection	o	o	+	Supporting	[16]
Adaptive cruise control	o	+	+	Assisting	[92]
Lane keeping assist	o	+	+	Assisting	[35,47,71]
Park assist	o	+	o	Assisting	[37,66]
Trailer assistant	o	o	o	Assisting	[45]

actuators. The functionality of distributed features is based on the connection of different domains. They often introduce and utilize exteroceptive sensors that provide information about the surroundings of the vehicle [14]. Of the 52 features included in the taxonomy 16 features belong to the distributed features class. The evaluation results of these features are presented in Table 3.

The advisory class contains features that utilize information of integrated features and exteroceptive sensors to provide additional information for safe and comfortable driving and potentially to influence the driver's behavior. Collision warning, lane departure warning, blind spot monitor and park distance warning depict advisory features to gain additional safety. Speed limit warning helps to stick to regulations and eco-drive advice intends to influence the driver's behavior to achieve a sustainable driving style. Range estimation and traffic adaptive routing support the driver's decisions regarding the selected route and stopovers.

The supporting class covers all features that actively influence the vehicle's state, but do not perform longitudinal or lateral vehicle motion control. It encompasses features such as adaptive high beam and adaptive bend light as well as automated start-stop. Rear impact protection represents an active safety feature that aims to decrease the damage induced to passengers during standstill, rear-end collisions. Bogenrieder [16] describes an approach that utilizes a backwards oriented radar sensor to detect an imminent rear-end collision.

The park assist and trailer assistant feature perform lateral control of the vehicle, while longitudinal control always remains with the driver. Therefore, these are automation level 1 features and part of the assisting features class. The Adaptive Cruise Control (ACC) feature performs longitudinal control and the LKA feature performs lateral control. While operated individually, both features represent automation level 1. If both systems are activated simultaneously, the feature combination represents automation level 2, "Partial Driving Automation". Consequentially, level 2 automation features are included in the assisting features class.

3.3   Cross-Linked Features

In the presented taxonomy, cross-linked features utilize sensor networks to derive information or to influence several actuators. These features span functional networks in distinction to the sequential functional chains of distributed features. They are based on the fusion of proprioceptive and exteroceptive sensor information to obtain a realistic and complete model of the vehicle's internal state and surroundings. Of the 52 features included in the taxonomy 12 features belong to the cross-linked features class. The evaluation results of these features are presented in Table 4.

Table 4. Evaluation results for the criteria hardware dependency (H), variety (V) and connectivity (C) of 12 features of the cross-linked feature class, their allocation to a specific automation sub-class and sources used in the evaluation.

Feature	H	V	C	Sub-class	Source
Road condition warning	-	+	++	Advisory	[8]
Green light speed advisory	--	+	++	Advisory	[8]
Range optimization	-	+	++	Supporting	[21, 78]
Recuperation control	-	+	+	Supporting	[93]
Emergency call	-	o	++	Supporting	[32]
Collision mitigation	-	+	+	Supporting	[20, 36]
Predictive cruise control	--	++	+	Assisting	[10, 85]
Lane change assist	--	++	+	Assisting	[22, 56]
Traffic jam pilot	--	++	++	Automating	[52, 57]
Highway pilot	--	++	++	Automating	[2]
Automated valet parking	--	++	++	Automating	[19, 59]
Full driving automation	--	++	++	Automating	[50, 98]

The road condition warning feature in the advisory class is described in the Car2Car communication consortium manifesto [8]. Severe road conditions are propagated via Car2Car communication or back end service between road users.

The green light speed advisory feature is also defined by the Car2Car consortium. It interacts with the road infrastructure and provides an optimal speed advice, averting an otherwise necessary red light stop. Both features require lane accurate positioning and access to various internal states and the communication platform of the vehicle. Therefore, they are classified into the cross-linked feature class.

The supporting class contains two energy management related features, the range optimization and the recuperation control. The range optimization calculates the remaining energy of the vehicle and predicts the required energy to reach the desired destination. If necessary, it shuts down power hungry comfort features like heating and air conditioning and limits the propulsion power. The recuperation control predicts the vehicle's energy flows and for example reduces battery load before long recuperation phases, to prevent waste of energy due to battery heat protection [93]. As these features influence various actuators and require predictive map data, traffic flow information and internal states for optimal performance, they are classified into the cross-linked category. Emergency call and collision mitigation round out the supporting feature class. These are active safety features that take action before an imminent collision and automatically call help after an accident.

Equivalent to the distributed features, the assisting class covers features of SAE automation level 1 and 2. The predictive cruise control feature controls the longitudinal motion of the vehicle [85]. It calculates an energy optimal velocity trajectory based on predictive map data and proprioceptive and exteroceptive sensor information. The included lane change assist feature guides the driver's lane change maneuver [22]. It requires various sensors and predicts the surrounding traffic to calculate a safe lane change trajectory [56]. Both feature's depend on several sensing, processing and acting primitive elements and, therefore, are classified as cross-linked features.

All features from SAE automation level 3 upwards belong to the automating class of cross-linked features. The example features traffic jam pilot, highway pilot, automated valet parking [59] and full driving automation [98] perform the complete DDT. The former three features are designed for a specific operational domain. Depending on their characteristics and implementation, all automating features utilize more or less comprehensive environmental perception and interpretation. Beside the longitudinal and lateral control of the vehicle, the features must control several actuators to perform the complete DDT. Automating features comprise the highest level of cross-linking.

4   Comprehensive Hierarchization for Logical System Architectures

As stated in Sect. 2.1, based on established system architecture modeling concepts, all functional behavior of the introduced features is modeled within one level of logical system architectures. Thereby, the differing character and integration depth of the individual functional elements is not considered. The representation resolves the complexity of the underlying functional dependencies and multiple usage scenarios of particular functional elements only to a limited degree. Hence, the systems engineering principles of modularization, abstraction and hierarchization are not employed to the full extent.

Section 2.2 outlines the approaches utilized by researchers in the field of automated driving. The utilization of psychological concepts offers a sound characterization for the functional components of automating features. This supports the structuring of fundamental sub-tasks of the DDT, but does not necessarily support the entire systems engineering process. Existing and established E/E systems were mostly neglected by the described architecture representations. For an holistic approach we need a hierarchical structure that supports a clear representation of the dependencies between functional elements and includes all automotive E/E features. It concurrently provides an abstraction that facilitates adaption and association of different shapes of systems engineering activities.

The aim of the proposed hierarchization of functional elements is to introduce a comprehensive domain-crossing functional architecture. The introduced hierachization is based on the integration level and the character of the processed information. This enables a flexible description of the existing chain of effects and their interaction with associated elements within one systematic approach and simplifies precise specification of interfaces. It facilitates the definition of tailored templates for activities, such as verification and validation, functional safety and release planning. These templates could guide developers, testers, project and quality managers during the configuration of function specific process implementations and the selection of a balanced set of suitable methods and tools.

Figure 6 depicts our newly introduced hierachization for the logical system architecture. The classified features of Sect. 3 were broken down into principal functional elements and arranged to represent a clockwise flow of information. The layered approach provided by Stiller [75] served as basis for the development of the logical system architecture.

The type of information that is processed by the respective element, is the major discrimination criterion we apply to assign the elements to a particular level. The physics level contains the functional elements to gain information from physical principles and vice versa to influence the physics. On the raw information level the derived raw information is filtered and actuation requests are processed. The filtered information of different functional elements is combined via information fusion techniques within the filtered information level and interpreted information is used to operate the actuators. On the highest level of the hierarchization, the interpreted information is used to predict and abstract the state and behavior of the system environment and the upcoming course and actions of the vehicle are planned. In the following, we explain these different levels, their characteristics and possible consequences for future systems engineering.

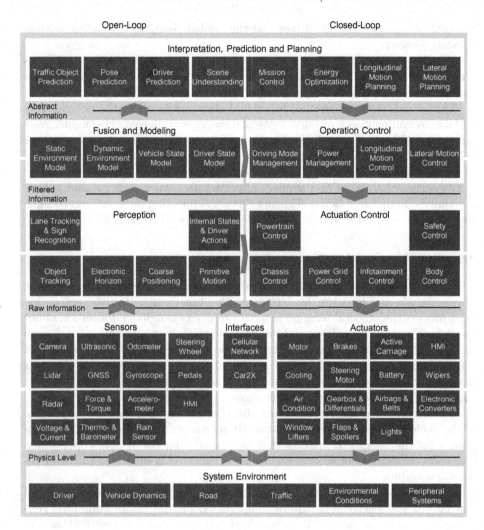

Fig. 6. The proposed holistic hierarchization approach for the logical system architecture in the automotive domain.

4.1 Physics Level

The physics level of the logical system architecture is composed of sensors, interfaces and actuators and comprises all interfaces to the system environment. Sensors utilize physical measurement principles and provide basis perception functions. They provide raw information in form of discrete, unfiltered sample data. The type of supplied information ranges from sampled physical quantities like force and torque, acceleration and velocity to the raw image provided by a camera and the point cloud of a lidar sensor. The sensors class also contains the control interfaces of the driver and the Global Navigation Satellite System (GNSS) receiver.

The interfaces class enables the interaction with affiliated technical systems. It provides access to cellular networks and communication entities such as Car2X, representing a bidirectional flow of information.

The actuators encompass all functional elements to affect the vehicle state and its environment as a physical system. The powertrain elements engine, gearbox and differentials influence the propulsion and the flow of energy of the vehicle. By application of steering torque, the steering motor affects the lateral movement of the vehicle, but also acts as an interface towards the driver. Active suspension, flaps and spoilers alter the properties of aero- and vehicle dynamics. Further functional elements serve a supporting purpose (e.g. cooling, wipers or lights) and to interact with the driver (e.g. HMI, speakers).

4.2   Raw Information Level

This level contains the functional elements required for filtering and processing of raw signals and to drive the actuators. The functions within the perception class process the physical sensor's raw data to derive tangible information about the vehicle's primitive motion and internal states. Coarse positioning is achieved by interpretation of the pseudoranges in the navigation satellite receiver and the electronic horizon provides information about the upcoming road segments from an internal data storage. Images and point clouds are processed to extract surrounding objects, lanes and traffic signs.

The actuation control class drives and controls the mechanical components of the vehicle via the physical actuators. It represents the basic functional components of the integrated features that are essential for the vehicle's operability. The software implementation of functions on this level is subjected to hard realtime constraints.

While the elements of the physics level represent the functional share of mechanical and electrical hardware components, the raw information level contains the functional part of the embedded software associated with those elements. Its development should be coupled with the processes of the physical level. On this level, the development of components is commonly carried out by Tier 1 suppliers. Validation and verification of the functional elements can mostly be done independent of other elements. The obtained information is commonly shared within the related domain of the vehicle's communication network.

4.3   Filtered Information Level

Functions on the filtered signal level perform fusion and abstraction of the various detached information sources and control the vehicle operation. The information of the proprioceptive and exteroceptive sensors is accumulated in the interpretation class. The static and dynamic environment model provide a condensed and consistent representation of the vehicles' surroundings. The vehicle state model consolidates all internal vehicle states and the driver state model describes the driver's features, such as level of attention and driving style.

The functions to control the lateral and the longitudinal motion of the vehicle are the most important items of the operation control class. Their task is to achieve the targeted velocity and vehicle pose within the operational limits. The driving mode management coordinates the underlying functions to attain a well-attuned driving experience. The power management approves and limits power consumption of the various components and coordinates the recuperation of electrified vehicles.

The functions of the filtered information level are not essential for the operability of the vehicle, but enable distributed features. The included longitudinal and lateral control elements are part of the assisting and automating features. The functions on this level are subjected to soft real-time constraints. Verification and validation of these functions is performed on the interface level. Simulation based techniques require modeling of not only the vehicle physics and environment, but also modeling of all underlying functional elements implemented in software.

4.4   Interpreted Information Level

The interpreted information level contains cognitive functions for interpretation, prediction and planning. Stochastic models enable the prediction of the behavior of traffic objects and driver intentions. The information of the vehicle state model facilitates the prediction of the vehicle's pose. The functional element scene understanding represents the interpretation of the aggregated information. The longitudinal and lateral motion planning functions are based on the interpreted information and act on the underlying control functions. A dedicated element for energy optimization enables the range optimization and the recuperation control features. The mission control function is an essential part of all automation features. It coordinates the individual elements to accomplish the driving task.

The functional elements of the interpreted information level are best suited for implementation on a centralized, high-performance control unit, as the amount of data necessary to provide the described information exceeds the capability of established communication networks. The functional elements of the interpreted information level resemble a service-oriented approach. Therefore, no guarantees for real-time constraints are given. Simulation models for verification and validation of these high level functions do not need detailed models of the vehicle mechanics or the physical background of the utilized sensors. Emulation of the model based environment representation and the control behavior of the filtered information level is sufficient.

5   Representation of Selected Features Within the Proposed Logical System Architecture

The elements within our proposed logical system architecture were derived from the analysis and taxonomy of existing and conceptual automotive features in

Sect. 3. In the following, we demonstrate the applicability by modeling selected features of all three main categories of the taxonomy. The modeling of established features shows the ability of our approach to maintain legacy content. The representation of research concepts proofs the ability to cope with future demands.

5.1   Integrated Features

Of the integrated feature class, the ESC and the EPS are modeled within our proposed logical architecture.

Electronic Power Steering. The EPS feature serves as an actuator to influence the lateral movement of the vehicle. Figure 7 depicts the logical system architecture of the ESC. It applies a torque to the steering wheel to support the driver actuation or to achieve a given target steering position. The power steering described by Kim [38] supports the driver's steering intention. It provides a detailed description of the architecture of a power steering control feature for driver support. The torque applied by the driver is sensed and amplified dependent on the vehicle velocity. Naranjo [54] describes a power steering feature for automated control of the vehicle. It applies steering torque to control the steering position. To obtain a satisfactory control behavior, the control is operated with a duty cycle of 10 ms.

Therefore, the power steering feature consists of the odometer and steering wheel elements of the sensors class, the power steering control function and the steering motor of the actuators class.

Electronic Stability Control. The ESC feature "is an active safety technology that assists the driver to keep the vehicle on the intended path and thereby helps to prevent accidents" [44]. The yaw movement of the vehicle is stabilized by

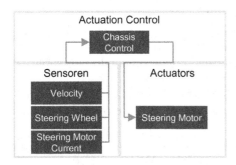

Fig. 7. The chain of effects of an Electronic Power Steering (EPS) feature described, using our newly introduced logical system architecture abstraction.

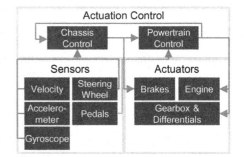

Fig. 8. The chain of effects of an Electronic Stability Control (ESC) feature described, using our newly introduced logical system architecture abstraction.

individually controlling the tire slip of each wheel. To avoid counteracting the driver, "it needs to accurately interpret what the driver intends for the vehicle motion in order to provide added directional control" [81].

Figure 8 depicts the logical system architecture of the ESC. The current yaw rate and vehicle movement is read in from a gyroscope, an odometer and an accelerometer. The driver intention is derived from the information of the steering wheel and the pedals. The stability control functional element calculates the individual tire slips necessary to obtain a stable movement. Actuation of the brakes is directly applied, the engine, gearbox and differentials are actuated via the powertrain control function.

5.2 Distributed Features

To represent the distributed features class, we selected the ACC feature as a longitudinal control feature and the LKA feature as a lateral control feature.

Adaptive Cruise Control. The ACC feature depicts an assisting feature that controls the vehicles longitudinal velocity and adapts it to the velocity of leading traffic. Winner [92] provides a comprehensive overview of the ACC feature. The radar based perception of the area in front of the vehicle is used to calculate and control the vehicle's velocity. The driver inputs are monitored to detect an override by throttle actuation and a deactivation by brake actuation. Moon [51] describes a two-level control structure, where the upper level controls the vehicles speed by requesting accelerations and the lower level controls the acceleration by throttle and brake actuation.

Figure 9 depicts the logical system architecture of an ACC feature. For comprehensibility, the elements for driver interaction, like activation and override,

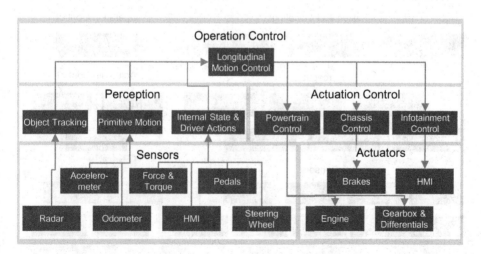

Fig. 9. The chain of effects of an Adaptive Cruise Control (ACC) feature described, using our newly introduced logical system architecture abstraction.

are removed and only the core elements are represented. The primitive motion of the vehicle is estimated based on internal sensor information. The radar signal is processed by the object tracking function and used to calculate and control the desired time-gap in the longitudinal motion control function. Actuation is performed via the stability control and the powertrain control elements.

Lane Keeping Assist. The lane keeping assist feature assists the driver in the lateral control task without assuming control of the complete DDT. Following Ishida [35], "The lane keeping assistance system consists of a camera-equipped lane recognition unit, the LKAS control unit, and the Electric Power Steering (EPS)." The lane tracking functional element extracts the lane markings in the camera image and calculates the lateral deviation, orientation and curvature. This information is used as control variable in the lateral motion control function. The actuation is a steering torque applied via the power steering control and the steering motor.

5.3   Cross-Linked Features

Of the cross-linked feature class we selected the PCC feature, which belongs to the assisting sub-class. Wahl [85] describes the PCC as a feature for optimal longitudinal control. The ACC is extended to adapt the velocity to the road topology and speed limits besides leading traffic. Figure 10 depicts the logical architecture of the PCC feature.

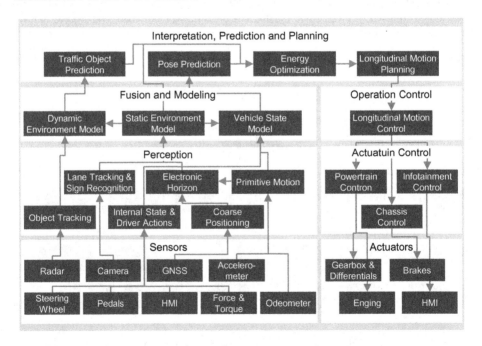

Fig. 10. The chain of effects of a predictive cruise control feature described, using our newly introduced logical system architecture abstraction.

The environmental perception of the ACC is extended by a camera system for lane tracking and traffic sign recognition. A GNSS receiver provides coarse positioning, which is used to provide the upcoming road topology via the electronic horizon function. A consistent model of the static environment, the vehicle state and the dynamic environment is formed on the interpretation level.

The feature implements a model predictive control strategy. Therefore, the pose of the vehicle and the movement of the traffic object are predicted and passed on. Bauer [11] splits up the control task of the PCC into two levels. This approach maps to the longitudinal motion planning element and the longitudinal motion control.

6 Conclusions

In this contribution, existing and upcoming automotive features were evaluated and classified in a taxonomy. We evaluated 52 current and upcoming automotive E/E features and presented the detailed results of the assessment. Our taxonomy presents a systematic and comprehensive overview of the current automotive E/E cosmos. This rehashed representation facilitates the discussion and delineation of the current challenges to the ASE practice.

Based on the taxonomy we developed a comprehensive logical system architecture for current and upcoming vehicles. An hierarchical structure was introduced to the logical system architecture to handle the increasing functional complexity. The utilized abstract functional elements were selected to cover all vehicle domains and enable the representation of functional chains and networks. The compact pattern provides a clear overview of the abstract representations of all E/E components of the vehicle and facilitates the attribution of properties to elements and interfaces. The abstraction enables the representation of new and legacy functional elements, whose combinations constitute the various features of the vehicle. Therefore, all feature dependencies and all stakeholders of specific functional elements are easily identifiable.

Future work will be necessary to evaluate the anticipated added value of the hierarchical structure of the logical system architecture. It is important to take a closer look on how the definition of beneficial guidelines for different architecture levels supports decision makers during the product development. For functions on different levels, varying process quality gates may be scheduled to consider the necessary effort for their hardware integration. With future software over the air capabilities, a feature ramp-up after the start of production is feasible for high-level features, while low-level features need to be integrated in the established fashion to ensure basic functionality. Also agile practices should be more or less beneficial on different levels. In conclusion, the definition of different meta-strategies for verification, validation and functional safety for the different levels of the logical system architecture should provide a substantial benefit.

References

1. Adler, N., Hillenbrand, M., Müller-Glaser, K.D., Metzker, E., Reichmann, C.: Graphically notated fault modeling and safety analysis in the context of electric and electronic architecture development and functional safety. In: 2012 23rd IEEE International Symposium on Rapid System Prototyping (RSP), pp. 36–42, August 2012

2. Aeberhard, M., et al.: Experience, results and lessons learned from automated driving on Germany's highways. IEEE Intell. Transp. Syst. Mag. **7**(1), 42–57 (2015)

3. Ahmed, M., Svaricek, F.: Preview optimal control of vehicle semi-active suspension based on partitioning of chassis acceleration and tire load spectra. In: 2014 European Control Conference (ECC), pp. 1669–1674, June 2014

4. ATESST2 Consortium: EAST-ADL Domain Model Specification, 2.1.12 (edn.) (2013)

5. Automotive-SIG: Automotive SPICE Process Assessment/Reference Model. VDA QMC, Berlin, Germany, 3.0 edn., July 2015. http://www.automotivespice.com/

6. AUTOSAR development cooperation: Specification of RTE. Munich, 4.2.1 edn., July 2015. https://www.autosar.org/specifications/release-42/

7. Bach, J., Otten, S., Sax, E.: A taxonomy and systematic approach for automotive system architectures - from functional chains to functional networks. In: 3rd International Conference on Vehicle Technology and Intelligent Transport Systems (VEHITS), Porto (2017)

8. Baldessari, R., et al.: CAR 2 CAR Communication Consortium Manifesto. CAR 2 CAR Communication Consortium, Brussels, 1.1 (edn.), August 2007

9. Bauer, E., et al.: PRORETA 3: an integrated approach to collision avoidance and vehicle automation. Automatisierungstechnik **60**, 755–765 (2012)

10. Bauer, K.L., Gauterin, F.: A two-layer approach for predictive optimal cruise control. In: SAE 2016 World Congress and Exhibition (2016)

11. Bauer, K.L., Gauterin, F.: A two-layer approach for predictive optimal cruise control. In: SAE Technical Paper 2016–01-0634 (2016)

12. Bechler, M., Makeschin, L.: Reichweitenschätzung für elektrofahrzeuge, September 2015. http://google.com/patents/DE102014204308A1?cl=de dE Patent App. DE201,410,204,308

13. Becker, J., Colas, M.-B.A., Nordbruch, S., Fausten, M.: Bosch's vision and roadmap toward fully autonomous driving. In: Meyer, G., Beiker, S. (eds.) Road Vehicle Automation. LNM, pp. 49–59. Springer, Cham (2014). https://doi.org/10.1007/978-3-319-05990-7_5

14. Bengler, K., Dietmayer, K., Färber, B., Maurer, M., Stiller, C., Winner, H.: Three decades of driver assistance systems. IEEE Intell. Transp. Syst. Mag. **6**(4), 6–22 (2014)

15. Bhave, A., Krogh, B.H., Garlan, D., Schmerl, B.: View consistency in architectures for cyber-physical systems. In: 2011 IEEE/ACM International Conference on Cyber-Physical Systems (ICCPS), pp. 151–160, April 2011

16. Bogenrieder, R., Fehring, M., Bachmann, R.: Pre-safe in rear-end collision situations. In: Proceedings 21st International Technical Conference on the Enhanced Safety of Vehicles, Stuttgart (2009)

17. Boll, B., Stinus, J., Salecker, M., Schneider, G.: Hill holder device for a motor vehicle, January 2004. https://www.google.com/patents/US6679810 US Patent 6,679,810

18. Broström, R., Engström, J., Agnvall, A., Markkula, G.: Towards the next generation intelligent driver information system (IDIS): the Volvo car interaction manager concept. In: Proceedings of the 2006 ITS World Congress, p. 32 (2006)
19. Buechel, M., et al.: An automated electric vehicle prototype showing new trends in automotive architectures. In: 2015 IEEE 18th International Conference on Intelligent Transportation Systems, pp. 1274–1279, September 2015
20. Coelingh, E., Eidehall, A., Bengtsson, M.: Collision warning with full auto brake and pedestrian detection-a practical example of automatic emergency braking. In: 2010 13th International IEEE Conference on Intelligent Transportation Systems (ITSC), pp. 155–160. IEEE (2010)
21. Conradi, P.: Reichweitenprognose für elektromobile. Vernetztes Automobil, pp. 179–186 (2014)
22. Cramer, S., Lange, A., Bengler, K.: Path planning and steering control concept for a cooperative lane change maneuver according to the h-mode concept. In: 7. Tagung Fahrerassistenzsysteme, November 2015
23. Dagan, E., Mano, O., Stein, G., Shashua, A.: Forward collision warning with a single camera. In: IEEE Intelligent Vehicles Symposium, pp. 37–42, June 2004
24. Daniel, J., Lauffenburger, J.P.: Fusing navigation and vision information with the transferable belief model: application to an intelligent speed limit assistant. Inf. Fusion 18, 62–77 (2014)
25. Dimig, S., et al.: Steering column lock apparatus and method, June 2003. https://www.google.com/patents/US6571587. US Patent 6,571,587
26. Dix, P., Bojarski, M.: Reprogrammable vehicle access control system, December 2004. https://www.google.com/patents/US20040263316. US Patent App. 10/602,750
27. Fawcett, J., Robinson, P.: Adaptive routing for road traffic. IEEE Comput. Graph. Appl. 20(3), 46–53 (2000)
28. Flemisch, F.O., Bengler, K., Bubb, H., Winner, H., Bruder, R.: Towards cooperative guidance and control of highly automated vehicles: H-mode and conduct-by-wire. Ergonomics 57(3), 343–360 (2014). https://doi.org/10.1080/00140139.2013.869355. pMID: 24559139
29. Fuerst, S.: AUTOSAR the next generation - the adaptive platform. In: CARS Critical Automotive applications: Robustness & Safety in 11th EDCC European Dependable Computing Conference (2015)
30. Gemeinschaftswerk, siehe Autorenliste: Bosch Automotive Electrics and Automotive Electronics, vol. 5. Robert Bosch GmbH, Plochingen (2007)
31. Golding, A.: Automobile navigation system with dynamic traffic data, August 1999. https://www.google.com/patents/US5933100. US Patent 5,933,100
32. Granier, E.: Device and method for emergency call, March 2004. https://www.google.com/patents/US6711399 US Patent 6,711,399
33. Haas, W., Langjahr, P.: Cross-domain vehicle control units in modern E/E architectures. In: 16. Internationales Stuttgarter Symposium, pp. 1619–1627 (2016)
34. Holder, S., Hoerwick, M., Gentner, H.: Funktionsübergreifende szeneninterpretation zur vernetzung von fahrerassistenzsystemen. In: AAET - Automatisiertes und vernetztes Fahren (2012)
35. Ishida, S., Gayko, J.E.: Development, evaluation and introduction of a lane keeping assistance system. In: Intelligent Vehicles Symposium, pp. 943–944. IEEE, June 2004
36. Kaempchen, N., Schiele, B., Dietmayer, K.: Situation assessment of an autonomous emergency brake for arbitrary vehicle-to-vehicle collision scenarios. IEEE Trans. Intell. Transp. Syst. 10(4), 678–687 (2009)

37. Katzwinkel, R., et al.: Handbuch Fahrerassistenzsysteme. Bremsenbasierte Assistenzfunktionen, pp. 471–477. Vieweg & Teubner Verlag, Springer Fachmedien, Wiesbaden (2012)
38. Kim, J.W., Lee, K.J., Ahn, H.S.: Development of software component architecture for motor-driven power steering control system using AUTOSAR methodology. In: 2015 15th International Conference on Control, Automation and Systems (ICCAS), pp. 1995–1998, October 2015
39. Kleine-Besten, T., Kersken, U., Pöchmüller, W., Schepers, H.: Handbuch Fahrerassistenzsysteme. Navigation und Telematik, pp. 599–624. Vieweg & Teubner Verlag, Springer Fachmedien, Wiesbaden (2012)
40. Korsiakoff, A., Sweet, W.N., Seymour, S.J., Biemer, S.M.: Systems Engineering Principles and Practice. Wiley, Hoboken (2011)
41. Kulkarni, M., Shim, T., Zhang, Y.: Shift dynamics and control of dual-clutch transmissions. Mech. Mach. Theory **42**(2), 168–182 (2007). http://www.sciencedirect.com/science/article/pii/S0094114X06000565
42. Kuroda, S., Kiyomiya, T., Matsubara, A., Kitajima, S.: Engine automatic start stop control apparatus, January 2003. https://www.google.com/patents/US6504259. US Patent 6,504,259
43. Leen, G., Heffernan, D.: Expanding automotive electronic systems. Computer **35**(1), 88–93 (2002)
44. Liebemann, E.K., Meder, K., Schuh, J., Nenninger, G.: Safety and performance enhancement: the bosch electronic stability control (ESP). SAE Paper 2004, 21–0060 (2004)
45. Lundquist, C., Reinelt, W., Enqvist, O.: Back driving assistant for passenger cars with trailer. In: SAE 2006 World Congress & Exhibition (2006)
46. Majeed, K.N.: On/off semi-active suspension control, September 1991. https://www.google.com/patents/US5062657. US Patent 5,062,657
47. Matsumoto, S., Kimura, T., Takahama, T., Toyota, H.: Lane keep control for vehicle, April 2003. https://www.google.com/patents/US6556909. US Patent 6,556,909
48. Matthaei, R., Maurer, M.: Autonomous driving - a top-down-approach. Automatisierungstechnik **63**(3), 155–167 (2015)
49. Miller, R.H., Tascillo, A.L.: Blind spot warning system for an automotive vehicle (2005). https://www.google.com/patents/US6859148. US Patent 6,859,148
50. Montemerlo, M., et al.: Junior: the stanford entry in the urban challenge. J. Field Robot. **25**(9), 569–597 (2008). https://doi.org/10.1002/rob.20258
51. Moon, S., Yi, K., Moon, I.: Design, tuning and evaluation of integrated ACC/CA systems. In: 17th World Congress of the International Federation of Automatic Control (IFAC 2008). IFAC Proceedings Volumes, vol. 41, pp. 8546–8551, July 2008
52. Müller, T.: Chancen und risiken auf dem weg zum ppilotierte fahren. In: Internationaler Automobil Kongress, October 2016
53. Muñoz-Organero, M., Magaña, V.: Validating the impact on reducing fuel consumption by using an EcoDriving assistant based on traffic sign detection and optimal deceleration patterns. IEEE Trans. Intell. Transp. Syst. **14**(2), 1023–1028 (2013)
54. Naranjo, J.E., Gonzalez, C., Garcia, R., de Pedro, T., Haber, R.E.: Power-steering control architecture for automatic driving. IEEE Trans. Intell. Transp. Syst. **6**(4), 406–415 (2005)
55. Navale, V.M., Williams, K., Lagospiris, A., Schaffert, M., Schweiker, M.A.: (R)evolution of E/E architectures. SAE Int. J. Passeng. Cars Electron. Electr. Syst. **8**(2), 282–288 (2015)

56. Nilsson, J., Brännström, M., Coelingh, E., Fredriksson, J.: Lane change maneuvers for automated vehicles. IEEE Trans. Intell. Transp. Syst. **PP**(99), 1–10 (2016)
57. Neural Network: Traffic jam assist. http://products.bosch-mobility-solutions. com/en/de/driving_comfort/driving_comfort_systems_for_passenger_cars_1/driver_ assistance_systems_4/driver_assistance_systems_5.html
58. N.N.: Predictive efficiency assistant (2012). http://www.audi-technology-portal. de/en/mobility-for-the-future/audi-future-lab-mobility_en/audi-future-engines_ en/predictive-efficiency-assistant
59. Nordbruch, S., Quast, G., Nicodemus, R., Scheiger, R.: Automated valet parking. In: 7. Tagung Fahrerassistenzsysteme, November 2015
60. Park, J.H., Kim, C.Y.: Wheel slip control in traction control system for vehicle stability. Veh. Syst. Dyn. **31**(4), 263–278 (1999)
61. Pohl, K., Hoenninger, H., Achatz, R., Broy, M.: Model-Based Engineering of Embedded Systems - The SPES 2020 Methodology. Springer, Heidelberg (2012). https://doi.org/10.1007/978-3-642-34614-9
62. Prabhu, S.M., Mosterman, P.J.: Model-based design of a power window system: modeling, simulation and validation. In: Proceedings of IMAC-XXII: A Conference on Structural Dynamics, Society for Experimental Mechanics Inc., Dearborn (2004)
63. Pretschner, A., Broy, M., Krueger, I.H., Stauner, T.: Software engineering for automotive systems: a roadmap. In: FOSE Future of Software Engineering (2007)
64. Rabinovich, V., Alexandrov, N., Alkhateeb, B.: Automotive Antenna Design and Applications. CRC Press, Boca Raton (2010)
65. Rahimzei, E., Sann, K., Vogel, M.: Kompendium: Li-ionen-batterien. Technical report, VDE Verband der Elektrotechnik (2015)
66. Reif, K.: Automobilelektronik, 5th edn. Springer, Heidelberg (2014). https://doi. org/10.1007/978-3-658-05048-1
67. Reinhardt, D., Kucera, M.: Domain controlled architecture - a new approach for large scale software integrated automotive systems. In: 3rd International Conference on Pervasive Embedded Computing and Communication Systems, pp. 221– 226 (2013)
68. Renner, I.: Methodische Unterstützung funktionsorientierter Baukastenentwicklung am Beispiel Automobil. Ph.D. thesis, Technische Universität München, München, May 2007
69. SAE international: Taxonomy and definitions for terms related to driving automation systems for on-road motor vehicles, September 2016
70. Schäuffele, J., Zurawka, T.: Automotive Software Engineering - Grundlagen, Prozesse, Methoden und Werkzeuge effizient einsetzen, 5 (edn.). Springer Fachmedien Wiesbaden GmbH (2012)
71. Schmitz, C.: Method and apparatus for driver assistance, April 2011. https://www. google.com/patents/US8031063. US Patent 8,031,063
72. Schofield, K., Larson, M., Vadas, K.: Vehicle headlight control using imaging sensor, August 1998. https://www.google.com/patents/US5796094. US Patent 5,796,094
73. Seidel, W.: Process for controlling front or rear spoilers, September 2006. https:// www.google.com/patents/US7113855. US Patent 7,113,855
74. Stämpfle, M., Branz, W., et al.: Kollisionsvermeidung im längsverkehr-die vision vom unfallfreien fahren rückt näher. 3. Tagung Aktive Sicherheit durch Fahrerassistenz (2008)
75. Stiller, C., Färber, G., Kammel, S.: Cooperative cognitive automobiles. In: Proceedings of the 2007 IEEE Intelligent Vehicles Symposium, pp. 215–220, June 2007

76. Stolz, W., Kornhaas, R., Sommer, T.: Domain control units - the solution for future E/E architectures? In: SAE Technical Paper 2010–01-0686, pp. 221–226 (2010)
77. Streichert, T., Traub, M.: Elektrik/Elektronik-Architekturen im Kraftfahrzeug - Modellierung und Bewertung von Echtzeitsystemen. Springer, Heidelberg (2012). https://doi.org/10.1007/978-3-642-25478-9
78. Strobel, S., Rösinger, K., Bröcker, M.: Radikale Innovtionen in derMobilität. Fuzzy-Logik basiertes Energiemanagement fürElektrofahrzeuge, pp. 211–224. Proff, H. (2014)
79. Takagi, M., Asano, T., Yamada, T.: Automotive vehicle with adjustable aerodynamic accessory and control therefor, March 1989. https://www.google.com/patents/US4810022. US Patent 4,810,022
80. Tas, Ö.S., Kuhnt, F., Zöllner, J.M., Stiller, C.: Functional system architectures towards fully automated driving. In: 2016 IEEE Intelligent Vehicles Symposium (IV) (2016)
81. Tseng, H.E., Ashrafi, B., Madau, D., Brown, T.A., Recker, D.: The development of vehicle stability control at ford. IEEE/ASME Trans. Mechatron. 4(3), 223–234 (1999)
82. van Zanten, A., Kost, F.: Handbuch Fahrerassistenzsysteme. Bremsenbasierte Assistenzfunktionen, pp. 356–394. Vieweg & Teubner Verlag, Springer Fachmedien, Wiesbaden (2012)
83. Vector Informatik GmbH: PREEvision User Manual Version 8.0. Stuttgart (2016)
84. van der Voort, M., Dougherty, M.S., van Maarseveen, M.: A prototype fuelefficiency support tool. Transp. Res. Part C: Emerg. Technol. 9(4), 279–296 (2001). http://www.sciencedirect.com/science/article/pii/S0968090X00000383
85. Wahl, H.G.: Optimale Regelung eines prädiktiven Energiemanagements von Hybridfahrzeugen. Ph.D. thesis, Karlsruher Institut für Technologie (2015)
86. Walker, P.D., Zhang, N., Tamba, R.: Control of gear shifts in dual clutch transmission powertrains. Mech. Syst. Signal Process. 25(6), 1923–1936 (2011)
87. Walter, M., Fechner, T., Hellmann, W., Thiel, R.: Handbuch Fahrerassistenzsysteme. Lane Departure Warning, pp. 543–553. Vieweg & Teubner Verlag, Springer Fachmedien, Wiesbaden (2012)
88. Weber, J.: Automotive Development Process. Springer, Heidelberg (2009). https://doi.org/10.1007/978-3-642-01253-2
89. Weber, M., Weisbrod, J.: Requirements engineering in automotive development - experience and challenges. In: IEEE Joint International Conference on Requirements Engineering (RE 2002) (2002)
90. Wiesenthal, M., Collenberg, H., Krimmel, H.: Aktive hinterachsinematik akc - ein beitrag zu fahrdynamik, sicherheit und komfort. In: 17. Aachener Kolloquium Fahrzeug- und Motorentechnik (2008)
91. Willats, R., et al.: Vehicle access control and start system, December 2003. https://www.google.com/patents/US20030222758. US Patent App. 10/348,233
92. Winner, H., Danner, B., Steinle, J.: Handbuch Fahrerassistenzsysteme. Adaptive Cruise Control, pp. 478–521. Vieweg+Teubner Verlag, Wiesbaden (2012). https://doi.org/10.1007/978-3-8348-8619-4_33
93. Woestman, J., Patil, P., Stunz, R., Pilutti, T.: Strategy to use an on-board navigation system for electric and hybrid electric vehicle energy management, 26 2002. https://www.google.com/patents/US6487477. US Patent 6,487,477
94. Wolf, D., Hess, G., Twichel, J.: Automatic start/stop system and method for locomotive engines, September 2005. https://www.google.com/patents/US6941218. US Patent 6,941,218

95. Zechmann, J., Irion, A.: Method and apparatus for controlling the brake system of a vehicle, January 2000. https://www.google.com/patents/US6009984. US Patent 6,009,984
96. Zhang, R., Krishnan, A.: Using delta model for collaborative work of industrial large-scaled E/E architecture models. In: Whittle, J., Clark, T., Kühne, T. (eds.) MODELS 2011. LNCS, vol. 6981, pp. 714–728. Springer, Heidelberg (2011). https://doi.org/10.1007/978-3-642-24485-8_52
97. Zhang, Y., Wang, W., Kobayashi, Y., Shirai, K.: Remaining driving range estimation of electric vehicle. In: 2012 IEEE International Electric Vehicle Conference, pp. 1–7, March 2012
98. Ziegler, J., et al.: Making bertha drive - an autonomous journey on a historic route. IEEE Intell. Transp. Syst. Mag. **6**(2), 8–20 (2014)

User-Based Relocation of Stackable Car Sharing

Haitam Laarabi[1(✉)], Chiara Boldrini[1], Raffaele Bruno[1], Helen Porter[2], and Peter Davidson[2]

[1] Institute for Informatics and Telematics (IIT),
Italian National Research Council (CNR), Pisa, Italy
haitam.laarabi@iit.cnr.it
[2] Peter Davidson Consultancy (PDC), Berkhamsted, England

Abstract. The relocation of carsharing vehicles is one of the main challenges facing its economic viability, in addition to the operational costs and infrastructure deployment. In this paper, we take advantage of an innovative technological proposal of a one-way carsharing system, to test and validate a user-based relocation strategy. The new technology allows vehicles to be driven in a road train by either an operator (up until eight vehicles) or a customer (up to two). The proposed strategy encourages a customer to take a second vehicle along the way, when he/she happens to be moving from a station with excess of vehicles, to a deficient station. As a case study, we have considered a suburban area of the city of Lyon, of which we have a 2015 household travel survey to build a synthetic population undertaking various activities during a day. Then, we inject this population in a detailed multi-agent and multi-modal transport simulation model, to compare the relocation performance of a lower/upper-bound availability algorithm with three other naively intuitive algorithms. The study shows that: (i) relocation algorithm is very sensitive to the ratio of parking slots to fleet size, and (ii) with the right infrastructure we can relocate one vehicle and generate at least one additional trip.

Keywords: Carsharing · User-based relocation
Multi-agent traffic simulation · Stackable vehicles · Electric vehicles

1 Introduction

Carsharing systems are innovative mobility services that are increasingly becoming popular in urban and sub-urban areas and have the potential to solve real-world problems of urban transports [17]. The principle of a carsharing system is that customers can rent for limited period of times a car from a fleet of shared vehicle operated by a company or a public organisation. Although carsharing services have been proposed in the early 1970s, they have emerged as a worldwide phenomenon only in the last decade. This is due to the deployment of one-way carsharing systems in which the customers are allowed to leave the rented car

© Springer Nature Switzerland AG 2019
B. Donnellan et al. (Eds.): SMARTGREENS 2017/VEHITS 2017, CCIS 921, pp. 256–273, 2019.
https://doi.org/10.1007/978-3-030-02907-4_13

at a drop-off location different from the pickup location [3]. This provides an increased flexibility for the users compared to two-way systems.

Typically, one-way carsharing systems suffer from unbalance distribution of available vehicles in the service area. Specifically, some locations can be more popular than others at different times of the day (e.g., residential areas at night-time as opposed to industrial and commercial areas at peak hours). This imbalance of demand easily results into situations in which vehicles accumulates in areas where there is a lower number of rental requests, while at the same time there is shortage of vehicles where they are more needed [5]. When this happens, the operator can resort to rebalancing policies, i.e., relocation of vehicles from where they are not needed (taking into account the expected demand in the near future) with the objective of serving more effectively the travel demands. Clearly, this has a cost for the operator, thus relocation should be performed only when economically viable.

However, before the operator resorts to rebalancing, he needs to know the optimal solution for infrastructure planning, giving the high investments costs and travel demand. In other words, he needs to determine the number, size and location of parking stations to deploy in the area where the carsharing system is supposed to operate in. In the literature, this problem is generally solved considering a spatial-temporal formulation of a MILP [1,10]. In our previous work, we formulated a set-covering model coupled with queuing theory to guarantee certain level of service to customers [8].

Different approaches for vehicle relocation in carsharing systems exist [29]. Operator-based solutions require the use of dedicated staff for executing the redistribution tasks. On the contrary, user-based solutions rely on users willing to relocate vehicles to locations where they are needed, usually on the basis of an economic incentive. However, both approaches can be costly. Furthermore, it is still uncertain whether users are willing to accept incentives for deviations from their destinations. Finally, the design of optimisation frameworks for the decision of which vehicles to relocate to which location can become intractable due to the extremely large number of relocation variables [10].

To cope with the aforementioned issues, in this paper we suggest a user-based relocation algorithm that takes a conservative stance in order to predict the excess and deficiency of vehicles. When a customer queries the carsharing system about trip he/she desire to perform, the system reacts by verifying whether the origin station is in excess of vehicle and if the destination station is in deficiency of vehicles. In this case, the system will encourage the customer to take a second vehicle so to help at the rebalancing. The possibility to drive a second vehicle assume a new class of lightweight vehicles, called ESPRIT cars, which can be stacked, recharged and driven in a road train [13]. This is supposed caters for more efficient relocations since a single customer can relocated two vehicles at the same time.

To validate the performance of the proposed relocation strategy on a meaningful case we use the city of Lyon as case study. Specifically, we use a multi-agent simulation framework that we have previously designed [23]. It is based on MATSim, a popular open-source and agent-based traffic simulation platform, which supports dynamic traffic assignment, large scenarios and detailed modelling of transportation networks [2]. Then we set up a scenario using data from the 2015 Lyon conurbation household travel survey, which provides information about more than three million trips, and public data on the Lyon's public transit systems. Then, we analyse the impact of the infrastructure planning strategy on the user-based relocation in terms of number of rental trips and relocation trips.

The remainder of this paper is organised as follows. Section 2 provides an overview of related literature on infrastructure planning, vehicle relocation and carsharing performance evaluation. It also introduces the ESPRIT carsharing system and the user-based relocation in such a system. Section 3 sets the methodological ground of the relocation strategies on which is based this paper. Section 4 describes the Lyon case scenario, travel demand. Section 5 discusses the simulation results. Finally, the conclusion summarizes the paper and outlines future work.

2   Related Work

2.1   Models for Infrastructure Planning

Infrastructure planning tries to determine the number, size and location of parking stations in a carsharing system in order to maximise some performance measure, such as demand coverage or profit. From a general point of view, this is an instance of the facility location problem, which is an optimisation problem extensively studied in the field of logistics and transportation planning [14].

Existing planning frameworks typically rely on time-space optimisation approaches, which are models that assume a deterministic knowledge of the demand of vehicles at each time interval of the control period. For instance, A MILP formulation is used in [1] to maximise the profits of car-sharing system, which simultaneously optimises the location of parking stations and the fleet size under several trip fare schemes. The proposed model is then used to analyse a case study in Lisbon. A recent work [10] addresses the planning of an electric car-sharing system using a multi-objective MILP model that simultaneously determines the number, size and locations of stations, as well as the fleet size taking into account vehicle relocation and electric vehicle charging requirements. More recently, new modelling approaches (eg. queuing theory and fluid models) have been proposed to take into account that the demand process of customers is stochastic and exhibits seasonal effects. For instance, a closed queuing network modelling of a vehicle rental system is proposed in [16] to derive some basic principles for the design of system balancing methods. In our previous work [8], we formulated a set-covering model that minimises the cost of deployment (in terms of number of stations and their capacity) and leveraged on queuing theory to also guarantee a pre-defined level of service to the customers (in terms of probability of finding an available car/parking space).

2.2   Relocation: State of Art

Vehicle relocation strategies can be classified into the following two broad categories: (i) user-based schemes, which incentive customers to participate in the relocation program, and (ii) operator-based schemes, which leverage on dedicated staff for relocation activities.

In [20] two operator-based strategies are simulated. The shortest time strategy relocates vehicles to minimise the travel times of staff members. The inventory balancing strategy moves vehicles from over-supplied stations to stations with vehicle shortage. In [21] an inter-programming model is developed to minimise the costs associated to staff-based relocation. A similar model is developed in [19] to maximise the profit of the carsharing operator. In [25] a stochastic MIP model is formulated to optimise vehicle relocations, which has the advantage of considering demand uncertainty. A multi-objective MILP model for planning one-way car-sharing systems is developed in [10] taking into account vehicle relocation, station deployment and electric vehicle charging requirements. The design of optimal rebalancing algorithms with autonomous, self-driving vehicles has been recently addressed in [26] using a fluidic model, and [30] using a queueing-theoretical model. An alternative approach for operator-based relocation scheme consists in selecting trips so as to reduce vehicle imbalance, for instance by rejecting trips to stations with parking shortage [1, 27].

User-based relocation policies are typically considered more convenient for the carsharing operator as they do not require the use of a staff. However, it is still uncertain whether users would be willing to participate in a rebalancing program by accepting an alternative destination or a more distant vehicle [18]. For this reason, most of the studies in this field focus on designing pricing incentive policies for encouraging users to relocate the vehicles themselves [12,15]. Clearly, the effectiveness of these schemes highly depends on users' participation and their willingness to accept changes of their travel behaviours.

2.3   Relocation: Stackable Vehicles

The underlying design principles of cars are rapidly evolving and the design of innovative lightweight vehicles is coming to the fore of current academic and industrial research programs. The long-term vision is to reinvent urban mobility systems by leveraging on vehicles specifically designed for city use with significant smaller spatial use and carbon footprints, as well as considerably less expensive to own and operate [24]. For instance, several concept prototypes of stackable, and foldable two-seat urban electric cars are currently under development, such as the MIT BitCar [28], or EO Smart [7]. A step forward is take by the ESPRIT European Project that is designed and prototyping a new vehicle that is stackable with mechanical and electrical coupling, and it can be driven in road trains as shown in Fig. 1.

ESPRIT vehicles have the potential to facilitate the deployment of one-way carsharing by also supporting more efficient operational procedures. In particular, redistribution is made easier because the vehicles can be driven in a road

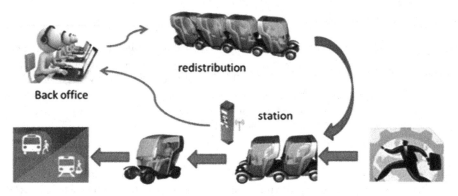

Fig. 1. The architecture of an ESPRIT-based car-sharing system [13].

train. As a consequence, a single staff can drive a road train of up to eight vehicles, or users may drive a road train of two vehicles with a conventional driving license. As discussed in the previous section, one of the main hurdles for user-based relocation strategies is to encourage the users to change their destination to perform a relocation task.

With ESPRIT, we can afford a different way of user-based relocation, where operator can take advantage of actual trips and augmenting their relocation efficiency by delivering two vehicles instead of just one. However, this strategy has been proven, in the following paper, to have a low impact on the total number of carsharing trips.

2.4   Simulation of Carsharing Systems

In general, evaluating the performance of a carsharing system is a difficult task due to the complex and time-variant interplay between the demand and supply processes. Specifically, the availability of vehicles in a carsharing system is intrinsically dependent on trips that are demanded by the customers and vice-versa. In addition, there are several operational conditions that add uncertainties to the system about the future location of vehicles, such as the impact of pricing schemes impact on the decisions of individual users. Therefore, a simulation approach can be very useful to cope with operation complexities and to quickly evaluate the effectiveness of different planning and operation models.

Studies of micro-simulation for performance evaluation of carsharing system has been investigated as early as 1982 [9]. During that period, there was not yet the large panel of traffic simulation tools that are existing nowadays. Thus, the critics held by the author in [9] regarding the computational complexity and availability of data should be taken in moderation. In 1999, a queuing-based transport simulation has been proposed by [4] for the assessment of the performance of a shared one-way vehicle system. Different measures of efficiency were determined, such availability of vehicles, their distribution and energy consumption, while some relocation strategies were tested. However, the simulation model is exactly predictive and does not capture the inherent uncertainty of real world systems. A more detailed carsharing simulation model and open source was introduced by [11], where it is based on multi-modal agent-based traffic simulator, such that each agent seeks to fulfils its daily plan as a set of activities connected by legs. In our previous work, we designed a similar but more sophisticated carsharing simulator [23], in such a way to separate the carsharing mobility simulation model from the operational and demand model. The purpose is to allow users test different operational models and strategies using the same tool. We have, therefore, used this simulation model to study the performance of a new carsharing system deployed in a suburban area of Lyon.

3   Relocation Strategies

The need for vehicles relocation in carsharing systems stems from the unbalance of availability of vehicles that naturally emerges at different moments of the day. A station manifesting an excess of availability of vehicles should be leveraged to provide additional vehicles to stations manifesting a deficiency of vehicles. In this work we assume that customers are encouraged to relocate a second vehicle if they are planning to make a trip between stations with an excess and a shortage of vehicles, respectively.

Defining the metric that can be used to detect whether a station has an excess/deficiency of vehicles is not an easy task. In principle, availability might undergo large fluctuations due to a continuous stream of pickups and drop-offs of vehicles. In principle, relocating a vehicle from/to a highly variable station may negatively interfere with the natural flow of vehicles and cause a butterfly effect in the network. On the other hand, complex availability metrics would require to have additional knowledge about the carsharing system, e.g. to predict the minimum availability of vehicles over some time interval. In the following we explore both approaches. Specifically, we first propose a set of relocation policies that only rely on a simple characterisation of the carsharing dynamics based on the instantaneous number of vehicles that are available for rent at a station. Then, we descrive a more elaborated relocation heuristic, which assumes a knowledge of carsharing demand patterns.

Before describing our proposed scheme, it is also important to point out that are several business and operational factors that can affect the effectiveness of the relocation process. As a matter of fact, a relocation task is costly, since it consumes fuel and makes the vehicle unavailable during the trip period. Furthermore, a carsharing operator might want to ensure a high availability in a certain station by contrast to others following a specific marketing strategy. In this paper, we will not dive into all these complexities of real world systems, but we assume that a relocation strategy is effective if one relocated vehicle generates at least one additional trip i.e. the fraction of additional trips over the relocation trips should be superior or equal to one, as the minimum accepted performance.

Algorithm 1. Uniform relocation.

1: **procedure** RELOCATE(i, j, t)
2:     $rand \leftarrow$ uniform_random_generator(0,1)
3:     **if** $(v_i(t) > 2$ **AND** $p_j(t) > 2)$ **then**   ▷ Check relocation feasibility
4:         **if** $rand \leq \gamma$ **then**
5:             **return** TRUE   ▷ Yes
6:         **end if**
7:     **end if**
8:     **return** FALSE   ▷ No
9: **end procedure**

3.1   Policies Based on Current System State

First of all, we describe relocation policies that do not require to maintain the past system state, or to predict the future system state. This implies that these strategies are not influenced by the history of the carsharing system, and do not rely on a knowledge of the carsharing demand. The only information that is maintained by the carsharing operator is $v_i(t)$, defined as the number of available vehicles at station i at time t, and $p_i(t)$, defined as the number of available parking spaces at station i at time t. Clearly, $p_i(t) = c_i - v_i(t)$, where c_i is the capacity of station i.

The simplest relocation strategy is the *uniform* policy, illustrated in Algorithm 1, which incentivises each customer to take a second vehicle to his intended destination with a fixed probability γ. Specifically, let us assume that at time t a customer generates a request for a rental vehicle from location O to location D. The central controller of the carsharing system determines the station i that is the closest to location O with an available vehicle, and the station j that is the closest to location D with an available parking space. A relocation task is assigned to the customer with probability γ if and only if there are at least two vehicles available at station i and there is enough available parking space at station k to accomodate the train of two vehicles.

The second relocation strategy is illustrated in Algorithm 2, and it simply restricts the relocation task to destination stations that are empty. The rationale behind this strategy is to avoid situations in which two close booking requests can not be satisfied by a system.

Algorithm 2. Prioritise empty station.

1: **procedure** RELOCATE(i, j, t)
2:     **if** $(v_i(t) > 2$ **AND** $p_j(t)) == 0)$ **then**
3:         **return** TRUE   ▷ Yes
4:     **end if**
5:     **return** FALSE   ▷ No
6: **end procedure**

The third relocation strategy is the *balance* policy, illustrated in Algorithm 3, in which a customer takes a second vehicle to his intended destination only if this contributes to reduce the difference in the occupancy levels of the stations. The rationale behind this strategy is to use the redistribution to equalise as much as possible the utilisation of stations. This can be mathematically expressed computing the difference between the new occupancy levels that would be due to the movement of a single vehicle or a train of two vehicles. After standard algebraic manipulations it is straightforward to show that if $v_i(t) > v_j(t) + 4$ it is always beneficial to encourage a customer to take a second vehicle with him.

3.2 Policies Based on Predicted Minimum Availabilities

As noted before, the instantaneous car availability at a station is typically highly variable. We conjecture that a more reliable parameter for guiding the relocation decision is an estimates of the number of vehicles that will not be used because they are in excess with respect to the carsharing demand. This excess of vehicles can be estimated by measuring the minimum car availability over a period of time. Specifically, let us assume that the system time is divided into time intervals of duration τ. Then, let us denote with α_i^k the *minimum car availability* that is expected at station i during the time interval $[k\tau, (k + 1)\tau]$. The estimation of the minimum car availability of a carsharing system that does not perform relocation, say $\alpha_i^{k,nr}$, is straightforward as it is given by

$$\widehat{\alpha}_i^{k,nr}(t) = \min \{v_i(s) : s \in [k\tau, t]\}. \tag{1}$$

In Eq. (3), the function $v_i(t)$ is provided by historical information. On the other hand, the relocation process changes the system dynamics and observations from a system without relocation might be quite different from the ones of the system with relocation. Thus, we decide to also compute an *expected* minimum availability using forecasts of the carsharing demand. More precisely, let us denote with $\alpha_i^{k,r}$ the minimum availability in the time interval $[k\tau, (k + 1)\tau]$ based on the *estimated* number of vehicles that will be dropped off and picked up at station i during $[k\tau, (k + 1)\tau]$ according to the carsharing demand, and

---

Algorithm 3. Balanced offer.

1: **procedure** RELOCATE(i, j, t)
2:     **if** $(v_i(t) \geq 2$ **AND** $p_j(t) \geq 2)$ **then**   ▷ Check relocation feasibility
3:         $u_1 \leftarrow [(v_j(t) + 1)/c_j] - [(v_i(t) - 1)/c_i]$
4:         $u_2 \leftarrow [(v_j(t) + 2)/c_j] - [(v_i(t) - 2)/c_i]$
5:         **if** $|u_1| > |u_2|$ **then**
6:             **return** TRUE   ▷ Yes
7:         **end if**
8:     **end if**
9:     **return** FALSE   ▷ No
10: **end procedure**

---

taking into account that the initial number of vehicles is $v_i(k\tau)$ and the station has finite capacity c_i. To clarify the procedure used to estimate the minimum car availability Fig. 2 illustrates an example. As shown in the figure, we consider the sequence of expected pick-up and drop-off events to estimate the future evolution of the $v_i(t)$ function. Note that we discard pick-up and drop-off events that are not feasible, i.e. pick-ups that would occur when the estimated $v_i(t)$ function is equal to zero, and drop-offs that would occur when the estimated $v_i(t)$ function is equal to c_i.

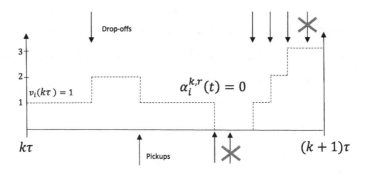

Fig. 2. Example of $\alpha_i^{k,r}$ estimation for a station with capacity equal to 3.

Finally, we have that

$$\alpha_i^k = \min\left\{\alpha_i^{k,nr}, \alpha_i^{k,r}\right\}. \qquad (2)$$

Different approaches can be used to estimate α_i^k. The simplest one is to use historical information about the car availability from a carsharing system in which relocation is not used. In this case, α_i^k would simply be equal to $\min\{v_i(t) : t \in [k\tau, (k+1)\tau]\}$. However, the shortcoming of this approach is that the relocation process changes the system dynamics and observations from a system without relocation are not representative of the system with relocation. In particular, rental requests that failed in the system without relocation can be successful in the system with relocation (and viceversa). Thus, we decide to use a combination of historical data and carsharing demand forecast. Specifically, let us assume that at time t, with $t \in [k\tau, (k+1)\tau]$, a customer wants to pick up a car at station i. Then, we split the computation of α_i^k into two components. The first one is $\widehat{\alpha}_i^k(t)$, which is given by:

$$\widehat{\alpha}_i^k(t) = \min\{v_i(s) : s \in [k\tau, t]\}. \qquad (3)$$

In other words, $\widehat{\alpha}_i^k(t)$ is the exact minimum availability of station i considering only the time interval $[k\tau, t]$ and the knowledge of the real car availability given by $v_i(t)$. The second one is $\overline{\alpha}_i^k(t)$, which represents the minimum availability in the time interval $[t, (k+1)\tau]$ based on the *estimated* number of vehicles that will be dropped off and picked up from station i during $[t, (k+1)\tau]$ according to the carsharing demand, and taking account that the initial number of vehicles is $v_i(t)$ and the station has finite capacity c_i. Finally, we have that

$$\alpha_i^k(t) = \min \left\{ \widehat{\alpha}_i^k(t), \overline{\alpha}_i^k(t) \right\}. \tag{4}$$

Following the same line of reasoning it is possible to also estimate β_i^k, defined as the *availability parking space availability* at station i during time interval $[k\tau, (k+1)\tau]$. Intuitively, β_i^k is the complement of the maximum car availability. We are now able to define a relocation policy that leverages the knowledge of the predicted minimum car and parking space availability, which is illustrated in Algorithm 4. Clearly, first the relocation strategy checks if the relocation tasks is feasible, i.e. there are at least two vehicles available at station i and there is enough available parking space at station k to accomodate the train of two vehicles. Then, the algorithm checks a similar but more restrictive condition, i.e. relocation is feasible if also minimum car and parking space availabilities are considered.

Algorithm 4. Minimum availabilities.

1: **procedure** RELOCATE(i, j, t)
2:     **if** ($v_i(t) \geq 2$ **AND** $p_j(t) \geq 2$) **then**   ▷ Check relocation feasibility
3:         **if** ($\alpha_i^k \geq 2$ **AND** $\beta_i^k \geq 2$) **then**   ▷ Check minimum availabilities
4:             **return** TRUE   ▷ Yes
5:         **end if**
6:     **end if**
7:     **return** FALSE   ▷ No
8: **end procedure**

---

4 Case Study

4.1 Scenario

Similarly to the work previously done in [22], we will test and validate the suggested user-based relocation strategies using the Lyon case study. The operating area of the simulated carsharing system is shown in Fig. 3, and corresponds to three suburban district of the city of Lyon. The road network is constructed from OpenStreetMap data and made of 141,795 links, not only limited to the study area with green background. Regarding the public transit systems, we used data publicly available from Grand Lyon Data platform[1] to define transit routes and modes (buses, tram, underground), transit stops, as well as schedules and vehicles capacities. One of the most important modelling task is to construct the travel demand for different transportation modes. Traditionally, travel demand

---

[1] http://data.grandlyon.com/.

data is organised as trip origin/destination (O/D) matrices, which simply contain the number of trips that are taken from an origin node to a destination node in a specific period of time. However, since we use a multi-agent modelling approach, the travel demands are constructed as individual daily plan dairies, which contain sequence of activities and the preferred transportation mode for trips between activities. As for [22], we created the demand for Lyon based on used census data from the INSEE website and uses data from the Lyon Travel Diary Survey 2015. The synthetic population representing the demand is of the order of 1.4 million agents, all of whom pass through the choice model to determine the destination and mode of trips. For this model, we considered five types of facilities: home, work, education, shopping and leisure. Such that home facility represents most of the facilities with 35,853 instance, while the remaining others make up 1549 instances. The Lyon Household and travel diary survey 2015 were used to estimate coefficients for generating the synthetic population. The records were split according to whether the synthetic person had both a driving licence and the household a car or not. The trip records were fitted to a nested mode (Car or PT) and destination choice model, and the coefficients at both levels of the nest were estimated simultaneously.

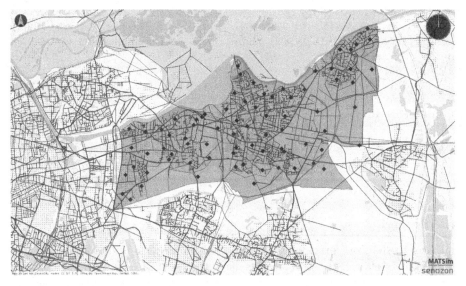

Fig. 3. Map of the simulated area, with blue diamond referring to the Esprit stations deployed within the study area (green background). The red and grey lines refer to, respectively, the PT and car networks [22]. (Color figure online)

The main novelty considered in this new version of the Lyon demand comparing to the one tested in [22], is at the level of the mode choice. We introduced a "walk" mode, as well as new combinations of modes, such:

1. Private car (car)
2. Park and Ride
3. Public Transport (pt)
4. ESPRIT (no Public Transport)
5. ESPRIT followed by Public Transport, "ESPRIT first"
6. Public Transport followed by ESPRIT, "ESPRIT last"
7. and Walk

ESPRIT first and ESPRIT last match to the concept of *first and last kilome-tre*. These trips are of particular interest. Individual agents make travel choices according to whether they have a driving licence, and whether there is car belong-ing to their household. If they can drive and have access to a car then all seven modes are available, but if the household does not have a car they may still choose to use ESPRIT. The choice nest takes account of their car availability status and directs the agent through the appropriate part of the choice nest accordingly. The demand contains a total number of agents of 80.740 agents and a customer base of 6.416 agents. While the modal share as shown in Fig. 4, shows Esprit share of represents 6.7% of the modal share, while private car is leading by a modal share of 64.3%, then public transport 17.3% and finally walking mode representing 11.7% of the modal split.

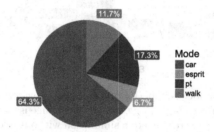

Fig. 4. Modal share of the base demand.

In this case study we considered a scenario where 350 vehicles were deployed in 77 stations, which are represented as diamonds in Fig. 3. The deployment was undertaken following the optimal deployment strategy introduced in [6]. However, we branched off two main variants of this scenario, so to compare the impact of the ratio of the parking slots to the vehicles on the relocation strategies. We assumed, therefore, in the first variant that each of the stations have reasonably very large parking space of 20 slots, in total 1540, i.e. a ratio slot:vehicle equal to 4.4. While in the second variant we assumed a reasonably smaller parking space of 10 slots per station, in total 770, i.e. a ratio equal to 2.2. We will see in the following why this ratio has a significant impact of the performance of the relocation strategies.

4.2   Results and Discussion

Considering the environment described above, we have executed multiple simulations with different set-ups. On one hand, we have executed two simulations with no relocation strategy activated on the two different parking slots: 1540 and 770. The goal is to obtain a reference line for deducing the performance of the relocation strategies. This reference line is represented with a straight solid line on both Figs. 5 and 8. It indicates the threshold of 100% of the number of trips without relocation, and any bar plot rising above it means that the Algorithm led to new successful bookings that have failed beforehand. Both variants of parking slots have produced roughly same number of trips: Successful bookings are dependent less on parking availability than fleet size [6].

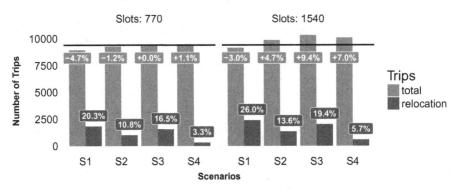

Fig. 5. Comparison of relocation performance when considering the four scenarios, with respect to the ratio vehicle/parking slot. Performance is measured in terms of (1) percentage of number of trips relative to a simulation without relocation (2) percentage of trips where a second vehicle has been offered to be relocated.

On the other hand, we run a first set-up using the different relocation strategies on the two variants of slot:vehicle ratios. The purpose is to compare the performance of each strategy and the impact of the ratio. The second set-up focused on the proposed sophisticated relocation strategy, *Algorithm* 4. Since this algorithm depends on the predefined bin of time per contra to other algorithms. Therefore, we have tested it on a set of time bins to observe their impact on its performance.

In order to compare between the performance of the different strategies and in different time interval, we have decided to use two metrics M_1 and M_2:

- M_1: the difference of trips obtained from the simulation without relocation and the one with relocation. A positive difference means that the relocation decisions allowed the booking success of new trips comparing to the simulation without relocation.

- M_2: The fraction of carsharing trips that actually served for transporting a second vehicles by the agent. The purpose is to know how many relocation trips have been required to get more successful booking trips.

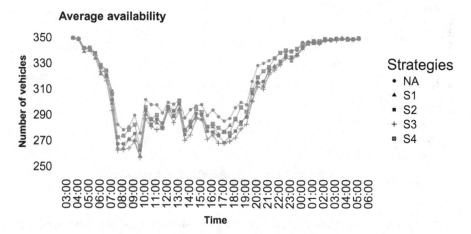

Fig. 6. Fluctuation of the sum of 30 min average availability of vehicles of all stations, for the different scenarios and solely for the case of 1540 parking slots.

Let us start with the graph result of first set-up shown in Fig. 5. The uniform relocation strategy $S1$, has scored negatively in both situations, respectively -4.7% and -3.0% for the metric M_1. While it led to the highest number of relocation such that $M_2 > 20\%$ of the shares of the total trips that served for relocating a second vehicle. This strategy based on uniformly distributed random numbers demonstrate that it cannot be at all a solution in dealing with the relocation problem. In addition to the fact that it can not be even used as a reference strategy with which we would measure how well our relocation strategy scores in comparison to a random behaviour.

Algorithm 2 scores quite well in the case of the large parking space variant: $M_1 = +4.7\%$ that is more than 400 additional trips. In contrast, it required $M_2 = 13.6\%$ of relocation trips, which is around 1300 trips. This strategy that consists in prioritising empty stations has led to a ratio of approximately 1:3. In other words, the decision maker will have to relocate 3 vehicles to ensure 1 new successful booking. This strategy is costly for the carsharing operator but it can be used as strategy with lower bound performance.

The balancing relocation strategy described in Algorithm 3 scores the highest number of additional trips in case of 1540 slots: $M_1 = +9.4\%$. To generate the additional 900 trips, the systems had to encourage around 2000 agents to relocate a second vehicle ($M_2 = 19.4\%$). This is equivalent to a ratio of 1:2, one additional vehicle for 2 relocation trips. In the case of 770 slots, the score is tied: $M_1 = +0.0\%$, while $M_2 = 16.5\%$ is still significantly high. We conclude that the

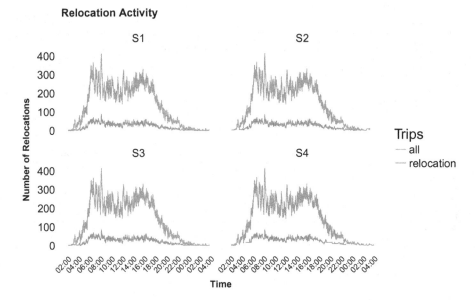

Fig. 7. Comparison of the ongoing activity of all carsharing trips with carsharing trips transporting a second vehicle to be relocated and solely for the case of 1540 parking slots.

three Algorithms 1, 2 and 3 all behave quite poorly when the ratio of slots to vehicles is low.

The proposed strategy based on minimum availability outperformed the other strategies in terms of M_1 to M_2 ratio. Even though it has scored less than strategy $S3$ in the case of 1540 slots: $M_1 = +7.0\%$. Yet the second metric has a score significantly lower than the other strategies: $M_2 = 16.5\%$, resulting in a ratio greater than 1:1. In other words, the carsharing system will require less relocation trips to generate more additional trips, in the case of a significantly large parking space availability. In the 770 slots variants, the algorithm offer poor performance, but still positive score and 1:3 ratio way better than the other algorithms.

In addition to Fig. 5, two other comparison plots were generated and depicted by Figs. 6 and 7. The average availability figure show the slight mitigation of the availability due to the relocation algorithms. While there is a drop of average availability of only 5 to 20 vehicles when comparing with the no relocation case, the Algorithm 4 remains the one with less mitigation availability relative to the three other algorithms. This observation led us to hypothesize that the poor performance of the other algorithms was due to the unavailability of vehicles due to excessive relocation decisions. This hypothesis is supported by the plots in Fig. 7, which shows that Algorithm 4 led to less relocation activities in comparison with the other algorithms.

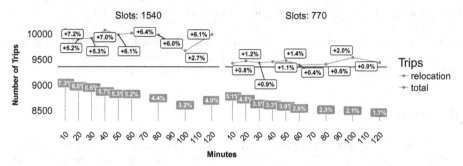

Fig. 8. Comparison of fluctuation of relocation performance when considering different time bins, with respect to the ratio vehicle/parking slot.

The results obtained with Algorithm 4 was in the case of predefined time bin of 40 min. At each start of the interval, the algorithm classifies the stations that are expected to be in excess and deficiency of vehicles. This list is not updated until the start of the next time interval. Since the minimum availability is sensitive to the time bin, we had to test the performance of the algorithm with different time bins.

The outcomes of these simulations have confirmed that (1) the ratio slots:vehicles is a sensitive factor on the performance of the relocation Algorithm 2) larger is the time interval, more conservative is the algorithm and less relocation trips were encouraged without degrading much the M_1 metric. Indeed, a quick calculation of the rate of change[2] in both cases led to a negative slope of -0.03%/min in terms of percentage of relocation trips, meanwhile the slope is no less than -0.01%/min (even positive in case of 770 slots) for the percentage of additional trips.

5 Conclusion

We have seen in this article how it is possible to achieve positive relocation performance, if the customer is encouraged to transport a second vehicle with him/her. This will be possible thanks to the ESPRIT model where it seeks to design stackable vehicles that can be driven in train of two by a customer with a car driving license. We have demonstrated that with a proper demand model and the right station deployment and parking slots to vehicles ratio, we can ensure a positive relocation performance with a ratio greater than 1:1, that is at least one additional trips is generated when relocating one vehicle, using the proposed lower/upper-bound availability algorithm. Still further studies are required to understand better the relationship between the relocation performance and the

---

[2] The rate of change was computed following the traditional formula: $\frac{1.7-5.1}{120-10} \approx \frac{4.0-7.3}{120-10} = -0.03\%$/min.

parking slots to vehicle ratio. While we aim for improving the proposed algorithm in such a way to always guarantee a ratio greater or equal to 1:1.

Acknowledgement. This work has been partially funded by the ESPRIT project. This project has received funding from the European Union's Horizon 2020 research and innovation programme under grant agreement No. 653395.

References

1. de Almeida Correia, G.H., Antunes, A.P.: Optimization approach to depot location and trip selection in one-way carsharing systems. Transp. Res. Part E: Logist. Transp. Rev. **48**(1), 233–247 (2012)
2. Balmer, M., Cetin, N., Nagel, K., Raney, B.: Towards truly agent-based traffic and mobility simulations. In: Proceedings of AAMS 2004, pp. 60–67. IEEE Computer Society (2004)
3. Barth, M., Shaheen, S.: Shared-use vehicle systems: framework for classifying carsharing, station cars, and combined approaches. Transp. Res. Rec.: J. Transp. Res. Board **1791**, 105–112 (2002)
4. Barth, M., Todd, M.: Simulation model performance analysis of a multiple station shared vehicle system. Transp. Res. Part C: Emerg. Technol. **7**(4), 237–259 (1999)
5. Barth, M., Todd, M., Xue, L.: User-based vehicle relocation techniques for multiple-station shared-use vehicle systems (2004)
6. Biondi, E., Boldrini, C., Bruno, R.: Optimal deployment of stations for a car sharing system with stochastic demands: a queueing theoretical perspective. In: The 19th IEEE Intelligent Transportation Systems Conference, pp. 1–7. IEEE (2016)
7. Birnschein, T., Kirchner, F., Girault, B., Yüksel, M., Machowinski, J.: An innovative, comprehensive concept for energy efficient electric mobility-EO smart connecting car. In: Proceedings of IEEE ENERGYCON 2012, pp. 1028–1033. IEEE (2012)
8. Boldrini, C., Bruno, R., Conti, M.: Characterising demand and usage patterns in a large station-based car sharing system. In: The 2nd IEEE INFOCOM Workshop on Smart Cities and Urban Computing (2016)
9. Bonsall, P.: Microsimulation: its application to car sharing. Transp. Res. Part A: Gen. **16**(5), 421–429 (1982)
10. Boyacı, B., Zografos, K.G., Geroliminis, N.: An optimization framework for the development of efficient one-way car-sharing systems. Eur. J. Oper. Res. **240**(3), 718–733 (2015)
11. Ciari, F., Schuessler, N., Axhausen, K.W.: Estimation of carsharing demand using an activity-based microsimulation approach: model discussion and some results. Int. J. Sustain. Transp. **7**(1), 70–84 (2013)
12. Clemente, M., Fanti, M.P., Mangini, A.M., Ukovich, W.: The vehicle relocation problem in car sharing systems: modeling and simulation in a petri net framework. In: Colom, J.-M., Desel, J. (eds.) PETRI NETS 2013. LNCS, vol. 7927, pp. 250–269. Springer, Heidelberg (2013). https://doi.org/10.1007/978-3-642-38697-8_14
13. ESPRIT: Esprit h2020 eu project - easily distributed personal rapid transit (2015). http://www.esprit-transport-system.eu/. Accessed 12 Dec 2016
14. Farahani, R.Z., Asgari, N., Heidari, N., Hosseininia, M., Goh, M.: Covering problems in facility location: a review. Comput. Indus. Eng. **62**, 368–407 (2012)

15. Febbraro, A., Sacco, N., Saeednia, M.: One-way carsharing: solving the relocation problem. Transp. Res. Rec.: J. Transp. Res. Board **2319**, 113–120 (2012)
16. George, D.K., Xia, C.H.: Fleet-sizing and service availability for a vehicle rental system via closed queueing networks. Eur. J. Oper. Res. **211**(1), 198–207 (2011)
17. Hampshire, R., Gaites, C.: Peer-to-peer carsharing: market analysis and potential growth. Transp. Res. Rec.: J. Transp. Res. Board **2217**, 119–126 (2011)
18. Herrmann, S., Schulte, F., Voß, S.: Increasing acceptance of free-floating car sharing systems using smart relocation strategies: a survey based study of car2go Hamburg. In: González-Ramírez, R.G., Schulte, F., Voß, S., Ceroni Díaz, J.A. (eds.) ICCL 2014. LNCS, vol. 8760, pp. 151–162. Springer, Cham (2014). https://doi.org/10. 1007/978-3-319-11421-7_10
19. Jorge, D., Correia, G.H., Barnhart, C.: Comparing optimal relocation operations with simulated relocation policies in one-way carsharing systems. IEEE Trans. Intell. Transp. Syst. **15**(4), 1667–1675 (2014)
20. Kek, A., Cheu, R., Chor, M.: Relocation simulation model for multiple-station shared-use vehicle systems. Transp. Res. Rec.: J. Transp. Res. Board **1986**, 81–88 (2006)
21. Kek, A.G., Cheu, R.L., Meng, Q., Fung, C.H.: A decision support system for vehicle relocation operations in carsharing systems. Transp. Res. Part E: Logist. Transp. Rev. **45**(1), 149–158 (2009)
22. Laarabi, H.M., Boldrini, C., Bruno, R., Davidson, H.P., Peter: on the performance of a one-way car sharing system in suburban areas: a real-world use case. In: 3rd International Conference on Vehicle Technology and Intelligent Transport Systems, vol. 1, pp. 102–110. Scitepress (2017)
23. Laarabi, M.H., Bruno, R.: A generic software framework for carsharing modelling based on a large-scale multi-agent traffic simulation platform. In: Namazi-Rad, M.-R., Padgham, L., Perez, P., Nagel, K., Bazzan, A. (eds.) ABMUS 2016. LNCS (LNAI), vol. 10051, pp. 88–111. Springer, Cham (2017). https://doi.org/10.1007/ 978-3-319-51957-9_6
24. Mitchell, W.J., Borroni-Bird, C.E., Burns, L.D.: Reinventing the Automobile: Personal Urban Mobility for the 21st Century. MIT Press, Cambridge (2010)
25. Nair, R., Miller-Hooks, E.: Fleet management for vehicle sharing operations. Transp. Sci. **45**(4), 524–540 (2011)
26. Pavone, M., Smith, S.L., Frazzoli, E., Rus, D.: Robotic load balancing for mobility-on-demand systems. Int. J. Robot. Res. **31**(7), 839–854 (2012)
27. Uesugi, K., Mukai, N., Watanabe, T.: Optimization of vehicle assignment for car sharing system. In: Apolloni, B., Howlett, R.J., Jain, L. (eds.) KES 2007. LNCS (LNAI), vol. 4693, pp. 1105–1111. Springer, Heidelberg (2007). https://doi.org/10. 1007/978-3-540-74827-4_138
28. Vairani, F.: bitCar: design concept for a collapsible stackable city car. Ph.D. thesis, Massachusetts Institute of Technology (2009)
29. Weikl, S., Bogenberger, K.: Relocation strategies and algorithms for free-floating car sharing systems. IEEE Intell. Transp. Syst. Mag. **5**(4), 100–111 (2013)
30. Zhang, R., Pavone, M.: Control of robotic mobility-on-demand systems: a queueing-theoretical perspective. Int. J. Robot. Res. **35**(1–3), 186–203 (2016)

A Maneuver Based Interaction Framework for External Users of an Automated Assistance Vehicle

Mohsen Sefati[(✉)], Denny Gert, Kai D. Kreiskoether,
and Achim Kampker

Chair of Production Engineering of E-Mobility Components,
Campus-Boulevard 30, 52074 Aachen, Germany
{m.sefati, d.gert, k.kreiskoether,
a.kampker}@pem.rwth-aachen.de

Abstract. Automated vehicles become gradually available for restricted environments. Fully Automated Vehicles (FAV) operate without a driver and need to cooperate and interact with other road users of any kind. This article illustrates an interaction framework, which allows a human user outside the car to interfere with the FAVs guidance. This is achieved by communicating a desired maneuver, where the external user is asked to choose among a set of possible maneuvers. This set of maneuvers is communicated by the FAV to the user and has been checked for execution feasibility by the FAV, based on its perception. To this end, the environment is represented as an occupancy grid and a path search without distinct goal is performed. A small set of paths will be selected and communicated to the external user in an abstract level. This article presents the planning framework, as well as basic implementations for suited path search algorithms. The conclusion addresses unsolved challenges.

Keywords: Automated driving · Assistance Vehicle
Human-machine-interaction · Cooperative automation · External user

1 Introduction

Vehicle guidance is a complex task for a human driver. The Automation of this task has therefore been a technical endeavor since the beginning of the automobile history. In past decades, the intelligence of automated vehicles has increased evolutionary, so that fully automated vehicles (FAV) are gradually available for defined roadways in restricted environments. FAVs address the highest automation level, in which no supervision of a human driver is required as it is defined in the SAE-Level 5 [1]. A variety of possible applications with different characteristics of automated driving can be found for FAVs. Wachenfeld et al. [2] have used distinctive features to distinguish between four main use-cases. While they don't claim, that the existence of further use-cases is excluded, they distinguish between the use-cases 'Fully Automated with available driver', 'Vehicle On-demand', 'Highway Pilot' and 'Automated Valet Parking'. In all the above use-cases, the user is mainly thought to function as a passenger and to use the driving time for other activities rather than driving in the vehicle

B. Donnellan et al. (Eds.): SMARTGREENS 2017/VEHITS 2017, CCIS 921, pp. 274–295, 2019.
https://doi.org/10.1007/978-3-030-02907-4_14

cabin. However, there are conceivable use-cases, in which the user is outside of the vehicle. In these use-cases, the user can be engaged with other tasks outside, while the driving task is overtaken by the system and no human supervision is required. The FAV will assist the user in the secondary task, by synchronizing its position and movement with the external user's activity. In this work, this use-case is named as 'Assistance Vehicle (ASV)'.

The application of an ASV is reasonable from an economic point of view, only if the accompanying user can pursue valuable activities during this time. Thus, ASVs' use case is more convincible for commercial and service vehicles rather than private ones. Some interesting examples are logistic and transport services such as delivery and pickup services (e.g. parcels or garbage transport), social services (e.g. care and insurance services), and maintenance services. Considering these services, the work flow can be described in three main phases with respect to the driving task, where the processes in the service central (e.g. depot or warehouse) is not considered as a phase (cp. Fig. 1).

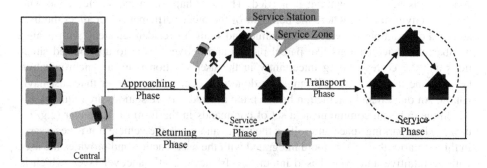

Fig. 1. Process phases for service vehicle with respect to the driving task.

Depicted process phases can typically be characterized as

- *Approaching/Returning Phase*: This phase consists of the outward and return rides between the central (e.g. depot) and the service zones. Service zones can be defined by clusters of service stations, in which the driver has to do the secondary task, e.g. delivery. Since the central might be placed of the city, this ride could include highways with a velocity up to 120 km/h.
- *Transport Phase*: The ride between service zones describes the transport phase. In most of the services, this phase happens mainly in urban areas with a maximum velocity of 50 km/h.
- *Service Phase*: Finally, the ride between the several service stations in the service zones describes the service phase. In conventional cases, this phase is a combination of walking and driving. Therefore, the maximum velocity of this phase is considered as 30 km/h.

The application of the ASV is mainly in these service phases. This also means, that the ASV will be transported to the service zone manually and the automation is realized only for this specific zone. The use of this concept can be very beneficial, if the service stations are within walking distance and therefore can be clustered as a zone. In such a scenario with a conventional vehicle, the service phase is usually executed very inefficiently, due to the high number of stop and go with short transport periods in between. More inefficiency could be caused when the user must move between the stopped vehicle and the next service station, which is not necessarily near to the actual position of the vehicle. This inefficiency can be highly improved with an ASV, where the user is discharged of the task of driving and can carry on with the work while the ASV finds a parking spot and can drive back when the user wants to move back into the vehicle. The vehicle will adapt its position and velocity, so that the availability of the vehicle in the operation field is always guaranteed. In this way, the user has on-demand access to the vehicle and the required tools at all times. The ASV can also increase the safety by decreasing the risk of accidents and protect the user by letting him be more focused on the service tasks. As a summary, an ASV provides the user with more comfort and safety and in turn increases the efficiency.

In more dynamic service phases, an ASV system must be more flexible and a spontaneous behavior planning is needed. This can happen in the service zone with dynamic environments, where the behavior of the other participants can affect the user action. For example, services in dynamic and built up residential areas with a high number of service stations like parcel pickup and delivery. Due to unexpected situations and the corresponding uncertainty in the user's action plan, a synchronization between the user and the ASV must be done in a deeper level, rather than just navigation. An on-demand interaction between the defined service stations are required, for which the user can communicate a set of commands in the form of maneuver (e.g. U-turn, finding parking space and follow the route) and guide the vehicle if it is necessary. In this scenario, the ASV is more integrated with the user's action and provides the user with an intuitive and event-based interaction framework at maneuver layer, which is not only limited to setting the navigation points. This interaction comprises of two communication ways: not only does the ASV understand the intention of the road participants based on its observation, but the user should also be able to communicate his goals and intentions to the ASV explicitly.

The main idea of this work is to introduce an interaction interface for the external user to interfere in the vehicle guidance explicitly and in a discrete form from a limited distance. The framework should not stress the user with the workload and must be generally applicable also in unknown traffic situations, also in absence of the digital map. The major challenge of this framework is to provide both the user and the FAV with a mutual understanding about the shared drivable area and possible actions. Since the user might have a restricted sight of the ASV, he might not be able to interpret the situation correctly as the ASV does. Furthermore, the ASV might not be able to map the desired user command to the available action due to the ambiguity in command understanding. Therefore, it is necessary to provide the user with a standard list of executable commands, which are extracted out of the current scene understanding of the ASV. Each command can be described as a simplification of a chain of actions, which can be clustered into a comprehensible single command for the user. The user commands (in form of the gesture or signal in case of use of smart devices) should be standardized and mapped directly to standard actions, which vanishes the ambiguity in communication.

1.1   Related Works

One of the most well-researched topics in the field of robotics is the concept of Human-Robot-Interaction (HRI). HRI basically allows human users to interact with robots in a more efficient way, meanwhile making it feel more natural. This is achieved by using techniques such as speech and gesture recognition. Due to its effectiveness it has found use in wide range of applications in various fields such as education, home and assistance for tele-operated and unmanned robots [3]. HRI involving the interactions between automobile driver and the vehicle, has gained huge popularity and has been extensively researched in the past few years. The H-Mode [4, 5] and Conduct-By-Wire (CbW) [6, 7] are some of the examples of such HRIs. Both of these concepts implement techniques to assist the driver in a semi-autonomous driving mode for both urban and highway scenarios (i.e. Automation Level 3, cp. [1]). This is achieved via an active maneuver-based interface providing continuous guidance to the driver. The CbW-Framework has been advanced further by adding a Gate-Concept [8] which intelligently splits the guidance task during execution by identifying key decision points. Lotz and Winner have suggested a similar maneuver-based guidance for special cases like lane changes or turns at intersections [9]. The focus of these techniques is maneuver-based guidance such as lane changes and turning maneuvers, which foremost needs the vehicle to move along a fixed navigation path. Unlike most concepts which deal with the interactions between the driver and the automated vehicle, only a handful of researches focuses on the interaction between the automated vehicle and an external user. The FAV of Google Car project is one such example which not only detects the presence of a cyclist but can also distinctively understand the hand signals [14]. Technical University of Munich's Tele-operated driving project assists the operator of the vehicle by both communicating and executing feasible paths that are obtained from the current scenario. The operator can then select one such path from the available set and use it for vehicle guidance [10].

2   Functional Concept of the Interaction Framework

2.1   Architecture and Working Principle

Inspired by [6], the here presented framework follows the approach of a maneuver-based interface for a cooperative vehicle guidance through interaction with the external user. The framework provides the user with a feasible set of maneuvers, where feasibility is checked on the current scene percept by the ASV. The user is asked to select one of the proposed maneuvers. To this end, the framework computes possible paths on the ASV static environment model. We call this step *Path Exploration* and address two different approaches in Subsects. 2.3 and 2.4 respectively. The output of this exploration is a set of feasible paths, from which a subset is chosen in the subsequent procedure, annotated as *Maneuver Extraction*. The respective output is a, generally smaller, set of feasible paths that determines the maneuver catalog which will be proposed to the user. This catalog is basically a bit vector, indicating which of the standard maneuvers are feasible for execution at computation time. It is communicated to the external user in an abstract way (cp. Fig. 2, right) in order to demand a minimum

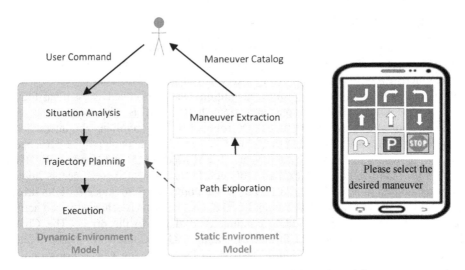

Fig. 2. Functional architecture of the framework (left) and illustration of the maneuver catalog presented to the external user (right). Parts of this illustration have been originally published in [11].

of interpretation. Calculated paths will be stored internally in order to serve as an initial solution for the actual trajectory planning that will happen once the external user has selected one maneuver for execution. During the trajectory planning, the initial solution is refined and a temporal planning is added, yielding velocity and acceleration set points for the execution. Here, we only address the procedure of generating the maneuver catalog (cp. Fig. 2, left).

Paths proposed to the external user as maneuver catalog should be drivable on the one hand – concerning vehicle dynamics and collisions with the static environment – and should mimic paths that a human driver would have chosen on the other. In example, a path should not end directly in front of a wall or be curvier than necessary. The problem of path exploring is closely related to the classical path planning in unstructured environments, that asks for a collision-free path from starting pose A to goal pose B through an arbitrary obstacle map. A basic path planning problem, the so called piano mover's problem, has been found to be PSPACE-hard [12]. Accordingly, path planners typically suffer from a rapid increase of computational complexity with growing problem size (e.g. size of the spatial domain and number of degrees of freedom of the robot). Unlike the classical problem setup, no goal state is given in case of the path exploration task. A suited algorithm not only needs to calculate a feasible path from A to B, but also needs to choose B considering the given environment representation. Moreover, in our case B is not restricted to be a single goal state, but generally is a set of "meaningful" goal states. "Meaningful" could be assessed by taking the goal B and the environment representation into viewpoint in order to evaluate if B is a suited target. Another possibility, more beneficial for the use case addressed here, is to additionally incorporate the path leading to B into the assessment. E.g. a goal that is not reachable from the ASV's position cannot be "meaningful". The path exploration method presented in Sect. 2.4 takes up the latter paradigm computing a high number of path prior to any evaluation, where the alternative method, presented in Sect. 2.3, takes up the first by choosing a set of goal states and subsequently performing a classical path planning to each of the determined goals.

From intuition, path exploration appears to be more computational demanding than a single path search towards a given goal. This is reasonable, if one assumes the same quality requirements on obtained path(s). Nevertheless, for the here addressed use case, a path obtained from the exploration is not supposed to be directly executed, respectively will not directly passed to the control loops of the ASV, but serves as an initial solution. Therefore, quality requirements can be relaxed, compared with common path planning tasks in autonomous driving (Fig. 3).

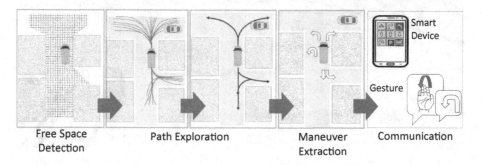

| Free Space Detection | Path Exploration | Maneuver Extraction | Communication |

Fig. 3. Illustration of the processing steps in the interaction framework. This illustration was originally published in [11].

2.2 Environment Model for Path Exploration

The environment model used for exploration of possible paths aims at capturing the static environment. The temporal planning dimension will be handled by the ASV without a possibility of user interaction after a user choice has been committed. In the following, we address the construction of the environment model in absence of radar sensors on basis of LiDAR sensors. This allows an evaluation of implemented algorithm on the public available Kitti dataset [13].

Occupancy Grid Map
The static environment is represented as a 2D occupancy grid, composed of quadratic grid-cells m_i. Each cell is updated according to new sensor data by individual Bayes Filter updates as described in [14]

$$l_{k,i} = l_{k-1,i} + \log \frac{p(m_i|z_k, x_k)}{1 - p(m_i|z_k, x_k)} - \log \frac{p(m_i)}{1 - p(m_i)} \tag{1}$$

with

$$l_{k,i} = \log \frac{p(m_i|z_{1:k}, x_{1:k})}{1 - p(m_i|z_{1:k}, x_{1:k})} \tag{2}$$

where $l_{k,i}$ is the log-odds representation of the occupancy probability at time step k. z_k denotes the range measurements and x_k denotes the FAV's position, which is assumed to be known by measurements (Fig. 4).

Fig. 4. Visualization of the free space detection in near range. Brighter color on the grid denotes a lower occupancy probability. Colored points represent the raw point cloud. The orange box represents a moving bicycle, causing a trace in the occupancy map. Raw data is taken from the Kitti dataset [13]. (Color figure online)

With obtaining range measurements x_k from LiDAR sensors, the algorithm cannot distinguish whether a measurement results from a dynamic or static object. Therefore, moving objects that travel through the free-space can cause a slight increase of occupancy probabilities in travelled-through areas. In order to prevent footprints of moving objects, an IMM-UK-PDA[1] filter can be deployed. Due to an extensive set of formulas, details shall be omitted here with referring to [15]. In each iteration 'newly-free' and 'newly-occupied' grid cells can be identified by observing a de- or increase of occupancy probability. It is beneficial to determine this change by observing the discrete derivative of the occupancy probability $p_{occ,i,k}$ in non-log form at a grid cell i at time step k

$$p_{occ,k,i} = \frac{1}{1 + e^{-l_{k,i}}}, \quad p_{occ,k,i} \in [0, 1] \tag{3}$$

$$\Delta p_{occ,k,i} = \frac{p_{occ,k,i} - p_{occ,k-1,i}}{T} \tag{4}$$

---

[1] Interacting-Multiple-Model Unscented Kalman Probabilistic Data Association.

where T denotes the sampling time. Equation (3) describes a sigmoid function, therefore the derivative is highest at $l_{k,i} = 0$ and decays to both the positive and negative direction. At this point, the occupancy probability is 0.5, therefore the occupancy is undecided. This way, an activation threshold can be defined, that damps noise effects for the detection. By making geometrical considerations, newly-changed cells can be merged and enclosed by a rectangular bounding box, which is treated as a potential dynamic object with states $(x_k, y_k, \theta_k, l_k, w_k)$ and gets tracked over time. Here x, y and θ define the pose of the bounding box w.r.t. the vehicle coordinate frame, and l, w denote length and width. The orientation of the bounding box is known by having cells classified as newly-free and newly-occupied. The dynamic object hypothesis is investigated by filtering the observations over time with help of the IMM-UK-PDA filter. This filter incorporates more than one motion model in underlying Kalman Filters. The respective predictions are combined with help of mode probability values $\mu_{j,k}$ for each deployed motion model $j \in [1, \ldots, N]$. These act as weights

$$\sum_{j=1}^{N} \mu_{j,k} = 1 \tag{5}$$

$$\widehat{x}_k = \sum_{j=1}^{N} \mu_{j,k} \widehat{x}_{j,k} \tag{6}$$

$$\widehat{P}_k = \sum_{j=1}^{N} \mu_{j,k} \left[\widehat{P}_{j,k} + \left(\widehat{x}_{j,k} - \widehat{x}_k \right) \left(\widehat{x}_{j,k} - \widehat{x}_k \right)^T \right] \tag{7}$$

where $\widehat{x}_{j,k}$ denotes predicted states of the Unscented Kalman Filter for motion model j and $\widehat{P}_{j,k}$ denotes the corresponding covariance matrices. $\widehat{x}_k$, $\widehat{P}_k$ denote the combined predicted states. The mode probabilities $\mu_{j,k}$ therefore indicate the relevance of each deployed motion model for the common filter output. Probabilities $\mu_{j,k}$ are updated in each filter step by likelihood considerations on measurements in comparison with predicted states. By deploying a noise model among the motion models, the corresponding mode probability can be used as an indicator to evaluate the dynamic object hypothesis and allow a classification of the tracked object as static or dynamic.

Obtaining the Euclidean Distance Transform and an Artificial Potential Field

Given the occupancy grid map cleared from dynamic influence, two additional representations can be efficiently obtained, that will be helpful for the path exploration. Given the occupancy map m, a binary representation $\widehat{m}$ can be constructed by thresholding the log odds. Subsequently, the Euclidean Distance Transform (EDT)

$$EDT_{\widehat{m}}(p) = \min_{q \in \mathbf{m}, \widehat{m}(q)=1} \|p - q_2\| \tag{8}$$

can be obtained, where $p, q \in m$ denote cell-centered grid points. The EDT stores the distance to the nearest occupied cell in meters for any grid point p. It can i.e. be used for efficient collision checks. A linear-time algorithm is given in [16]. Given the EDT, an artificial potential field $\boldsymbol{a}$ can be obtained as the logarithmic inverse distance transform

$$\boldsymbol{a}(p) = \log\left(\frac{1}{EDT_{\widehat{\boldsymbol{m}}}(p)}\right) \tag{9}$$

Different from the EDT, the potential field $\boldsymbol{a}$ does not need to be computed explicitly, but can be evaluated on query (Fig. 5).

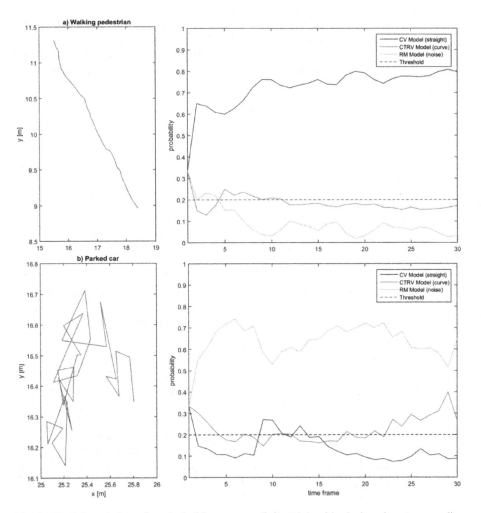

Fig. 5. Spatial evolution of tracked object centers (left). Right side depicts the corresponding time evolution of the mode probabilities. Besides the noise model two constant velocity models have been applied for straight and curved motion as described in [15].

2.3 Path Exploring by Modified RRT

The original RRT algorithm introduced in [17] as well as its extension to RRT*, yielding asymptotic optimality [18], is a randomized search that iteratively extends a search tree, where each node in the search tree corresponds to a state in the ASV's configuration space spanned by position x, y and orientation θ. Respectively, a path from the root to a leaf node corresponds to a drivable path. It is typical for randomized path search algorithms, that the quality of the solution path (e.g. the shortest path leading to a given goal position) grows with the number of iterations, due to a deeper exploration of the search space. A large variety of extensions and modifications have been developed since the original RRT algorithm has been proposed, which typically aim at increasing the quality of the solution path with respect to the runtime, often by goal-biased sampling techniques. Nevertheless, the RRT algorithm does not explicitly require a goal position to operate, which makes it particularly suitable for the here addressed use case. The modified RRT presented here aims at generating a large number feasible paths quickly, rather than aiming for a single high-quality solution path. The corresponding modifications to the sampling function, nearest node search and extend function of the original RRT are addressed in the following. New samples are generated in a polar region around the ASV as illustrated in Fig. 6. Once the environment model has been updated, several computing threads can work in parallel on the path exploration. Assuming five available workers, groups of two computing units can be deployed on separate search trees, depicted as green and red in Fig. 6. The purpose of the artificial initial position is to obtain forward drive paths from the green search area and reverse drive paths from the red search area. Within both computing groups, each of the two workers can work on expanding the same search tree. This setup is a combination of two schemes referred to as *AND-Parallelization* and *OR-Parallelization* [19]. The fifth remaining worker can simultaneously compute a turn-around maneuver (cp. Fig. 6, left). The target state of the turnaround is the artificial initial state of the red search tree. A successful search therefore allows additional maneuver proposals by combining the turnaround with the red search tree. The here presented modified RRT is not suited to compute the turnaround, which can be done by i.e. a Hybrid A^*-Search. Once a sample has been generated, the nearest node in the persistent search tree is taken into consideration, which can be efficiently identified by deploying a query on a three dimensional Kd-Tree. The respective dimensions are $\left(x, y, \frac{\vartheta}{\kappa}\right)$, where ϑ denotes the summated absolute steering effort that will be defined at a later point. A large value of ϑ indicates, that the path leading to the respective node requires a high steering effort. This setup causes the nearest node search to prefer nodes associated to lower steering effort if sufficiently close to the sample point in an Euclidean sense, where $\kappa \in \mathbb{R}$ is a scaling parameter. The modified extension step avoids solving the arising two-point boundary value problem of the RRT's extension function by only using the sample point for an Euclidean distance heuristics for a limited number of forward-simulations. The simulation deploys a half car model $\dot{x} = f(x, \delta)$ with state vector $x = [\beta \dot{\varphi} \varphi x y]$ denoting slip angle β, heading rate $\dot{\varphi}$, heading

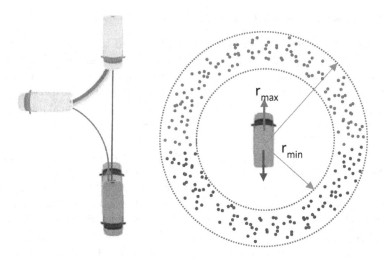

Fig. 6. Right: Illustration of the polar sampling region. The colors red and green indicate separate RRT threads for the current pose of the FAV (green) and an artificial pose rotated by π (red) for the second thread. *Left:* Illustration of a turnaround maneuver connecting the real and artificial pose of the FAV. The illustration was previously published in [11]. (Color figure online)

angle φ, position coordinates x, y and input variable δ representing the steering angle. The initial state vector for the extension x_0 and initial steering angle δ_0 are given as stored properties of the previously identified nearest node. (x_s, y_s) are coordinates of the sample point. Starting from the initial state, a predefined set of m steering angle increments $\Delta\delta_l$ is applied, each of them starting at the initial state and adding the increment on the steering angle in each integration step. The integration is aborted if a maximum number of n integration steps is exceeded, a sufficiently close neighborhood of the sample point is reached or if the respective extension path leads into a collision (Fig. 7).

Each integration state is evaluated by the cost function

$$J^i(x_k, x_s, y_s, k) = e(x_k, x_s, y_s) + \alpha \cdot a(\lfloor x_k \rfloor) + \beta \cdot \sigma^i(k) \tag{10}$$

with squared Euclidean distance

$$e(x_k, x_s, y_s) = (x_k - x_s)^2 + (y_k - y_s)^2 \tag{11}$$

the value of the artificial potential field a (cp. Eq. 4) evaluated for the grid point determined by rounding the position to the nearest integer coordinates

$$a(\lfloor x_k \rfloor) = a_i, \quad i \leftarrow (\lfloor x_k \rfloor, \lfloor y_k \rfloor) \tag{12}$$

```
              EXTEND (x_0, δ_0, x_s, y_s)
1             for l = 1:m
2                 x_{0,l} = x_0
3                 δ_{0,l} = δ_0
4                 for k = 1:n
5                     δ_{k,l} = δ_{k-1,l} + Δδ_l
6                     x_{k,l} ← f(x_{k-1,l}, δ_{k,l})
7                     if (collision(x_{k,l})) {
8                         C_{k,l} = inf
9                         break
10                    } else {
11                        C_{k,l} ← J^i(x_{k,l}, x_s, y_s, k)
12                    }
13                    if (reach(x_{k,l}, x_s, y_s, ε)) break;
14                end for
15            end for
16            i ← min(C_{k,l})
17            addpath(x_{1:k,l}, δ_{k,l})
```

Fig. 7. The extension function of the modified RRT formulated as pseudo-code. *Collision* incorporates the obstacle check on the EDT of the occupancy grid map (cp. Eq. 8). *Reach* evaluates if the current state reaches an ε-neighbourhood of the sample point. *Addpath* adds the state with lowest cost to the search tree.

and the cumulated absolute steering effort

$$\sigma^i(k) = k \cdot |\Delta\delta_i|. \tag{13}$$

$\alpha, \beta \in \mathbb{R}$ are weighting parameters. The earlier mentioned summated absolute steering effort ϑ used in the nearest node search is updated in each integration step as

$$\vartheta_k = \vartheta_0 + \sigma^i \tag{14}$$

with ϑ_0 the summated absolute steering effort at the start node of the extension. Updating ϑ according to Eq. (14) has the effect, that each node persistent in the search tree stores the steering effort that is needed to reach the respective node from the root.

2.4 Path Exploring by Skeletonization

The basic idea of the approach in this subchapter is to choose reasonable goal positions from junctions in the skeleton representation of the free-space, and to subsequently search for a path leading to this goal region. The skeleton of an image leaves a 'thin version' of the shape contained in the original image while representing certain geometric and topologic properties. In order to calculate the skeleton, the occupancy grid is converted into a binary image by thresholding, where unexplored areas are considered

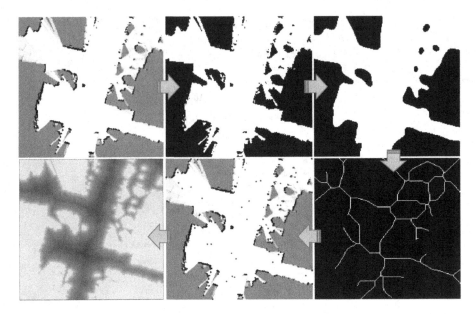

Fig. 8. The original occupancy map (top left), the map after thresholding (top mid) and the map after filtering (top right). Further, the resulting skeleton (bottom right) and the obtained goal points in red for junctions and blue for endpoints (bottom mid). The orange errors indicate the processing sequence of the goal point generation. The last step indicates an evaluation of each detected point on the artificial potential field (bottom left) by an evaluation of Eq. (9). (Color figure online)

as occupied, and subsequently blurred by a white Gaussian noise filtering. From the resulting skeleton representation, pixel coordinates of junctions and endpoints can be obtained by applying an image based edge detection. Subsequently, detected points are deleted if the associated value in the potential field (cp. Eq. 9) exceeds a threshold.

The skeletonization can alternatively applied to a binary representation of the artificial potential field by thresholding the cost value. This promises better results, but requires the potential field to be computed explicitly, rather than only evaluating on point-queries. A path search to each of the obtained goal positions results in too much computation effort. Therefore, an additional selection of points should be applied as will be addressed in the outlook. A path to each final selected goal point can be computed by i.e. applying a Hybrid A* approach. Hereby, a post-smoothing procedure can be omitted for the path exploration task.

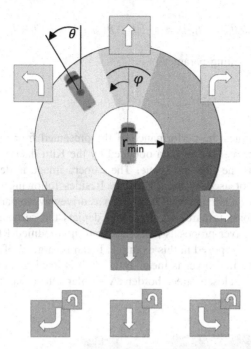

Fig. 9. Illustration of polar channels for classification of paths to maneuver classes. Path endpoints are mapped to the respective channel as illustrated (cp. red point). The best endpoint within a channel is determined by considering the deviation from an ideal orientation (cp. direction of illustrated arrows) and the summated absolute steering effort (cp. Eq. 14) associated to the path endpoint. The illustration was previously published in [11]. (Color figure online)

2.5   Maneuver Extraction

The goal of the maneuver extraction is to compare this set of candidate paths obtained from exploration to a predefined maneuver set in order to determine which of the maneuvers can be executed and therefore can be proposed to the external user in the maneuver catalogue (cp. Fig. 2). To this end, each path endpoint obtained from the path exploration is investigated. A candidate path endpoint (x, y, θ) is mapped to the corresponding channel index, which identifies the respective maneuver class. The endpoint is kept, if the orientation stays within channel associated limits. If more than one path is mapped to a maneuver class, the best path is determined according to a cost J for an endpoint i mapped to channel j, representing a trade-off between deviation from an ideal endpoint orientation (cp. Fig. 9), the summated absolute steering effort (cp. Eq. 14), and length of the path leading to the endpoint l_i

$$J(\theta_i, \vartheta_i, l_i, j) = \tau \cdot \left(\theta_i - \theta_{j,ideal}\right)^2 + \mu \cdot \vartheta_i + l_i \tag{15}$$

with $\tau,\ \mu \in \mathbb{R}$ as weighting parameters.

3 Results

This chapter investigates the performance of the presented framework by presenting simulation results for real sensor data obtained by the Kitti dataset [13].

Figure 10 depicts the first test-scene. The camera image is depicted for a better understanding, but not used by the algorithm. Besides following the road, the scene offers opportunities of a left and right-turn given as driveways to parking lots as well as enough free-space for a turnaround. Figure 11 depicts the corresponding occupancy grid, as well as maneuver proposals resulting from the modified RRT. All predefined maneuver classes are proposed in this example. It can be argued, if the proposal of the reverse-drive left turn maneuver is meaningful, as it is leading into a gap between to parking vehicles with closed street border. A similar effect can be observed in the second test-scene (Fig. 12).

Fig. 10. Camera image (top) and raw point cloud (bottom) of the first test-scene.

Fig. 11. Explored paths (red) and extracted paths (black). The grid map has 400×400 cells with a cell length of 0.25 m resulting in a search space of 100×100 m. The maximum radius for the polar sampling region of the modified RRT is 45 m. The ego-vehicle is centered. (Color figure online)

The second scene represents an urban city scenario. As in the previous scene, it can be argued if the proposal of the forward-drive left turn maneuver is meaningful. In case of the forward-drive right turn maneuver, a human would be able to predict from the camera image, that there will not be an opportunity to perform the right turn. From the perspective of laser scan data, or more precisely from the projection on a 2D grid, this is not clear. Due to the person right in front of the FAV, the area behind this person is considered unknown. The modified RRT explores a right turn, as unexplored space is considered as free-space. Note that this assumption was beneficial in case of the previous test-scene. A further interesting observation can be made concerning the reverse-drive right turn maneuver. It is of the same arguable type of previous mentioned proposals, but coincidently might appear reasonable due to the small alley branching

Fig. 12. Camera image (top) and raw point clouds with occupancy map (mid, bottom) of the second test-scene.

Fig. 13. Explored paths and maneuver proposals for the second test-scene.

the main street. The here applied modified RRT lacks probabilistic completeness, which results in missing maneuver opportunities even for infinite runtimes. This is a result of avoiding the solution of a two-point boundary value problem in the modified extension function of the RRT, resulting in reasonable runtimes despite the absence of a given goal. In this test-scene, the small alley is not being explored. It shall be investigated, if the path exploration by skeletonizing performs better (Fig. 13).

Figure 14 depicts goal points obtained from the skeleton. The grid-size has been reduced to 200×200 cells corresponding to 50×50 m, as the computational effort of a subsequent path search increases fast with growing size of search domain. Different from the RRT, the alley will be explored successfully. Also note that no goal point causing a right turn proposal has been determined. This is a result from considering

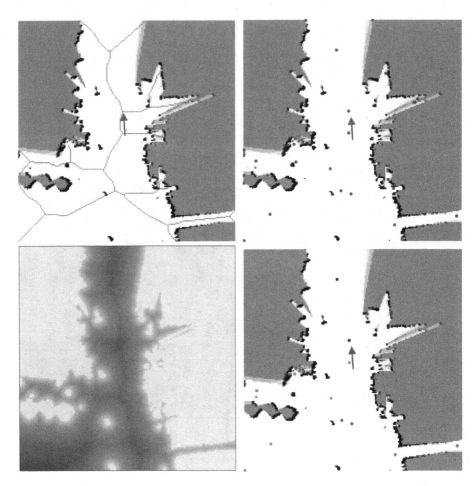

Fig. 14. Skeleton on the occupancy grid (top left) and goal points obtained from corner detection (top right). Goal points are evaluated on the artificial potential field (bottom left) and survive with a cost of −1.0 or lower. Note that 0 is the maximum value of the potential field (cp. Eqs. 8 and 9). Bottom right depicts final goal points. The colors denote association to a channel on the polar grid (cp. Sect. 2.5). (Color figure online)

unexplored areas as occupied (cp. Fig. 8). Further, no left turn will be considered. Nevertheless, the number of obtained goal points is quite high resulting in a high computational effort. The number should be further reduced, which will be further addressed in the outlook. Test-scene three depicts a situation, in which the road is blocked by a temporarily parking vehicle. The straight forward-drive maneuver is still proposed as an overtake maneuver. This indicates that the modified RRT is still beneficial compared to e.g. deploying motion primitives (Fig. 15).

Fig. 15. Point cloud of test-scene three (top) and maneuver proposals (bottom).

4 Conclusion and Outlook

A cooperative interaction framework has been presented, that allows a robot, such as
the ASV, to communicate feasible planning alternatives. The described approach is a
bottom-up approach, in which possible paths are computed prior to a decision. This
way, the decision can be made or negotiated by or with an external user. The modified
RRT approach deployed for the path exploration of the presented framework lacks
probabilistic completeness. Further, it cannot differentiate between meaningful goal
regions, that a human driver would have chosen, and target destinations that are not
worth to propose, at least cannot up to a satisfying extend. The skeletonizing approach

suffers from a similar property, especially being vulnerable against sharp shapes of free-space boundaries. Further, the number of goal points to be checked by individual path search queries might be high. In order to further reduce the number, a construction of a road graph based on the skeleton might be helpful. This way, rule based selection rules could be applied on the graph structure, aiming at a selection of most relevant targets. Both methods highly depend on parametrization. In order to gain more parameter independence, more robustness in meaningful maneuver proposals especially for urban scenarios, as well as adopting more maneuver proposals such as parking opportunities, the algorithm framework should incorporate semantic environment information. This information could be extracted from a digital map, or directly from the perception of the ASV. Addressing the latter, we aim to incorporate e.g. a pixel-wise semantic segmentation of the camera image, in order to obtain a more distinct detection of the drivable space. Besides that, we pursue the idea of deploying a convolutional neuronal network for proposing meaningful goal points.

References

1. SAE International: Taxonomy and definitions for terms (2014)
2. Wachenfeld, W., et al.: Use-cases des autonomen Fahrens. In: Maurer, M., Gerdes, J.C., Lenz, B., Winner, H. (eds.) Autonomes Fahren, pp. 9–37. Springer, Heidelberg (2015). https://doi.org/10.1007/978-3-662-45854-9_2
3. Tsai, C.-C., Hsieh, S.-M., Hsu, Y.-P., Wang, Y.-S.: Human-robot interaction of an active mobile robotic assistant in intelligent space environments. In: International Conference on Systems, Man and Cybernetics, pp. 1953–1958. IEEE (2009). 978-1-4244-2794-9/09/$25.00 ©2009
4. Flemisch, F., Adams, C., Conwary, S., Goodrich, K., Palmer, M., Shutte, P.: The H-Metaphor as a guideline for vehicle automation and interaction, Virginia, NASA/TM-2003-212672 (2003)
5. Kienle, M., Damböck, D., Kelsch, J., Flemisch, F., Bengler, K.: Towards an H-Mode for highly automated vehicles: Driving with side sticks. In: 1st International Conference on Automotive User Interfaces and Interactive Vehicular Applications Automotive, UI 2009 (2009)
6. Hakuli, S., Kluin, M., Geyer, S., Winner, H.: Development and validation of manoeuvre-based driver assistance functions for Conduct-by-Wire with IPG CarMaker, Budapest. In: FISITA 2010 World Automotive Congress (2010)
7. Geyer, S.: Maneuver-based vehicle guidance based on the Conduct-by-Wire principle. In: Maurer, M., Winner, H. (eds.) Automotive Systems Engineering, pp. 111–132. Springer, Heidelberg (2013). https://doi.org/10.1007/978-3-642-36455-6_6
8. Geyer, S.: Entwicklung und Evaluierung eines kooperativen Interaktionskonzepts an Entscheidungspunkten für die teilautomatisierte, manöverbasierte Fahrzeugführung, Dissertation. Institute for Automotive Engineering, Technische Universität Darmstadt (2015)
9. Lotz, F., Winner, H.: Maneuver delegation and planning for automated vehicles at multi-lane road intersections. In: 17th International Conference on Intelligent Transportation Systems (ITSC). IEEE (2014)
10. Hosseini, A., Wiedemann, T., Lienkamp, M.: Interactive path planning for teleoperated road vehicles in urban environments. In: IEEE 17th International Conference on Intelligent Transportation Systems (ITSC) (2014)

11. Sefati, M., Gert, D., Kreiskoether, K.D., Kampker, A.: An interaction framework for a cooperation between fully automated vehicles and external users in semi-stationary urban scenarios. In: 3rd International Conference on Vehicle Technology and Intelligent Transport Systems (2017)
12. Reif, J.H.: Complexity of the mover's problem and generalizations. In: Proceedings of the IEEE Symposium on Foundations of Computer Science (1979)
13. Geiger, A., Lenz, P., Stiller, C., Urtasun, R.: Vision meets robotics: the KITTI dataset. Int. J. Robot. Res. (IJRR) **32**, 1231–1237 (2013)
14. Thurn, S., Burghard, W., Fox, D.: Probabilistic Robotics. MIT Press, Cambridge (2006)
15. Schreier, M., Willert, V., Adamy, J.: Compact representation of dynamic driving environments for ADAS by parametric free space and dynamic object maps. Trans. Intell. Transp. Syst. **17**(2), 367–384 (2016)
16. Felzenszwalb, F., Huttenlocher, P.: Distance transforms of sampled functions. Theory Comput. **8**, 415–428 (2012)
17. Lavalle, S.M.: Rapidly-exploring random trees: a new tool for path planning (1999)
18. Karaman, S., Frazzoli, E.: Sampling-based algorithms for optimal motion planning. Int. J. Robot. Res. **30**(7), 267–274 (2011)
19. Carpin, S., Pagello, E.: On parallel RRTs for multi-robot systems. In: Proceedings of the 8th conference of the Italian Association for Artificial Intelligence, pp. 834–841 (2002)

A Personal Robot as an Improvement to the Customers' In-store Experience

António J. R. Neves[1,2]([✉]), Daniel Campos[1], Fábio Duarte[1], Filipe Pereira[1], Inês Domingues[1], Joana Santos[1], João Leão[1], José Xavier[1], Luís de Matos[1], Manuel Camarneiro[1], Marcelo Penas[1], Maria Miranda[1], Ricardo Silva[1], and Tiago Esteves[1]

[1] FollowInspiration, S.A., MACB, Zona Industrial do Fundão, Lote 154, 6230-348 Fundão, Portugal
[2] IEETA/DETI, University of Aveiro, Campus Universitário de Santiago, 3810-193 Aveiro, Portugal
an@ua.pt

Abstract. Robotics is a growing industry with applications in numerous markets, including retail, transportation, manufacturing, and even as personal assistants. Consumers have evolved to expect more from the buying experience, and retailers are looking at technology to keep consumers engaged. There are currently many interesting initiatives that explore how robots can be used in retail. In today's highly competitive business climate, being able to attract, serve, and satisfy more customers is a key to success. A happy customer is more likely to be a loyal one, who comes back and often to the store. It is our belief that smart robots will play a significant role in physical retail in the future. One successful example is wGO, a robotic shopping assistant developed by FollowInspiration. The wGO is an autonomous and self-driven shopping cart, designed to follow people with reduced mobility in commercial environments. With the Retail Robot, the user can control the shopping cart without the need to push it. This brings numerous advantages and a higher level of comfort since the user does not need to worry about carrying the groceries or pushing the shopping cart. The wGO operates under a vision-guided approach based on user-following with no need for any external device. Its integrated architecture of control, navigation, perception, planning, and awareness is designed to enable the robot to successfully perform personal assistance while the user is shopping. This paper presents the wGOs functionalities and requirements to enable the robot to successfully perform personal assistance while the user is shopping in a safe way. It also presents the details about the robot's behaviour, hardware and software technical characteristics. Experiments conducted in real scenarios were very encouraging and a high user satisfaction was observed. Based on these results, some conclusions and guidelines towards the future full deployment of the wGO in commercial environments are drawn.

Keywords: Robotics · Retail · Reduced mobility
Requirements · Functionalities

© Springer Nature Switzerland AG 2019
B. Donnellan et al. (Eds.): SMARTGREENS 2017/VEHITS 2017, CCIS 921, pp. 296–317, 2019.
https://doi.org/10.1007/978-3-030-02907-4_15

1 Introduction

In recent years, a high concern with user satisfaction has been observed in the retail industry. This is particularly accentuated with the rise in on-line shopping which pushes retailers to provide a better in-person shopping experience to attract customers. Among customers in the public, one of the main groups of interest are people with disabilities (the elderly, people in wheelchair, pregnant women, those with temporary reduced mobility, etc.). This is visible not only in the marketing strategies but also at the political level, where accessibility for disabled people is becoming the topic of regulation and legislation.

It is estimated that in Portugal about 8% to 10% of the population has some form of disability [1], and that in Europe alone there are about 50 million people with disabilities and 134 million people with reduced mobility. Apart from people using wheelchairs, there are other cases in which people are temporarily or permanently disabled, these include: an elderly person using a cane, or someone with a foot or leg injury who requires the use of crutches, pregnant ladies and parents with prams.

In fact, if we add the disabled, the elderly, pregnant women, and couples with children, we find that between 30% to 40% of all Europeans could benefit from improved accessibility. In addition to those people with reduced mobility due to disability or injury, there are many people without mobility issues who could benefit from assistance in carrying heavy bags. Shopping environments are highly heterogeneous and give rise to a high frequency of dynamic interactions that trigger various senses and emotions in humans. This often causes a high level of stress in people, and those with mobility limitations. Some of the identified difficulties include [2]:

- People who use wheelchairs;
- no adequate forward reach at basins, counters and tables;
- surfaces that do not provide sufficient traction (e.g. polished surfaces);
- People who have trouble walking;
- no seating in waiting areas, at counters and along lengthy walkways;
- access hazards associated with doors, including the need to manipulate a handle while using a walking aid;
- surface finishes that are not slip-resistant or are unevenly laid.

Besides the difficulties brought by the shopping environment itself, conventional shopping carts, which can carry many products and which are provided with wheels so that the shoppers can push them, also have serious drawbacks. One of them being their considerable size. This is simultaneously an important asset and a significant drawback, as although shopping carts can hold large and bulky products, the increased mass complicates manoeuvrability and handling.

Manoeuvrability is particularly compromised when making turns in supermarket aisles or when avoiding other carts, shelves, and indeed other shoppers [3]. Smaller baskets appeared on the market to overcome the traditional shopping cart's drawbacks. These baskets were developed to hold a set of items while at the same time being easy to move. They contain wheels or rolling elements incorporated into the bases which allow them to be moved when parallel to the floor or when inclined. However, even though these baskets improve manoeuvrability due to their reduced size and capacity, they also have drawbacks typical of their morphology, such the need for the user to bend down for placing or removing items, among others.

Furthermore, such baskets can have drawbacks typical of the way they are stored, since stacking them vertically can entail a problem for elderly shoppers or shoppers with any type of physical limitation [3].

With these described difficulties in mind, this paper presents a new robotic concept to help and assist people (giving special emphasis to people with reduced mobility) in these type of environments, through a user-following scenario. It describes also the design concerns and decisions taken into consideration during the development of the robot.

The wGO, presented in Fig. 1, is an autonomous and self-driven shopping cart, designed to follow people with reduced mobility (elderly, people in wheelchairs, pregnant women, temporary reduced mobility, etc.) in commercial environments [4]. With the robot, the user can control the shopping cart without the need to push it. This brings numerous advantages and a higher level of comfort, since the user does not need to worry about carrying the groceries or pushing the shopping cart.

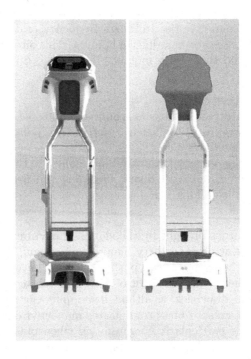

Fig. 1. wGO: on the left the front view; on the right the back view [4].

Internally, the robot is divided into several modules: Sensors, Vision, Behaviour, Executing, and Control system. The Sensors module receives data from the sensors and verifies the existence of obstacles. The Vision module acquires and processes RGB and Depth information (people detection, identification, and false positive reduction). The Behaviour module includes the tracking of the user and the generation of the path for the robot to follow. The Executing system receives the generated path and the obstacle detection information and, depending on the desired action for the robot, sends commands to the Control module that moves the robot along the defined path.

The wGO is designed to have an ergonomic shape, friendly both to the target users (people with reduced mobility) and the environment (e.g. a commercial retail environment). Its operation is guided by three different types of sensors: ultra-sound, Laser Range Finder (LRF), and active vision. This combination was selected due to their complementary features.

Experiments conducted in real scenarios were very encouraging and a high user satisfaction was observed. Comments like "My shopping was very fast!", "In fact, it was a precious help!" and "I think it's awesome, I will certainly use and recommend it!" were made by participants.

The paper starts by reviewing the existing solutions for people with reduced mobility to shop (Sect. 2) and some of the relevant legislation related with technology equipments (Sect. 3). Design related requirements are given in Sect. 4, as well as a description of the robot's behaviour, hardware and software details. Next, in Sect. 5 the wGO design evolution and justifications for the changes made are given. Section 6 identifies the main risks related with the wGO usage in a real scenario. Section 7 analyses some technical results and a user satisfaction study made in a relevant, unconstrained scenario. Conclusions, future work and other applications of this technology are given in Sect. 8.

2   Existing Solutions

Looking at the commercial market, the most obvious existing solutions are those provided by shopping cart producers[1]. These providers typically have products targeted for customers in wheelchairs, but not products for other types of users with reduced mobility (e.g. pregnant women). A different type of solution is the adapted system. Some examples are the "amigo mobility" scooter[2] and adapted wheelchairs[3], etc. These products are, however, not particularly user friendly. The user needs to first move into the mobility auxiliary device and then to learn how to use it (which may be particularly hard for the scooter case). In the case of wheelchair users, the user also needs to leave their own personal chair, which may cause discomfort and unnecessary stress. Another problem with these solutions is that the user is visibly distinguishable from the other supermarket clients, which may discourage some people from using it [5].

---

[1] E.g. wanzl (www.wanzl.com).

[2] www.myamigo.com.

[3] E.g. meyra (www.meyra.de), promoted by Egiro (www.egiro.pt).

While this topic of assisted shopping using robotics has received very little attention in the academic research community, several systems exist where robots are used to help people with reduced mobility.

In [6], an anticipative shared control for robotic wheelchairs, targeted at people with disabilities is presented. The same idea, of intelligent wheelchairs, is also the focus of the work in [7] where a data analysis system which provides an adapted command language is presented. A smart companion robot for elderly people, capable of carrying out surveillance and tele-presence tasks, is described in [8]. The work in [9] presents an analysis of the implementation of a system for navigating a wheelchair with automation, based on facial expressions, especially eyes closed using a Haar cascade classifier, aimed at people with locomotor disability of the upper and lower limbs.

A smart companion robot for elderly people, capable of carrying out surveillance and tele-presence tasks, is described in [8]. Also, with the aim of helping elderly people through tele-presence, a low-cost platform capable of providing augmented reality for pill dose management was developed in [10]. In [11] an approach based on the Dynamical System Approach for obstacle avoidance of a Smart Walker device to help navigation of elderly people is presented.

Perhaps the closest application to the focus of this paper is presented in [12] where a product locator application is proposed. The application runs on heterogeneous personal mobile devices keeping the user private information safe on them, and it locates the desired products over each supermarket's map. We believe that such a system could be complementary to the wGO and could be used in combination to further improve customers' shopping experiences.

3   Existing Legislation

In a joint effort started in 1995, the United Nations Economic Commission for Europe (UNECE) and IFR, engaged in working out a preliminary service robot definition and classification scheme, which has been absorbed by the current ISO Technical Committee 184/Subcommittee 2 resulting in a novel ISO-Standard 8373 which became effective in 2012 [13].

There, a **robot** is an actuated mechanism programmable in two or more axes with a degree of autonomy, moving within its environment, to perform intended tasks. Autonomy in this context means the ability to perform intended tasks based on current state and sensing, without human intervention.

A **service robot** is a robot that performs useful tasks for humans or equipment excluding industrial automation application.

A robot system is a system comprising robot(s), end-effector(s) and any machinery, equipment, devices, or sensors supporting the robot performing its task [14].

Being an autonomous and self-driven shopping cart, designed to follow people with or without reduced mobility in commercial surfaces, wGO follows under the **service robot** category. Thanks to the sensors (RGBD cameras and LRF) wGO detects and identifies its user in less than 2 seconds, he just needs to push the "start" button and the wGO will start following him. Furthermore, distance sensors, RGBD cameras and LRF allow wGO to identify and avoid any obstacle along the way.

The international standard IEC 60950-1 gives the general requirements for the safety of Information technology equipments [15]. As stated, it is essential that designers understand the underlying principles of safety requirements in order that they can engineer safe equipment. Designers should take into account not only normal operating conditions but also likely fault conditions, consequential faults, foreseeable misuse and external influences.

The standard also assumes that users will not intentionally create a hazardous situation. The priorities depicted in Fig. 2 should be observed in determining what design measures to adopt.

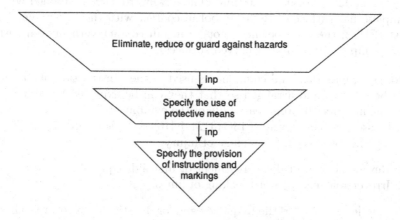

Fig. 2. Priorities for design measures to adopt [15]. inp stands for "if not possible".

The application of a safety standard is intended to reduce the risk of injury or damage due to: electric shock, energy related hazards, fire, heat related hazards, mechanical hazards, radiation or chemical hazards.

Here, we are mostly interested in the mechanical hazards, whose injuries may result from: sharp edges and corners, moving parts, equipment instability, flying particles.

Suggestions of measures to reduce risks include: rounding of sharp edges and corners, guarding, provision of safety interlocks, providing sufficient stability to free standing equipment. May also comprise selecting cathode ray tubes and high pressure lamps that are resistant to implosion and explosion respectively, and provision of markings to warn users where access is unavoidable.

Concerning Stability, the standard [15] states that under conditions of normal use, units and equipment shall not become physically unstable to the degree that they could become a hazard to an operator or to a service person. Compliance is checked by the following tests, where relevant. Each test is conducted separately.

During the tests, containers are to hold the amount of substance within their rated capacity producing the most disadvantageous condition.

- A unit having a mass of 7 kg or more shall not fall over when tilted to an angle of 10° from its normal upright position;
- A floor-standing unit having a mass of 25 kg or more shall not fall over when a force equal to 20% of the weight of the unit, but not more than 250 N, is applied in any direction except upwards, at a height not exceeding 2 m from the floor;
- A floor-standing unit shall not fall over when a constant downward force of 800 N is applied at the point of maximum moment to any horizontal surface of at least 125 mm by at least 200 mm, at a height up to 1 m from the floor. The 800 N force is applied by means of a suitable test tool having a flat surface of approximately 125 mm by 200 mm. The downward force is applied with the complete flat surface of the test tool in contact with the Equipment Under Test (EUT); the test tool need not be in full contact with uneven surfaces (for example, corrugated or curved surfaces).

Moreover, materials and components used in the construction of equipment should be selected and arranged so that they can be expected to perform in a reliable manner for the anticipated life of the equipment.

Risk assessment as given in ISO14121-1 [16] is depicted in Fig. 3. The first level concerns the severity of damage of injury:

- S1: Reversible, e.g. medical treatment or first aid required;
- S2: Irreversible, e.g. loss or breaking of limbs.

The second level describes the frequency and/or duration of exposure to hazard:

- F1: 1 day up to 2 weeks; 2 weeks up to 1 year;
- F2: Less than 1 h; 1 h up to 1 day.

The last level is the possibility of avoiding the hazard

- A1: Possible, probable;
- A2: Impossible.

4   Design Requirements and Specifications

The wGO is designed to have an ergonomic shape, friendly both to the target users (people with reduced mobility) and the environment (commercial retail environment). Several of the robot desired functionalities and requirements were taken into consideration:

- The robot should not have sharp edges;
- The bag should be accessible to every type of user;
- Product placement in the bag should be easy;

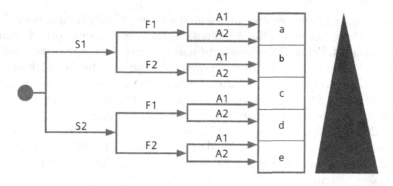

Fig. 3. Risk assessment as per [16].

- The robot should be able to carry at least 20 kg;
- The Start/Stop button should be well identified and accessible;
- The Emergency button should be well identified and accessible;
- There should be redundancy in the sensors;
- The robot should detect and follow people with 1.3 m height or more;
- The robot should be easy to clean.

Since not every type of user would be able to reach a touch screen (e.g. users in wheel chairs), a further requirement was specified that the wGOs operation should not depend on a touch interface. An illustration of the robot's hardware is shown in Fig. 4.

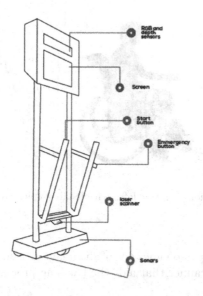

Fig. 4. wGO hardware illustration.

Its main internal sensors are: ultra-sound sensors, a Laser Rangefinder (LRF), and active vision sensors. This combination was selected due to their complementary features. While the ultra-sound sensor detects any type of material that is not sound absorbing, it has as main drawback the wide beam width and echo problems. LRF provides 270° information and its precision is high. It is, however, sensitive to dust. Active vision provides very rich information (image + 3D), but it has a relatively small field of view and low precision.

The ultra-sound sensors have a minimum and maximum ranges of 3 cm and 4 m respectively, and an estimated field of view of 60°. The LRF has a maximum range of 6 meters and preforms 270° laser scanning. The specifications of the active vision systems are:

- Range: 0.6 to 8 m (Optimal 0.6 to 5.0 m)
- Colour camera: 1280 × 960 at 10 FPS
- Depth camera: 640 × 480 (VGA) 16bit at 30 FPS
- Horizontal Field of view: 60°
- Vertical Field of view: 49.5°

4.1   Behaviour Description

The system is initialized when the user presses the start button. At this moment, the user is typically facing the wGO (Fig. 5). After initialization, the user starts shopping and the wGO follows him or her.

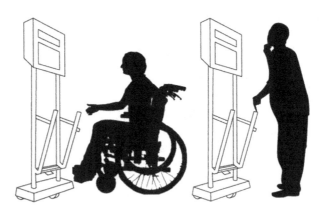

Fig. 5. wGO initialization: on the left, wheelchair typical case; on the right, non-wheelchair typical case.

In cases where the person goes out of the image sensors' field of view (Fig. 6), there is a 270° laser scanner that aids the tracking process.

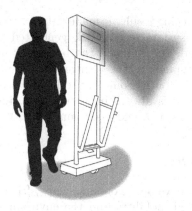

Fig. 6. Illustration of a case where the user is out of the field of view of the RGB and depth sensors, but visible by the 270° laser scanner. (Color figure online)

4.2 System Architecture

Figure 7 depicts the functional diagram of the application that is embedded in the robot, and is responsible for gathering information from the sensors, for example, the encoders and the RGB and depth cameras as well as controlling the movement of the robot. Therefore, this figure depicts a high-level representation of how perceptual data can be combined and used to enable a robot to follow a user in a realistic environment.

Internally, the application is divided into several modules: Vision, Sensors, Behaviour, Executing and Control system. The Vision module grabs and processes RGB and Depth information. In addition, the same module performs people detection [17], false positive reduction, and identification tasks. The Sensors module grabs data received from the sensors and verifies the existence of obstacles. The Behaviour module includes the tracking [18] of the detected person and the generation of the path [19] for the robot to follow.

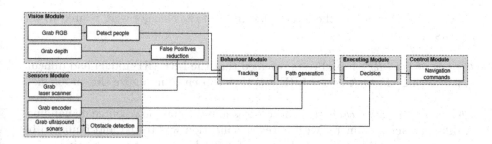

Fig. 7. wGO software flowchart.

In path generation, a local localization method based on odometry is used to retrieve the position of the robot and a route planning is performed. The fusion of the vision and laser tracking results is made [20] in order to perceives the user position. The Executing system receives the generated path and the obstacle detection information and, according to the behaviour and the desired action for the robot, sends commands to the Control [21,22] module that moves the robot along the pre-defined path. Some of the low-level navigation procedures, that ensure that the robot is always in a safe state, include: hardware fault identification, obstacle detection, and maximum velocity limitation.

5   Design Evolution

The wGO is designed to have an ergonomic shape, friendly both to the target users (people with reduced mobility) and the environment (commercial retail environment).

It has, however, undergone an evolution, as can be seen in Fig. 8.

The design on the left of Fig. 8 was the initial proposal (v0.1). From that version to the next (v0.2) several alterations were implemented, namely:

- LRF position was changed: It was noted that the different height would make visible more of the obstacles typically present in a commercial site. Moreover, stability of the support assembly easiness was increased.
- Smaller base: A smaller increases maneuverability. It cannot, however, be too small due to stability issues which might jeopardize safety.
- Emergency button was repositioned: The emergency button needs to be visible and of easy access. It cannot, however, be positioned in a place where it can be accidentally triggered. A new location was found that complies with these requirements.
- Changes to the electronic box: Some changes to the electronic box were performed in order to increase safety. For example, the box was electrically isolated.
- Redesign of the bag holder support: Angular vertices were smoothed in order to make the product less prone to accidents.

From that version to the next (v0.3) most of the changes were related to the sensors:

Change of the LRF: Initial tests revealed that the LRF (RP Lidar) had difficulties in seeing black. A better LRF was thus chosen to minimize this problem.

Change of the active camera: Kinect 2 was found to be too resource demanding. A different active camera, with lower specification was tested and found to be enough for our application, at the gain of being less resources consuming.

Inclusion of a Pan and Tilt system: A pan and tilt system was included so that the active camera responsible for performing tracking would better follow the user.

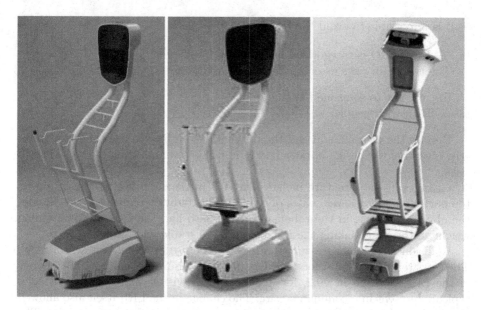

Fig. 8. wGO evolution [4].

Inclusion of two additional active cameras: Two active cameras were added in order to decrease the dead area and have a better obstacle avoidance performance.

Changes in the structure: The middle main structure was redesigned in order to be thinner (a sufficiently wide diameter to allow the required cables to pass through it) and less heavy.

Reorganization of the interior of the base: The base was redesigned in order to accommodate two batteries and thus increase the energetic efficiency of the robot.

Head redesign: The inclusion of the additional active cameras forced a complete redesign of the head of the robot in order to accommodate them.

6   The Risks

Safety is a critical characteristic for robots designed to operate in human environments. Looking back at the risk assessment as given in ISO14121-1 [16] and depicted in Fig. 3, we observe that:

– the severity of damage of injury is irreversible (e.g. breaking of limbs) - **S2**;
– the frequency and duration of exposure to hazard is in the order of 30 min, twice a week (typical duration and frequency of a shopping experience) - **F1**;

– it is impossible to avoid the hazard - **A2**.

In this way, wGO is categorized in the **d** level.
    The two main risks associated with the wGO are:

– To turn due to someone hanging on it;
– Collision with people.

Besides the user manual and the instructions on how to use the robot properly, several of the design options had in mind the minimization of the identified risks.

Changes in the structure included in v0.3 made the robot less heavy on its top part, giving it a lower centre of gravity. The addition of one battery, besides the obvious increase in the robot autonomy, also had the side effect of adding weight to the base making the robot even more stable and harder to turn.

Collision avoidance is assured mostly by the use of redundant sensors [23]. Their existence, type and location has underwent several changes. The change in the LRF position in v0.2, for instance, made visible more of the obstacles typically present in a commercial site. The change in the LRF itself also allowed the detection of darker obstacles (people wearing a black suit for instance). The addition of two active cameras in v0.3 is another example of redundancy of sensors to detect obstacles.

Finally, there is an emergency button, clearly visible and positioned in a place with easy access for every type of user in case an unpredicted situation happens. It is important to note that this button has not yet been used in any of the internal, external, controlled or uncontrolled tests.

7   Results

In the first part of this section, some technical results on a real scenario are shown. A formal, quantitative, real-world evaluation is highly complicated due to many complex factors, such as the need to test in multiple different environments, testing with several user groups (including those with reduced mobility), the lack of any accepted standard evaluation protocol for the objective measurement of robotic assistance in a retail environment, etc. Therefore, only initial qualitative results based on realistic experimentation are shown in this paper. A formal evaluation which addresses each of these issues will be performed in future work.

The second part describes a user satisfaction inquiry made on a real retail scenario on a population of 143 clients and its results.

7.1   Technical Results

Starting with the detection process, the top left part of Fig. 9 shows the original RGB capture from a typical user following scenario in a commercial shopping environment. In the bottom left, the initial detection gives rise to two false positives - corresponding to the two ladies in the back, while the intended target is the men with his back to the robot. By using the RGB and depth information, shown in the top, these false positives can be removed, with the result shown in the bottom right.

Fig. 9. People detection example. Top left: RGB information; top right: depth information; bottom left: original detections; bottom right: result after removal of false positives. (Color figure online)

The tracking process is illustrated next. In Fig. 10, the person is visible both by the vision and by the 270° laser scanner, while in Fig. 11 the person is only visible by the 270° laser scanner. In both cases, the tracking is not lost and the wGO can follow the person.

Path generation is used to decide about the navigation strategy of the wGO. Since the tracking module can in general return results from multiple sources of information (e.g. vision and laser), it is necessary to merge (fuse) them into one. An example of this results combination is provided in Fig. 12. This fusion step makes the system more robust to errors in either one of the sensors and helps in producing more stable trajectories.

Having one estimation of the person's localization, it is now necessary to decide where to send the robot (path generation). Moreover, it is important to keep some consistency in the results. Inaccuracies produced by the sensors, can lead to highly unstable paths, which is not desirable. An example of a path made by the wGO is shown in the left part of Fig. 13. An increase in the smoothness of the final route, when compared with a traditional approach, is observed.

Fig. 10. Tracking example where the person is visible both by the vision and by the 270° laser scanner. Purple dot in the map image corresponds to the person as localized by the image module, while the blue dot corresponds to the person as localized by the 270° laser scanner. (Color figure online)

Fig. 11. Tracking example where the person is only visible by the 270° laser scanner. Blue dot in the map image corresponds to the person as localized by the image module. (Color figure online)

Finally, a sample of the control results is given in the right part of Fig. 13. It can be observed that the trajectory is stable while avoiding all the present obstacles.

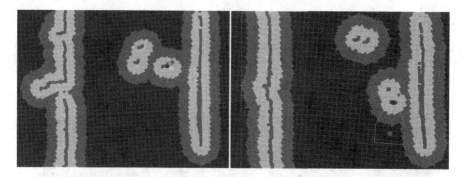

Fig. 12. Fusion examples. On the left is a case where the person is visible both by the vision and by the 270° laser scanner, while on the right the person is only visible by the 270° laser scanner. Purple dot in the map image of the left corresponds to the person as localized by the image module, while the blue dot corresponds to the person as localized by the 270° laser scanner, and green dots are the fusion result. Blue dot in the map image on the right corresponds to the person as localized by the image module and is also the final result. (Color figure online)

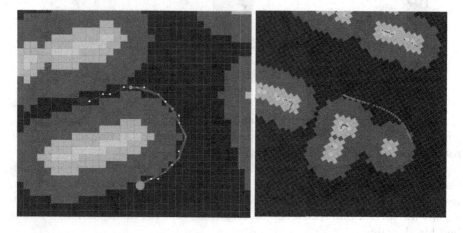

Fig. 13. Path generation and effective path illustrations. In these figures, the purple dots are the effective path done by the wGO, blue circle represents the wGO, large green circle the target destination, white circles the waypoints generated by a traditional path generation approach, smaller green circles the waypoints given by the technique present in the wGO, red dashed lines the trajectory given by the traditional algorithm, purple dashed lines the final trajectory produced by the wGO's system. In the background map, yellow and cyan areas are obstacles, red areas are security zones (although it is not advisable, the wGO can still use them if strictly necessary) and grey areas are free zones. (Color figure online)

7.2   User Satisfaction Survey

This section summarizes the demonstration of wGO in a relevant, unconstrained scenario (Fig. 14). Two wGOs were available for the tests and only users with reduced mobility were asked to participate.

Fig. 14. wGO performing in supermarket, one of the target scenarios [4].

Transportation of the wGOs from their production site to the destination where the demonstration (more than 1800 km) was made by truck with the robots accommodated in disposable plywood boxes.

Concerning the location description, the site had enough space for circulation, and there were no areas where the wGO did not fit. Its use was, however, restricted to days where the store was too crowded. The hypermarket had locations with several different levels of brightness and the main corridor's ceiling had big skylights.

143 clients tested the wGO during two weeks of demonstration. An average of 14 users a day used the wGO with the exception, as already mentioned, of days where the store was too crowded.

An average of 35 minutes was spent per trip, with most users being female (Fig. 15).

Most of the users had ages between 25 and 36, with the second most represented class people with 55 or more years (Fig. 16).

The categorization of users is shown in Fig. 17. As can be seen most of the volunteers presented some type of reduced mobility. Among those, parents with a baby stroller are the most represented followed by people in wheelchair and the elderly.

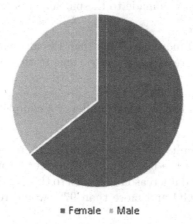

Female   Male

Fig. 15. Analysis of the population gender that used the wGO [4].

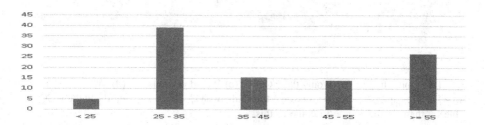

Fig. 16. Analysis of the population age that used the wGO [4].

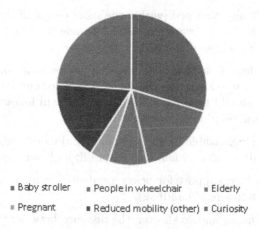

Baby stroller   People in wheelchair   Elderly

Pregnant   Reduced mobility (other)   Curiosity

Fig. 17. Categorization of the wGO users [4].

A small questionnaire was made to the participants with the following questions:

- Do you find the wGO more agreeable than the alternatives?
- Would the wGO be a reason for you to come to this store?
- Would you reuse the wGO?
- Would you recommend the wGO?
- From a scale of 0 to 4 (being 0 not al all and 4 completely), how satisfied are you with the wGO?

The questionnaire results are presented in Fig. 18, respectively. It can be seen that the vast majority of the users find wGO better than existing alternatives. More than 64% would find it a reason to return to this particular store. More than 90% would reuse the wGO and more than 97% would recommend it. Average satisfaction was 3.5 out of 4:

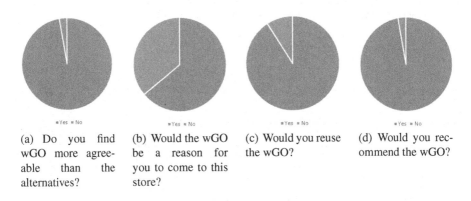

(a) Do you find wGO more agreeable than the alternatives?

(b) Would the wGO be a reason for you to come to this store?

(c) Would you reuse the wGO?

(d) Would you recommend the wGO?

Fig. 18. Questionnaire results [4].

7 out of the 143 have returned (within the short period of the demonstration) to use the wGO. Of these, 5 were male and 2 female.

Comments from the users included:

Detection and Identification: at times, wGO follows someone else; the wGO should have the ability to follow the customer in crowded environments; a passive bracelet or plotter should be provided so that wGO will follow one person; and the wGO should function in exterior light.

Visibility: To be recognizable by the customer, and more visible to others (e.g. beacon, light signal); to show when it is on or off; and increase the volume.

Capacity: Larger bag; support for heavy products, such as water bottles; and hook for personal items (e.g. handbag).

Usability: To go faster and to shorten the distance between the wGO and the user.

Movement: More fluidity, especially in narrow passages and angles and reactivity in crowded environments.

Usage restrictions: To go outside and to go to the car's trunk.

Other Features: To call a collaborator; to be able to scan products, know the price and remaining bag capacity; system for fetching products; speech recognition (dialogue with wGO); guide the customer in the store.

Fig. 19. wGO Technology applied in an industrial scenario.

8   Conclusion

The wGO, an autonomous shopping cart, has been introduced in this paper. A high focus was given to security concerns and its implication the design of the robot. The hardware and software details of the robot have been presented, as well as a presentation of the robot's evolution.

Experiments made in real scenarios are very encouraging and a high user satisfaction was observed. The participants on the user study demonstrated a comfortable behaviour during the experiments as well as a very easy understanding of the robot's operating system (especially, related with the perception and navigation). Comments like "My shopping was very fast!", "In fact, it was a precious help!" and "I think it's awesome, I will certainly use and recommend it!" were made by participants.

Some problems, however, remain to be solved. One of them is the limited space for the shopping items. Concerning the robot's behaviour, low velocity, identification errors and difficulties to move in crowded environments were mentioned. Moreover, there also some interesting features to be included in future versions like the WGO be able to work in outdoor environments and follow the user to the car, to point an example.

As a future work to attack these problems, improvements on the identification algorithm are being developed. When identifying better the user, the wGO will not start following a different person and will behave better in crowded environments by, again, always following the same user.

Concerning the exterior use, new sensors are being tested that are able to acquire depth information in the exterior. At the same time, new algorithms that do not make use of depth information and only the RGB are being studied.

Although the short-term application for the wGO is for commercial environments usage scenarios, several other applications are foreseen, for instance, at the shop-floor of the manufacturing industry and logistics (Fig. 19). As consequence, a new project were started in 2015 in order to develop a robot capable of performing tasks in an autonomous way through an industrial environment (NORTE-01-0247-FEDER-011109). New developments have been made in mapping, localization [24], navigation [25] and multi-robot cooperation algorithms [26]. The application of autonomous vehicles in these scenarios has several advantages in the LEAN process offering flexibility, reducing times and therefore the optimization of the operational costs.

Acknowledgements. The authors would like to thank all the institutional supporters, including TIC RISCO, Portugal Capital Ventures, SGPS, S.A, CEiiA, Fundão City Hall. We would also like to thank to R&D entities including, Tente, CEiiA, Centimfe and Alma Design. These developments were undertaken under several EU cofunding programs, namely: QREN (projects: CENTRO-07-0201-FEDER-023962 and CENTRO-07-0202-FEDER-024692); Portugal 2020 (projects: PFOCI-01-0247-FEDER006398 and NORTE-01-0247-FEDER-011109).

References

1. CRPG, ISCTE: Elementos de caracterizaccão das pessoas com deficiências e incapacidades em Portugal. In: Programa Operacional de Assistência Técnica ao QCA III: eixo FSE (2007)
2. Australia, G.o.W.: Section five - access and inclusion. Disability access and inclusion plan - training package, disability Services Comission (2016). Accessed 18 Nov 2016
3. Rodriguez, V.J.L., et al.: Shopping cart. Justia Patent (2014)
4. Neves, A., et al.: Functionalities and requirements of an autonomous shopping vehicle for people with reduced mobility. In: Proceedings of the 3rd International Conference on Vehicle Technology and Intelligent Transport Systems (VEHITS 2017), pp. 373–380 (2017)
5. Iezzoni, L.I., McCarthy, E.P., Davis, R.B., Siebens, H.: Mobility difficulties are not only a problem of old age. J. Gen. Intern. Med. **16**, 235–243 (2001)
6. Pinheiro, P.G., Cardozo, E., Pinheiro, C.G.: Anticipative shared control for robotic wheelchairs used by people with disabilities. In: IEEE International Conference on Autonomous Robot Systems and Competitions, pp. 91–96 (2015)
7. Faria, B., Reis, L., Lau, N.: User modeling and command language adapted for driving an intelligent wheelchair. In: IEEE International Conference on Autonomous Robot Systems and Competitions, pp. 158–163 (2014)
8. Pavon-Pulido, N., Lopez-Riquelme, J., Pinuaga-Cascales, J., Ferruz-Melero, J., Morais Dos Santos, R.: CYBI: a smart companion robot for elderly people: improving teleoperation and telepresence skills by combining cloud computing technologies and fuzzy logic. In: IEEE International Conference on Autonomous Robot Systems and Competitions, pp. 198–203 (2015)
9. Figueredo, M., Nascimento, A., Monteiro, R.L., Moret, M.A.: Analysis of a sorter cascade applied to control a wheelchair. In: Robot Control InTech (2016)

10. Rico, F.M., Rodriguez Lera, F., Matellan Olivera, V.: Myrabot+: A feasible robotic system for interaction challenges. In: IEEE International Conference on Autonomous Robot Systems and Competitions, pp. 273–278 (2014)
11. Faria, V., Silva, J., Martins, M., Santos, C.: Dynamical system approach for obstacle avoidance in a smart walker device. In: IEEE International Conference on Autonomous Robot Systems and Competitions, pp. 261–266 (2014)
12. Gómez-Goiri, A., Castillejo, E., Orduña, P., Laiseca, X., López-de-Ipiña, D., Fínez, S.: Easing the mobility of disabled people in supermarkets using a distributed solution. In: Bravo, J., Hervás, R., Villarreal, V. (eds.) IWAAL 2011. LNCS, vol. 6693, pp. 41–48. Springer, Heidelberg (2011). https://doi.org/10.1007/978-3-642-21303-8_6
13. For Standardization, I.O.: Robots and robotic devices - vocabulary. Standard ISO, March 2012
14. of Robotics, I.F.: Definition of service robots. http://www.ifr.org/service-robots/ (2017). Accessed 23 Jan 2017
15. Commission, I.E.: Part 1: general requirements. Information technology equipment - Safety (2005)
16. for Standardization, I.O.: Safety of machinery - risk assessment - part 1: principles. Standard, ISO (2007)
17. Jafari, O.H., Mitzel, D., Leibe, B.: Real-time RGB-D based people detection and tracking for mobile robots and head-worn cameras. In: 2014 IEEE International Conference on Robotics and Automation (ICRA), pp. 5636–5643 IEEE (2014)
18. Nebehay, G., Pflugfelder, R.: Clustering of static-adaptive correspondences for deformable object tracking. In: Proceedings of the IEEE Conference on Computer Vision and Pattern Recognition, pp. 2784–2791 (2015)
19. Haro, F., Torres, M.: A comparison of path planning algorithms for OMNI-directional robots in dynamic environments. In: 2006 IEEE 3rd Latin American Robotics Symposium, LARS 2006, pp. 18–25 IEEE (2006)
20. Abdulhafiz, W.A., Khamis, A.: Bayesian approach to multisensor data fusion with pre-and post-filtering. In: 2013 10th IEEE International Conference on Networking, Sensing and Control (ICNSC), pp. 373–378. IEEE (2013)
21. Palmieri, L., Arras, K.O.: A novel RRT extend function for efficient and smooth mobile robot motion planning. In: 2014 IEEE/RSJ International Conference on Intelligent Robots and Systems (IROS 2014), pp. 205–211. IEEE (2014)
22. Park, J.J., Kuipers, B.: A smooth control law for graceful motion of differential wheeled mobile robots in 2D environment. In: 2011 IEEE International Conference on Robotics and Automation (ICRA), pp. 4896–4902. IEEE (2011)
23. Ponz, A., Rodríguez-Garavito, C.H., García, F., Lenz, P., Stiller, C., Armingol, J.M.: Laser scanner and camera fusion for automatic obstacle classification in ADAS application. In: Helfert, M., Krempels, K.-H., Klein, C., Donnellan, B., Gusikhin, O. (eds.) Smart Cities, Green Technologies, and Intelligent Transport Systems. CCIS, vol. 579, pp. 237–249. Springer, Cham (2015). https://doi.org/10.1007/978-3-319-27753-0_13
24. Pinto, A.M., Costa, P.G., Moreira, A.P.: Architecture for visual motion perception of a surveillance-based autonomous robot. In: IEEE International Conference on Autonomous Robot Systems and Competitions (ICARSC) (2014)
25. Costa, P.J., Moreira, N., Campos, D.: Localization and navigation of an omnidirectional mobile robot: the robot@factory case study. In: IEEE Revista Iberoamericana de Tecnologias del Aprendizaje, pp. 1–9 (2016)
26. Santos, J., Costa, P., Rocha, L.F., Moreira, A.P., Veiga, G.: Time enhanced A*: towards the development of a new approach for multi-robot coordination. In: IEEE International Conference on Industrial Technology (ICIT) (2015)

Road Safety: Human Factors Aspects of Intelligent Vehicle Technologies

Cristina Olaverri-Monreal[✉]

Chair for Sustainable Transport Logistics 4.0,
Johannes Kepler University Linz, Altenbergerstraße 69, 4040 Linz, Austria
cristina.olaverri-monreal@jku.at

Abstract. The design of road-vehicle systems has a crucial impact on the driver's user experience. A post-market trial-and-error approach of the product is not acceptable, as the cost of failure may be fatal. Therefore, to design a suitable system in the automotive context that supports the driver during their journey in an unobtrusive way, a thorough survey of human factors is essential. This article elucidates the broad issues involved in the interaction of road users with intelligent vehicle technologies and summaries of previous work, detailing interaction-design concepts and metrics while focusing on road safety.

Keywords: Human factors · Driving task · User interfaces
interaction · Intelligent vehicles

1 Introduction

The International Ergonomics Association [11] defines ergonomics or human factors (HF) as "the scientific discipline concerned with the understanding of interactions among humans and other elements of a system, and the profession that applies theory, principles, data and methods to design in order to optimize human well-being and overall system performance", a definition that has also been adopted by the Human Factors and Ergonomics Society [12]. According to the International Standardization Organization (ISO) standard 9241-210, user experience (UX) stands for an individual's perception and responses resulting from the use of a product, system, or service [27]. The user's emotions, beliefs, preferences, physical and psychological responses, behaviors and accomplishments that occur before, during and after use all figure into the overall user experience. Relying on these definitions and the holistic approach depicted in Fig. 1, HF aims at meeting user needs generating products for a positive UX that are a joy to own and to use [44]. To this end it aligns knowledge and methods from multiple disciplines (i.e. software engineering, psychology, statisticians, designers, etc.) based on empirical data collection and evaluation. For example, graphical user interfaces (GUI) for pedestrian navigation and routing systems might enhance road safety and provide an optimal UX if they are developed in a user friendly manner. They can include aspects that connote a positive feeling

© Springer Nature Switzerland AG 2019
B. Donnellan et al. (Eds.): SMARTGREENS 2017/VEHITS 2017, CCIS 921, pp. 318–332, 2019.
https://doi.org/10.1007/978-3-030-02907-4_16

and a pleasurable experience, such as a shaded path in hot and sunny summer days [48].

According to the ISO 9241-210:2010 – "Ergonomics of human-system interaction - Human-centered design for interactive systems" user, system (consisting of task to be performed and product interacted with) and context of use affect UX [27]. Figure 2 shows how vehicular content as context of use can be arranged into this scheme, where the system is traffic in which the primary driving task (and also sometimes secondary and tertiary) takes place, the vehicle is the product interacted with and the driver is the user.

70% to 80% of new product development failure is due to a lack of understanding of user needs rather than a lack of advanced technology [24]. User centered design (UCD) entails consideration of the user's requirements within the design phase of a service or good being produced. As a consequence, the intended context of use will be reflected in the representation of the system's mental model of how such system should work [18]. Therefore, user mental models need to be considered to effectively design systems that reflect user expectations. This is particularly important for systems that are critical to decision making processes in cognitively demanding scenarios [42].

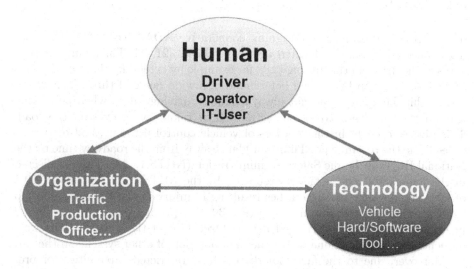

Fig. 1. Holistic approach to the scientific discipline of human factors (adapted from [13]).

The overall diffusion of the application of digital technologies in homes, buildings and cities presents the possibility of designing systems whose functioning is based on intelligent technologies that simultaneously reside in multiple, interconnected applications [37]. As a consequence, the development of intelligent road-vehicle systems such as advanced driving assistance systems (ADAS) is rapidly increasing [17,35]. Many of these systems rely on sensors that collect certain data, for example to identify the distance to the preceding vehicle or the information shown on traffic signs. Vehicle-to-Vehicle (V2V) communication

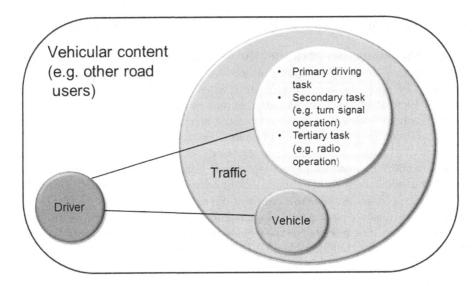

Fig. 2. Representation for a vehicular content of the elements that affect UX according to the ISO 9241-210:2010.

opens up the possibility of designing cooperative ADAS (co-ADAS) that use data collected by sensors located in other vehicles [21,38]. Their purpose is to support the driver in the driving task, for example by extending the driver's field-of-view so they can be warned beforehand of a wide range of threats. However, these vehicular systems do not always enhance driving safety when perceiving this information leads to visual distraction or taking one's eyes off the road. This diversion could instigate a loss of vehicle control if the eyes-off-road time exceeds 2 s, the recommended limit for glance away from the roadway time by the National Highway Traffic Safety Administration (NHTSA) [14]. This is reflected in the number of road accidents reported by the NHTSA: in 2011 alone, 10% of fatal crashes and 17% of crashes resulting in injury in the US were reported as distraction-affected [36], meaning that 390.331 people were killed or injured in crashes involving a distracted driver. Over 17% of these distractions were influenced by mobile phone use or the manipulation of other systems in the car.

Moreover, due to the fact that drivers have restricted capabilities for processing multiple sources of information, attentional demand can cause driver overload of attentional capacity when their processing capabilities are already near the upper threshold (e.g., when the traffic is demanding). Therefore road-vehicle systems intended to increase the driver's awareness of the surrounding environment need to be designed to ensure high usability, acceptance, efficiency and understanding on the part of the driver.

This article elucidates the broad issues involved in the interaction of road users with intelligent vehicle technologies, including summaries of previous work, and will focus on the information flow to which we are exposed while driving. It will include input and output modalities and detail the visual demand required in the vehicle. It will finish with a close look at cooperative systems and automation.

2   Vehicular Interaction

2.1   Information Flow

The overall driving task consists of operating a vehicle by performing many smaller, uncomplicated tasks concurrently. Complexity of vehicular interaction increases with the number of vehicles in traffic and higher traveling speed, as these factors make it more difficult to maintain awareness of the surrounding traffic environment and react to unexpected events and driving maneuvers from other road users. Some of the tasks related to driving are as follows:

- control of speed
- observance of the distance to the leading car
- steering
- traffic observation and action prediction
- navigation
- interaction with car controls
- awareness of the in-vehicle information output
- traffic signs and rules awareness

In addition, interaction in a vehicular environment can include lane shifts involving a combination of vehicle speeds and sizes, an ample spectrum of driving styles and a variety of illumination, weather and road conditions [53]. Furthermore, a transfer of information occurs during the driving process from driver to vehicle, driver to traffic environment and driver to co-driver or passengers and vice versa (Fig. 3).

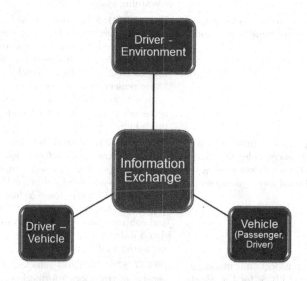

Fig. 3. Information flow in a vehicular context.

According to multiple resource theory, people are supposed to have a variety of resources (visual, auditory, cognitive, and psychomotor) that are dynamically allocated to tasks based on their characteristics [55]. If sources of information are augmented by the automobile industry by adopting technologies that can be found in other mobile environments, such as smart phones and tablets that stream personalized content into the car, the capacity of attentional resources

Table 1. Most relevant standards for the design of in-vehicle interfaces.

Targeted HMI system	Relevant standards
Human centered design principles and activities for computer-based interactive systems	ISO 9241-210:2010 "Ergonomics of human-system interaction - Human- design for interactive systems" [27]
Specifics for elliptical models in three dimensions to represent location of driver's eyes and determine field of view	ISO 4513:2010 "Road vehicles - Visibility - Method for establishment of "eyellipses" for drivers eye location" [5]
Warning Messages and Signals - to clearly perceive and differentiate alarms, warnings and information signals while taking into account different degrees of urgency and combining modalities of warnings	ISO 11429:1996 "Ergonomics - System of auditory and visual danger and information signals" [1] ISO/TR 12204:2012 "Road Vehicles - Ergonomic aspects of transport information and control systems - Introduction to integrating safety-critical and time-critical warning signals" [6] ISO/TR 16352:2005 "Road vehicles - Ergonomic aspects of in-vehicle presentation for transport information and control systems - Warning systems" [3]
Driver's Visual Behavior - Assessment of impact of human-machine interaction	ISO 15007-1:2014 "Road vehicles - Measurement of driver visual behavior with respect to transport information and control systems - Part 1: Definitions and parameters" [7] ISO 15007-2:2014 "Road vehicles - Measurement of driver visual behavior with respect to transport information and control systems - Part 2: Equipment and procedures" [8]
In-Vehicle Displays, e.g. image quality, legibility of characters, color recognition, etc. and procedures for determining the priority of on-board messages presented to drivers	ISO 15008:2009 "Road vehicles - Ergonomic aspects of transport information and control systems - Specifications and compliance procedures for in-vehicle visual presentation" [4] ISO/TS 16951:2004 "Road Vehicles - Ergonomic aspects of transport information and control systems - Procedures for determining priority of on-board messages presented to drivers" [2]
Suitability of Transport Information and Control Systems (TICS) for Use While Driving	ISO 17287:2003 "Road vehicles - Ergonomic aspects of transport information and control systems - Procedure for assessing suitability for use while driving" [28]

or the total amount of information the human brain is capable of retaining at any particular moment [29] can be affected.

As people spend a considerable amount of time driving, the latest implemented vehicular technologies intend to improve the travel experience by increasing safety and comfort as well as by enhancing entertainment possibilities. Research related to the human processing capabilities particularly regarding how much information the human being can process under driving conditions is therefore essential to ensure road safety.

The ISO is responsible for a vast number of standards which could be used either during the development of a particular human machine interface (HMI) or for its subsequent testing. The requirements, specifications, guidelines and characteristics compiled in Table 1 can be used consistently to ensure that a system is suitable for use in a vehicular context.

They focus on an in-vehicle presentation of information that improves the user experience and eases the learning process of road-vehicle systems. This was reflected in the results of several studies that examined the display modality and location in a central versus distributed display [30]. Additional works analyzed location preferences for driver information systems (DIS) and ADAS relative to the information they provided [39] and subsequently validated the results in a driving simulator [45].

2.2 Input Modalities

Today's cars have become more complex regarding driver interaction, due in part to a digital era that has changed people's habits and perception of content. An increase in novel interactions on mobile devices and a growing reliance on the provided content of developed applications has stimulated our perception and acceptance of interactive digital technology, thereby creating many opportunities for interacting with useful and attractive in-vehicle user interfaces devoted to safety, comfort and infotainment. As a consequence, input modalities in the form of mechanical parts of HMIs are increasingly being replaced by digital substitutes.

Interaction in the vehicle takes place while driving therefore affecting control of the vehicle, and the resulting driving performance while performing secondary and tertiary tasks can be measured by comparing individual performance with standard values that have been gathered as average from participants.

In this context, the selection of the most appropriate driving performance metrics depends on the objective of the experiment and the system to be tested. The following parameters are commonly used to monitor driving behavior [49]:

- speed-related parameters for measuring visual distraction while performing secondary tasks. For example, reduced speed is an indicator of road visual distraction that involves taking one's eyes off the road [25].
- lateral position parameters for measuring driving under cognitive load.
- parameters which describe headway performance changes measure visual or cognitive distraction.

Research on the best input modality for in-vehicle device interaction is being undertaken and investigating for example buttons and sensors in the steering wheel or the use of handwriting recognition. The ideal interaction should be eyes and hands free and one which causes no distraction. Natural language processing makes it possible to intuitively interact with a natural user interface through speech. Dialog systems based on spoken language, allow the interaction with computer-based applications through the use of voice [32]. A distraction from traffic as brief as powering on a car radio or answering a call on a mobile phone represents an increased risk of accident, therefore the potential of using dialog systems based on spoken language in road-vehicle applications is very large.

2.3 Display of Information

To design a suitable user interface in the automotive context the amount of information to display needs to be considered. As stated in [46] a large quantity of messages can be conveyed, expanding the range of possible display areas through non-conventional locations. Particularly the use of windshields as HMI could provide benefits for a low penetration rate of connected or automated vehicles, as they would provide an effective way to convey important safety-related information [43].

Figure 4 illustrates an approach in which visual data related to safety distance is provided to the rear vehicle in real-time, independently of the communication capabilities of the following vehicle and which relies on an asynchronous collaborative process [40].

Fig. 4. Windshield-conveyed message related to safety distance, independent of the communication capabilities of the trailing vehicle (adapted from [43]).

Regarding in-vehicle information location, a distance warning system based on vehicle-to-vehicle (V2V) communication would be able to additionally convey the time-gap between the connected vehicles, issuing a warning inside the following vehicle depending on an established threshold [33].

Regarding the preferences for in-vehicle location of information in the study by [41], 4 fairly homogeneous groups of functions were distributed in 4 different display locations, the location's proximity to the field of view of the driver

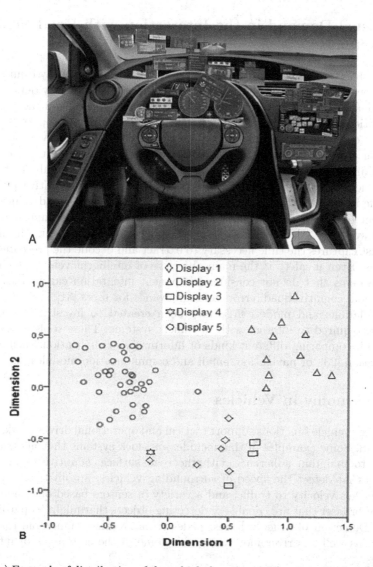

Fig. 5. (a) Example of distribution of the vehicle functions in the given displays [47] and (b) Perceptual map with the relative positioning of all functions in the multidimensional scaling diagram [41].

increasing according to the urgency of the messages that they conveyed. An example of this principle being that drivers preferred having entertainment-related functions displayed outside their visual field, and social media and apps integration in a vehicular context were not considered essential to be displayed at all. Figure 5 shows (a) an example of distribution of the vehicle functions in given displays [47] and (b) the perceptual map with the relative positioning of all functions in the multidimensional scaling diagram [41].

3   Visual Demand in the Interaction with Road-Vehicle Systems

Driving distraction implies diversion of the driver's attention away from the road to focus on another activity instead [50]. Furthermore, not every demand loads attentional resources equally. Each task has its own demands on the processing capabilities of the driver. The environment exhibits different properties, such as static or dynamic and familiar or unfamiliar, and it represents varying degrees of visual complexity. As indicated in [45], multiple glances between in-vehicle devices and the road can affect driver attention, reducing the ability to maintain vehicle control and delaying and/or interrupting the cognitive processing of traffic information. As stated in the ISO 15007-1:2014 - "Road vehicles measurement of driver visual behavior with respect to transport information and control systems", glance duration is usually thought of as being the measure that best captures the time necessary to extract and decode the presented visual stimulus. Even if safety is the main objective of intelligent vehicles technology, some systems that do not consider component integration can increase visual distraction, cognitive load, errors and annoyance for users [31].

Simulations and models [51] have been created to investigate the visual demand required to interact with in-vehicle systems. These tools assess visual workload, comparing different kinds of information visualization, such as route information [23], or navigation, email and communication modules [20].

4   Autonomy in Vehicles

Intelligent vehicle functions support tactical and operational driving tasks as part of ADAS. Some examples of this include: anti-lock systems that allow a motor vehicle to maintain adherence with the road surface; adaptive cruise control systems that detect the speed of surrounding vehicles, automatically adapting the vehicle's velocity to traffic; and a variety of sensors based on vision, radar, infrared or laser that are capable of detecting objects that might jeopardize road safety. Detection of other vehicles, pedestrians or other vulnerable road users (VRU), as well as driver monitoring, are some of the core areas of intelligent vehicles.

As previously mentioned, some co-ADAS rely on cooperative messages broadcasted using vehicle-to-vehicle (V2V) and vehicle-to-infrastructure communication (V2I) (V2X, collectively) to sense the surroundings. According to the US

Department of Transportation (USDOT), many connected vehicle (CV) application concepts have been developed through prototyping and demonstration in recent years. A classification of these connected vehicle applications within the CV Pilot Deployment Program is listed in Table 2.

A system to augment the visual perception of the road in an overtaking maneuver based on V2X technologies was presented in [38]. The system used Vehicular Ad Hoc Network (VANET) technology to provide a video-stream of the view from the front of a leading vehicle to a rear vehicle and was particularly helpful in cases of long and vision obstructive front vehicles. The visual display was located inside the vehicle. A projection through a head-up-display (HUD) [21] conveyed the information in a later version of the system and which took into account the length of the vehicle ahead. This idea to enhance road safety was adopted by Samsung in 2015 and posted in the blog Safety Truck [10].

Table 2. Classification of the most recent connected vehicle applications sponsored by the USDOT CV Pilot Deployment Program (adapted from [52]).

V2V safety	V2I safety
Red light violation warning	Emergency Electronic Brake Lights (EEBL)
Curve speed warning	Forward Collision Warning (FCW)
Stop sign gap assist	Intersection Movement Assist (IMA)
Spot weather impact warning	Left Turn Assist (LTA)
Reduced speed/work zone warning	Blind Spot/Lane Change Warning (BSW/LCW)
Pedestrian in signalized crosswalk warning (transit)	Do Not Pass Warning (DNPW)
	Vehicle Turning Right in Front of Bus Warning (Transit)

Research on vehicles with conditional automation [9] that are equipped with automated functions is advancing and more and more reports about vehicles that are able to operate autonomously for some portions of the trip are being released. In conditional automation the vehicle's control is relayed back to humans through a Take Over Request (TOR) in situations that the automation is not able to handle. To this end, it is essential to assess the driver's state, their capabilities and the driving environment at the time of a TOR, as the potential boredom and road monotony associated with higher automatism of vehicles might lead to a reduction in driver situational awareness [37].

Attempting to address this need, the use of continuous, in-vehicle visual stimulus to reduce driver reaction time after a period of hypovigilance was studied in [19]. Relying on peripheral vision, the authors implemented and tested an unobtrusive method based on luminescence. They showed a tendency among drivers to respond faster to a TOR when their peripheral vision detected the stimulus.

To monitor the driver and the road conditions a mobile application can be used as backup system and issue a warning to avoid inattentive driving in the event of a TOR as suggested in [15] and latter extended in [16].

How other road users are going to interact with fully autonomous vehicles in different scenarios is not known yet. Most of the research concerning this question has been based on simulated scenarios, including working technical systems operated by a human operator (Wizard of Oz (WOZ)) or survey studies (i.e. [22,34,54]. However, the authors in [26] performed a field test with driverless, fully autonomous vehicles through a smartphone application that indicated to pedestrians that an autonomous vehicle was approaching. The application was intended to develop a trust in the autonomous vehicle technology. Results showed that the application supported the pedestrians in the verification process of trusting autonomous vehicles as a reliable, safe technology. Figure 6 shows the evaluation process of the application as a means to trusting autonomous vehicles.

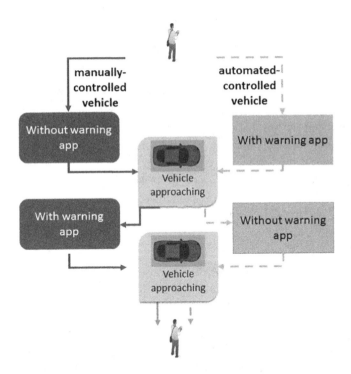

Fig. 6. Evaluation process of the application intended to develop a trust in autonomous vehicle technology [26].

5   Conclusion

Human factors in an automotive context have been researched extensively in recent years as they are decisive in road safety. This paper provides a concise introduction to the most important human factors related to road safety and addresses interaction design principles that heavily emphasize UCD, problems of distraction and workload and visual perception.

Some of the most valuable metrics for testing ADAS have been introduced as well. The information presented here provides the reader with the background, a variety of approaches and an example of applications of the most current and important concepts in the interaction with intelligent vehicle technologies.

References

1. ISO 11429:1996 - Ergonomics - System of auditory and visual danger and information signals (1996). http://www.iso.org/iso/catalogue_detail.htm?csnumber=19369
2. ISO/TS 16951:2004 - Road vehicles - Ergonomic aspects of transport information and control systems (TICS) - Procedures for determining priority of on-board messages presented to drivers (2004). http://www.iso.org/iso/catalogue_detail.htm?csnumber=29024
3. ISO/TR 16352:2005 - Road vehicles - Ergonomic aspects of in-vehicle presentation for transport information and control systems - Warning systems (2005). http://www.iso.org/iso/iso_catalogue/catalogue_tc/catalogue_detail.htm?csnumber=37859
4. ISO 15008:2009 - Road vehicles - Ergonomic aspects of transport information and control systems - Specifications and test procedures for in-vehicle visual presentation (2009). http://www.iso.org/iso/catalogue_detail.htm?csnumber=50805
5. ISO 4513:2010 - Road vehicles - Visibility - Method for establishment of eyellipses for driver eye location (2010). http://www.iso.org/iso/catalogue_detail.htm?csnumber=45845
6. ISO/TR 12204:2012 - Road vehicles - Ergonomic aspects of transport information and control systems - Introduction to integrating safety critical and time critical warning signals (2012). http://www.iso.org/iso/catalogue_detail.htm?csnumber=51275
7. ISO 15007-1:2014 - Road vehicles measurement of driver visual behaviour with respect to transport information and control systems. Part 1: Definitions and parameters (2014). http://www.iso.org/iso/catalogue_detail.htm?csnumber=56621
8. ISO 15007-2:2014 - Road vehicles measurement of driver visual behaviour with respect to transport information and control systems. Part 2: Equipment and procedures (2014). http://www.iso.org/iso/catalogue_detail.htm?csnumber=56622
9. SAE International's Levels of Driving Automation for On-Road Vehicles (2014). http://www.sae.org/misc/pdfs/automated_driving.pdf
10. Safety Truck (2015). http://uk.businessinsider.com/samsung-safety-truck-makes-driving-safe-2015-6?r=US#ixzz3jA0HkU1I
11. Definition and Domains of Ergonomics (2017). http://www.iea.cc/
12. Definitions of Human Factors and Ergonomics (2017). http://www.iea.cc/ergonomics/
13. Human factors. What is it? (2017). http://www.arpansa.gov.au/regulation/Holistic/humanfactors.cfm

14. National Highway Traffic Safety Administration: Visual-manual NHTSA driver distraction guidelines for in-vehicle electronic devices. Washington, DC, National Highway Traffic Safety Administration (NHTSA), Department of Transportation (DOT) (2012)
15. Allamehzadeh, A., Olaverri-Monreal, C.: Automatic and manual driving paradigms: cost-efficient mobile application for the assessment of driver inattentiveness and detection of road conditions. In: Intelligent Vehicles Symposium (IV), pp. 26–31. IEEE (2016)
16. Allamehzadeh, A., de la Parra, J.U., Hussein, A., Garcia, F., Olaverri-Monreal, C.: Cost-efficient driver state and road conditions monitoring system for conditional automation. In: Intelligent Vehicles Symposium (IV), pp. 1497–1502. IEEE (2017)
17. Brookhuis, K.A., De Waard, D., Janssen, W.H.: Behavioural impacts of advanced driver assistance systems-an overview. Eur. J. Transp. Infrastruct. Res. 1(3), 245–253 (2001)
18. Carroll, J.M., Anderson, N.S., National Research Council, et al.: Mental models in human-computer interaction: research issues about what the user of software knows. No. 12, National Academies (1987)
19. Çapalar, J., Olaverri-Monreal, C.: Hypovigilance in limited self-driving automation: peripheral visual stimulus for a balanced level of automation and cognitive workload. In: Proceedings 20th Intelligent Transportation Systems Conference, Yokohama, Japan (ITSC). IEEE (2017)
20. Gellatly, A.W., Kleiss, J.A.: Visual attention demand evaluation of conventional and multifunction invehicle information systems. In: Proceedings of the Human Factors and Ergonomics Society Annual Meeting, vol. 44, pp. 3–282. SAGE Publications (2000)
21. Gomes, P., Olaverri-Monreal, C., Ferreira, M.: Making vehicles transparent through V2V video streaming. IEEE Trans. Intell. Transp. Syst. 13(2), 930–938 (2012)
22. Habibovic, A., Andersson, J., Nilsson, M., Lundgren, V.M., Nilsson, J.: Evaluating interactions with non-existing automated vehicles: three wizard of Oz approaches. In: Intelligent Vehicles Symposium (IV), pp. 32–37. IEEE (2016)
23. Hanowski, R.J., Perez, M.A., Dingus, T.A.: Driver distraction in long-haul truck drivers. Transp. Res. Part F: Traffic Psychol. Behav. 8(6), 441–458 (2005)
24. von Hippel, E.: An emerging hotbed of user-centered innovation. Harv. Bus. Rev. 85(2), 27–28 (2007)
25. Hosking, S.G., Young, K.L., Regan, M.A.: The effects of text messaging on young drivers. Hum. Factors 51(4), 582–592 (2009)
26. Hussein, A., García, F., Armingol, J.M., Olaverri-Monreal, C.: P2V and V2P communication for pedestrian warning on the basis of autonomous vehicles. In: 19th International Conference on Intelligent Transportation Systems (ITSC), pp. 2034–2039. IEEE (2016)
27. ISO: 9241–210: 2010. Ergonomics of human system interaction-Part 210: Human-centred design for interactive systems. International Standardization Organization (ISO). Switzerland (2010)
28. ISO, E.: 17287 (2003). Road vehicles - Ergonomic aspects of transport information and control systems - Procedure for assessing suitability for use while driving (2003)
29. Kahneman, D.: Attention and Effort. Princeton, Citeseer (1973)
30. Lee, J.D., Gore, B.F., Campbell, J.L.: Display alternatives for in-vehicle warning and sign information: message style, location, and modality. Transp. Hum. Factors 1(4), 347–375 (1999)
31. Lee, J.D., Kantowitz, B.H.: Network analysis of information flows to integrate in-vehicle information systems. Int. J. Veh. Inf. Commun. Syst. 1(1–2), 24–43 (2005)

32. McTear, M.F.: Spoken dialogue technology: enabling the conversational user interface. ACM Comput. Surv. (CSUR) **34**(1), 90–169 (2002)
33. Michaeler, F., Olaverri-Monreal, C.: 3D driving simulator with VANET capabilities to assess cooperative systems: 3DSimVanet. In: Intelligent Vehicles Symposium (IV), pp. 999–1004. IEEE (2017)
34. Mok, B.K.J., Sirkin, D., Sibi, S., Miller, D.B., Ju, W.: Understanding driver-automated vehicle interactions through Wizard of Oz design improvisation. In: Proceedings of the International Driving Symposium on Human Factors in Driver Assessment, Training and Vehicle Design, pp. 386–392 (2015)
35. Molin, E., Marchau, V.: User perceptions and preferences of advanced driver assistance systems. Transp. Res. Rec.: J. Transp. Res. Board **1886**, 119–125 (2004)
36. NHTSA: Traffic Safety Facts Research Note Distracted Driving 2011. Technical report (2013). https://crashstats.nhtsa.dot.gov/Api/Public/ViewPublication/811737
37. Olaverri-Monreal, C.: Autonomous vehicles and smart mobility related technologies. Infocommunications J. **8**(2), 17–24 (2016)
38. Olaverri-Monreal, C., Gomes, P., Fernandes, R., Vieira, F., Ferreira, M.: The see-through system: a VANET-enabled assistant for overtaking maneuvers. In: Proceedings of the IEEE Intelligent Vehicles Symposium (IV), pp. 123–128 (2010)
39. Olaverri-Monreal, C., Lehsing, C., Trübswetter, N., Schepp, C.A., Bengler, K.: In-vehicle displays: driving information prioritization and visualization. In: Proceedings of the IEEE Intelligent Vehicles Symposium (IV), pp. 660–665 (2013)
40. Olaverri-Monreal, C., Lorenz, R., Michaeler, F., Krizek, G., Pichler, M.: Tailigator: cooperative system for safety distance observance. In: Proceedings 2016 International Conference on Collaboration Technologies and Systems, Orlando, Florida, USA (2016)
41. Olaverri-Monreal, C., Bengler, K.J.: Impact of cultural diversity on the menu structure design of driver information systems: a cross-cultural study. In: 2011 IEEE Intelligent Vehicles Symposium (IV), pp. 107–112. IEEE (2011)
42. Olaverri-Monreal, C., Gonçalves, J.: Capturing mental models to meet users expectations. In: 2014 9th Iberian Conference on Information Systems and Technologies (CISTI), pp. 1–5. IEEE (2014)
43. Olaverri-Monreal, C., Gvozdic, M., Bharathiraja, M.: Effect on driving performance of two visualization paradigms for rear-end collision avoidance. In: 20th International Conference on Intelligent Transportation Systems (ITSC). IEEE (2017)
44. Olaverri-Monreal, C., Hasan, A.E., Bengler, K.: Intelligent agent (IA) systems to generate user stories for a positive user experience. Int. J. Hum. Cap. Inf. Technol. Prof. (IJHCITP) **5**(1), 26–40 (2014)
45. Olaverri-Monreal, C., Hasan, A.E., Bulut, J., Körber, M., Bengler, K.: Impact of in-vehicle displays location preferences on drivers' performance and gaze. IEEE Trans. Intell. Transp. Syst. **15**(4), 1770–1780 (2014)
46. Olaverri-Monreal, C., Jizba, T.: Human factors in the design of human-machine interaction: an overview emphasizing V2X communication. IEEE Trans. Intell. Veh. **1**(4), 302–313 (2016)
47. Olaverri-Monreal, C., Lehsing, C., Trübswetter, N., Schepp, C.A., Bengler, K.: In-vehicle displays: driving information prioritization and visualization. In: Intelligent Vehicles Symposium (IV), pp. 660–665. IEEE (2013)
48. Olaverri-Monreal, C., Pichler, M., Krizek, G., Naumann, S.: Shadow as route quality parameter in a pedestrian-tailored mobile application. IEEE Intell. Transp. Syst. Mag. **8**(4), 15–27 (2016)

49. Östlund, J., et al.: Driving performance assessment-methods and metrics (2005)
50. Ranney, T.A., Mazzae, E., Garrott, R., Goodman, M.J.: NHTSA driver distraction research: past, present, and future. Driver Distraction Internet Forum (2000)
51. Stevens, A., Burnett, G., Horberry, T.: A reference level for assessing the acceptable visual demand of in-vehicle information systems. Behav. Inf. Technol. **29**(5), 527–540 (2010)
52. USDOT: Connected vehicles pilot deployment program (2017). https://www.its.dot.gov/pilots/cv_pilot_apps.htm
53. Vanderbilt, T.: Traffic. Vintage, New York City (2008)
54. Waytz, A., Heafner, J., Epley, N.: The mind in the machine: anthropomorphism increases trust in an autonomous vehicle. J. Exp. Soc. Psychol. **52**, 113–117 (2014)
55. Wickens, C.D.: Processing resources and attention. Mult.-Task Perform. **1991**, 3–34 (1991)

Evaluation Methodology for Cooperative ADAS Utilizing Simulation and Experiments

Sebastian Bittl[1]([✉]), Dominique Seydel[2], Jakob Pfeiffer[2], and Josef Jiru[2]

[1] HU Berlin, Berlin, Germany
sebastian.bittl@mytum.de
[2] Fraunhofer ESK, Munich, Germany
{dominique.seydel,jakob.pfeiffer,josef.jiru}@esk.fraunhofer.de

Abstract. Wireless vehicular networks are to be deployed in both Europe and the USA within upcoming years. Such networks introduce a new promising source of information about vehicular environments to be used by cooperative advanced driver assistance systems (ADAS). However, development and evaluation of such cooperative ADAS is still challenging. Hence, we introduce a novel methodology for their development and evaluation processes. It is applied to evaluate the fulfillment of requirements on position accuracy information within exchanged messages. Such requirements are only roughly defined and not sufficiently evaluated in field tests. This holds especially for Global Navigation Satellite Systems (GNSS) optimized for maximum integrity of obtained positions. Such configuration is required to increase robustness and reliability of safety critical ADAS. We find that pure GNSS-based positioning cannot fulfill position accuracy requirements of studied ADAS in most test cases.

Keywords: VANET · ETSI ITS · ADAS · Positioning · Evaluation

1 Introduction

Vehicular ad-hoc networks (VANETs) are an active research area. They promise to increase traffic safety leading to high interest of both automotive industry and governments. Hence, mass deployment can be expected within the next years enabling the introduction of novel cooperative advanced driver assistance systems (ADAS) [1,23]. A well known concept for the development of ADAS is to include simulation based testing and evaluation early in the development process [19,22]. Moreover, the simulation environment gets adapted in each development phase, e.g., for Vehicle in the Loop tests [45]. Such concepts are especially needed for VANET applications, due to a wide range of traffic scenarios with

---

S. Bittl—During work on the presented results the author was with Fraunhofer ESK.

B. Donnellan et al. (Eds.): SMARTGREENS 2017/VEHITS 2017, CCIS 921, pp. 333–353, 2019.
https://doi.org/10.1007/978-3-030-02907-4_17

many involved vehicles. Hence, it is hard to realize a complex testing setup in a reproducible way during field tests [39].

Development and testing of ADAS are complex tasks, and errors within the implementation of an ADAS can have a severe impact on the results of a conducted evaluation as well as on performance in real world use cases. This is especially important for cooperative ADAS, as the considered use cases are safety critical [9]. Hence, the focus of this work is on a novel methodology lowering the effort for cooperative ADAS development and evaluation processes, by close integration of both simulations and field tests. A first look at this issue is provided in [37], and this work provides an extension of the one given in [37].

The first VANET use cases, e.g., Road Works Warning (RWW), require only a positioning accuracy that allows to assign a vehicle to a dedicated lane. In regard to the longitudinal direction, demands on the position accuracy are even lower [2]. Our evaluation concentrates on two important VANET use cases, which are Intersection Collision Risk Warning (ICRW) [3] and Longitudinal Collision Risk Warning (LCRW) [4]. A comparison of the different early use cases of VANETs [9] shows that these two ADAS have advanced requirements on position accuracy and communication latency. A core reason of using ICRW and LCRW for the conducted evaluation of positioning requirements is the availability of comparable reference systems realized with conventional sensors, like radar sensors. The results of the VANET based implementation should resemble the ones of the reference system as good as possible, to allow regarding VANETs as a reliable sensor for ADAS. Thus, we propose a development environment, which allows to compare intermediate results of the VANET based ADAS with a radar based reference system, following our proposed methodology.

The basic question of our evaluation is whether pure Global Navigation Satellite Systems (GNSS), like the Global Positioning System (GPS), yield a sufficiently high position accuracy for ADAS, while using a configuration optimized for maximum integrity of obtained positions. This also answers the question whether vehicles being already on the road can be equipped with VANET technology by plug-in devices, which need nothing except a power supply from the vehicles energy system, like navigation systems.

The remainder of this work is outlined as follows. A review of related work is provided in Sect. 2. Afterwards, Sect. 3 introduces the proposed evaluation methodology, which is applied in Sect. 4 to ICRW and LCRW realizations. Finally, a conclusion about achieved results and possible topics of future work are given in Sect. 5.

2   Related Work

Basic requirements of VANET based ADAS have been determined and some realizations, based on early versions of VANET standards, were tested during field tests in Europe (e.g., DRIVE C2X, simTD) and the US [23]. These tests identified accurate positioning of vehicles to be a main issue in VANETs.

In general, current VANET approaches use wireless cyclic broadcast of beacon messages for basic information distribution among nodes (e.g., vehicles or road side units (RSUs)). Two similar approaches are followed within the European Telecommunications Standards Institute (ETSI) Intelligent Transport Systems (ITS) and US Wireless Access in Vehicular Environments (WAVE) standardization frameworks [1,23]. Communication within VANETs is often referred to as Vehicle-to-X (V2X) communication. In the following, we stick to the ETSI ITS nomenclature, but porting to the WAVE system is straight forward.

Within ETSI ITS, beacon messages are called Cooperative Awareness Messages (CAMs). These are sent and received by a so called Cooperative Awareness Basic Service within the facility layer. The facility layer combines all functionality from layers five upwards within the ISO/OSI model [7].

2.1   Evaluation Environments

A coupled simulation environment including network and traffic simulation is used in [26], to show the impact of VANETs for an intersection collision detection application. Within this simulation the effectiveness of vehicular networks for cooperative collision detection is evaluated, especially on packet level.

Currently it is not possible to use other popular VANET standard implementations, which are often used for research and application evaluation, like VSimRTI [36] or iTETRIS for ETSI ITS [30], or Veins for WAVE [39]. This is caused by close coupling of such implementations to their simulation environments (ns-3 resp. OMNet++), which makes porting of this approaches to real hardware hard. Moreover, protocol implementations dedicated for real hardware typically make direct use of operating system functionality, e.g., to obtain time information. Thus, such information is hard to be replaced by the one provided by a simulation environment. Moreover, parallel and coordinated usage of many on-board units (OBUs) within a test setup leads to very high complexity within the setup and testing process. Therefore, rigid usage of an abstraction layer for all data input sources is required, which is implemented within the ezCar2X framework [34], as described in Sect. 4.1.

2.2   Development of ADAS

Development of ADAS is a complex process. Much work has been dedicated to strategies limiting required effort within the development process, and ensuring testability of obtained ADAS [15,19,22,45]. As a main subject, many aspects from the field of software engineering have been adapted to specific needs of the automotive domain. An approach for an integrated testing and simulation framework is given in [44]. Our implementation in Sect. 4 adapts some of the concepts from [44] to the needs of cooperative ADAS.

However, development of cooperative ADAS shows even higher complexity in comparison to such ADAS using only information from within a single vehicle. Especially, testability is a significant challenge, due to a massive increase in the variety of data sets within the newly known vehicular environment. Thus, we propose an extension to well known ADAS development methods to enable their usage in the development process of future VANET based ADAS.

In [43], a cooperative vehicle collision warning system is proposed based on communicated node positions obtained with differential GPS, and the feasibility of trajectory prediction based on the communicated positions is examined. An architecture is proposed implementing a "Future Trajectory Estimator" for received vehicle data. The following main weaknesses of VANET based applications were identified: (a) The prediction accuracy decreases when the position error increases. (b) Applications are vulnerable to long-period GPS blockage and communication drop-outs. (c) There is limited tolerance of changes in the driver's intention, due to a slow update frequency of the determined positions.

Evaluation of ADAS performance is an important part of the development process. Thus, it is described in more detail in Sect. 2.3.

2.3   Sensor Based Reference System

In [46] security gains from the combination of data from VANET messages with radar measurements are studied. The aim is to validate position information from received messages. In contrast, our aim is to evaluate if the position information is accurate enough for robust usage within ADAS.

Different types of distance measurement sensors have been proposed for usage in the automotive domain. These include laser scanners, radar sensors, photonic mixing devices (PMDs) and cameras, which are common sensors for current ADAS equipped vehicles [45]. The measured distance, relative velocity and heading of the detected objects are passed to the ADAS for further processing. Basic requirements of a reference system for ADAS are given in [12].

OBUs typically use a GNSS based positioning system, e.g., Cohda Wireless MK4a [21]. We use a GNSS software receiver and compare its real live measurements to the ones of an automotive radar system. Details about the satellite based position estimation are given in Sect. 2.4. The input data for an ADAS based on a VANET approach are messages sent by vehicles. CAMs mainly contain the vehicle's position, including the positions confidence information (optional), speed, heading and generation time of the position [7].

Further processing of the different input data sets for evaluation of the VANET based ADAS is described in Sect. 3.2.

2.4   Satellite Based Positioning

GNSS have deeply entered the consumer market and are widely used for car navigation systems. GNSS based service applications offer a number of opportunities to the transportation market. Professional and reliability critical applications like road tolling, anti-theft systems and dangerous goods tracking have been implemented in vehicles. These applications are highly sensitive to the appearance of jamming devices, one account of weaknesses of current satellite navigation systems: the extremely low signal power. A close look on security implications of VANET dependency on GNSS input is given in [16,17].

Basically, GNSS like US American GPS, European Galileo, Russian Glonass, and Chinese Beidou are based on the same concept: Navigation satellites are placed in a medium earth orbits in a height of about 20.000 km to 25.000 km. These satellites emit highly precise ranging signals with up to 50 W for the estimation of the signal travel time from the satellite to the receiver on the earth. If the receiver is able to track the signal of 4 satellites or more, it can calculate it's actual position and time by use of the signal travel time. If the receiver would be perfectly aligned to the common GNSS clock, 3 satellites in view would be enough. However, most receivers are not perfectly aligned to the system time of GNSS and the receiver's clock offset to the system time has to be estimated. Beside the actual time information valid for the transmitting satellite, the satellite signal also carries information to determine the position of all satellites of the corresponding system, corrections for ionospheric and tropospheric effects and offset information for the satellite's unique clocks.

Most current commercial GNSS receivers use the satellite signals of the GPS L1 civil navigation signal. Galileo and Beidou provide open positioning signals at the same center frequency of 1575.42 MHz. Thus, modern civil GNSS receivers will also be able to use multiple of these systems in parallel. From the time the receiver estimated the timing offset between these satellite navigation systems, all corresponding satellites can be used for a joint positioning.

The GPS L1 C/A (coarse/acquisition) civil navigation signal offers a performance of about 5 to 10 m to the civil user. In case of scattering or multi path effects due to vegetation, buildings or other surrounding objects, the position accuracy can decrease to several tens (or in worst case hundreds) of meters. In case of more precise requirements on the position, one might also use either differential correction services, like DGNSS, or more sophisticated dual frequency receivers. The usage of two frequencies allows the receiver to estimate the ionospheric signal distortions, which effects the distance estimation between receivers and satellite the most. Thus, the usage of two frequencies improves the accuracy up to 2 to 9 m and in combination with differential corrections down to tens of centimeters. For additional information, see [10,24,27,32].

3   Evaluation Methodology

A central aspect of our development and evaluation methodology is to combine pure software based simulation, a real world reference system and a real world VANET. Thereby, we try to keep the amount of changing code as low as possible, when moving between the simulation environment to real world experiments. This is done due to two major reasons. At first, it avoids the effort for implementing a lot of wrapper code, which speeds up the development process. Secondly, it avoids errors introduced by changing behavior of the ADAS between the different evaluation environments.

The proposed development process of ADAS includes a simulation environment into the testing process in an early development phase. This methodology is similar to well known test-driven development [13] leading to an enhanced and extended version of simulation-driven development, which has shown good results used within a single vehicle [19]. Each implemented feature is tested with simulated data inputs and its performance is evaluated as early as possible.

In contrast to prior work, we do not use a newly implemented feature only on a single entity in the simulation environment. Instead, after the feature showed the expected behavior on a single vehicle it is deployed on all entities (i.e., vehicles, road side units) within the simulation environment. Thereby, also the correct interaction of each component regarding multiple communicating entities can be verified early in the development process.

A central requirement of the proposed methodology is the availability of a set of traffic scenarios, which are characteristic for the use case(s) of the ADAS to be implemented. To obtain such scenarios we use two complementary approaches.

Firstly, the standardized requirements of the ADAS are used to identify relevant road topologies. Afterwards, various traffic flows for these road topologies are generated. We use deterministic traffics flows as well as such from random trip generation. This can be done by various mechanisms, as such implemented within the Simulation for Urban Mobility (SUMO) framework [14].

Secondly, traffic scenarios which are identified as challenging for ADAS by real world field tests are included. To identify and characterize such scenarios, a possibility to obtain detailed traces from test drives and to re-run the entire drive with consistent timing of data inputs within the simulation environment is required. Requirements for such playback functionality can be found in [11], our implementation is described in detail in Sect. 4.1.

As already mentioned, we do not only use simulations to evaluate an ADAS. We also apply (multiple) vehicle-mounted sensors. As our target ADAS's aim at collision avoidance, distance measurement sensors are used. Thereby, we address the issue that it is hard to perfectly resemble traffic scenarios from simulation in practice, because of many cooperating entities.

The distance sensors are used to realize a reference ADAS serving as ground truth for the evaluation of the cooperative ADAS based on V2X data. To obtain a well usable reference system, we recommend to follow best practices for such single vehicle applications proposed in prior work [19,22]. The reference system is designed to obtain the maximum performance achievable by an ADAS using only local sensor information. Field tests are used to compare the reference system and cooperative ADAS. In a first testing step, side effects caused by the implementation are identified and removed. Afterwards, system limitations are determined during a second evaluation step.

In the following, a software architecture, which allows to implement a framework for the above described evaluation methodology, is described in Sect. 3.1. Afterwards, Sect. 3.2 describes a set of information processing steps which is common to many VANET based ADAS. Thus, these processing steps can be regarded as an extension to the standardized information handling steps within current ETSI ITS and WAVE frameworks.

3.1   Software Architecture

To switch from simulation to a real world environment with little effort we facilitate input interfaces that can be used for real sensors as well as for the input data from a simulation environment. The received data at the interfaces

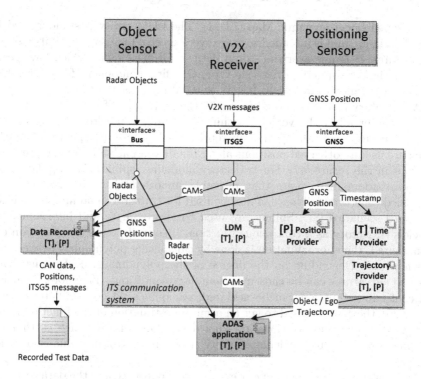

Fig. 1. Software architecture of the evaluation system.

is then processed in the exact same manner for both types of environments. The resulting software architecture is depicted in Fig. 1. Required functionality from the advanced ITS implementation are described in the next paragraphs.

The time provider is the central component to provide the current time, which can be requested by all the other components of the system. It can be based on different time systems, like the local system time, a GPS time or the timestamps used in the simulation environment. The main reason for this single time provider is a synchronized time base for the whole evaluation system. Another advantageous feature is the ability to change the speed of the systems time lapse. Thus, in a simulation or playback environment (which is described later in Sect. 4.1) development time is reduced significantly. The position provider is the time provider's equivalent component for position data. Within the system other entities can request the current position of the ego vehicle. The position provider obtains its input data from a position sensor connected to the system, like a GNSS receiver or from a traffic simulator.

After a VANET message has been received, it is propagated through the protocol stack until it arrives at the application layer. Within ETSI ITS parts of the application layer are located in the so called facility layer. It contains a dedicated facility entity for each message type, e.g., the basic service for CAMs. After a message has been processed by its respective facility entity, it is handed over to the so called Local Dynamic Map (LDM). The LDM entity is standardized by ETSI ITS in [6], and is intended to store every received message until its validity time expires. It hands over the stored data to all data sinks, which have registered for the corresponding message type. Our first component for information processing is called data fusion. It acts as a data sink of the LDM.

A trajectory provider is used to model the trajectory of a vehicle, either the ego vehicle, or all of the vehicles within its vicinity. Trajectories can be modeled in various ways. Most common representations use a sequence of points, which are usually observed positions along the driven path and for the predicted behavior in the future [47]. For trajectory modeling, splines are popular in the automotive domain, because they resemble smooth trajectories satisfying the demanded constraints. They can also be used for trajectory planning in collision avoidance scenarios [31]. Splines are used to approximate and interpolate the vehicle's trajectory as a continuous function. If several points of a function can be measured, the approximation is done by calculating polynomials between the known points. Depending on the degree of the polynomials, continuous gradients and curvatures can be chosen as boundary conditions. Therefore, a smooth and continuous function can be achieved. As splines do not oscillate at their boundaries they can be used to get an accurate estimation of the function in the near future, where no positions are known yet. One disadvantage is, that the interpolation has to be recalculated every time a new position value is obtained [25].

An important component of our evaluation environment is the data recorder. It connects to all the defined interfaces during initialization phase of the framework. At runtime, it records all the input data streams together with a time stamp of the recording time for each entry. This synchronized input data of a driving scenario is required for the offline evaluation of the ADAS.

3.2   General Purpose Information Processing

Most safety critical applications of VANETs account for collision avoidance systems with a low amount of exceptions, like broken down vehicle warning [9]. They all are based on the VANET protocol stack. The analysis of requirements for such collision avoidance systems shows that advanced information processing is very similar for many of them. Thus, we define a basic set of common information processing blocks to be used by all these ADAS.

The information processing chain consisting of the commonly used processing blocks is shown in Fig. 2. It is described in the following. Details about the actually chosen methods for our dedicated evaluation are given in Sect. 4.1.

The object sensors input is pre-processed by the corresponding component. Likewise, received CAMs are handled by the V2X data pre-processing component. Both components provide local object lists, D^R and D^V with the same format. These object lists are the input for the data fusion and vehicle tracking function generating a global object list D^G with all detected vehicles in the vicinity of the ego vehicle. To compare the evaluation results of the object sensor based ADAS and the VANET based one, the data fusion component can

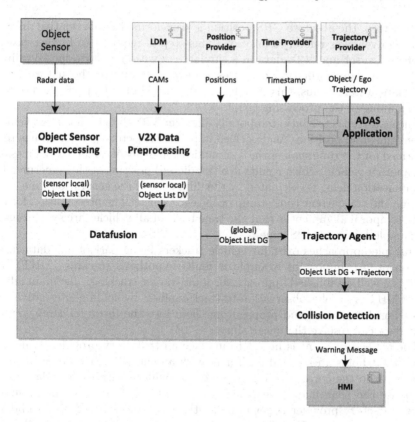

Fig. 2. Information processing chain for an ADAS.

also handle each class of input data separately. Then, for each object in D^G the trajectory agent creates a new trajectory provider or updates the existing one with the new position data for the dedicated object. The trajectory agent itself offers an interface to subscribe for the object list D^G that also delivers the associated trajectory provider for each object. Finally, the collision detection entity, which subscribes to D^G, receives every update of the list. If necessary, a warning message is issued and displayed on the human machine interface (HMI).

The object sensor data pre-processing has two main tasks. The first one is to filter all object sensor messages and only pass those with detected objects. The second task is to transform the object's position data so that it relates to the vehicle's reference position (in the centre of the front bumper) and not to the object sensor's mounting position any more. The relative position of each object sensor in relation to the reference position has to be measured very accurately and is statically configured for every test vehicle.

Within the LDM, CAMs are already filtered according to their relevance for the ego vehicle and their validity time. In addition, the purpose of the V2X data pre-processing differs from the object data in the case of position transformation.

In a first step, the absolute GPS position is transformed into a relative distance to the ego vehicle's reference position. This step is required to further process the object sensor and V2X data in a similar way within a Cartesian coordinate system. This transformation produces a deviation, due to the nature of the coordinate system. Thus, this transformation should only be performed once at the beginning of the processing chain.

The next step is to spatio-temporally align the V2X data with the ego vehicles current time and position. As described in [40], a trajectory prediction has to be performed for the time span of the communication delay of each received message from another vehicle, which results in a (predicted) position at this juncture. The communication delay is calculated by the difference of the generation time of the message and the current time stamp of the ego vehicle. The precondition for this spatio-temporal alignment is that the time basis of all vehicles are synchronized, for example by using GPS time.

Multiple approaches exist for vehicle trackers being part of our data fusion component. An important example is multi hypothesis tracking (MHT) [18], which is a well established approach for multi-target tracking. The main advantage of MHT is to solve observation-to-track conflicts by holding alternative data association hypotheses and propagating them into the future. Thereby, subsequent data can resolve the uncertainty.

The trajectory agent is meant to manage all trajectory providers generated from D^G. It holds an internal hash map with an entry for each object together with a link to the corresponding trajectory provider. With each update of the object list the hash map is searched for all objects. If one can not be found, a new trajectory provider is generated. Otherwise, the existing one is updated with the new position. For all entries of the hash map that are not available in D^G any more the corresponding trajectory providers are destructed.

Collision detection can be done by using motion prediction models like trajectory prediction, state-space models, structured environment approximation, classification models or neural networks. An overview on existing approaches for collision risk estimation is given in [20].

As depicted in Fig. 2, processing of the radar based reference ADAS and the V2X based ADAS only differs in the pre-processing steps. The reason for this architectural decision is to achieve comparability of both ADAS during the validation phase. More precisely, the object sensor and the V2X receiver have the same tasks. They decode and filter the incoming messages. One delaying factor is data access within the LDM. Hence, it is compensated by the spatio-temporal alignment during the V2X data pre-processing. Only the transformation algorithm used in the object sensor and the spatio-temporal alignment within the V2X data pre-processing have a varying impact on the precision of the objects position. Since, the transformation is based on the exactly determined distance of the mounted radar and the ego vehicles reference position it is accurate to one millimeter. Thus, the inaccuracy is negligible. Hence, the impact of the processing steps on the evaluation results can be reduced to the accuracy of the algorithm used for the spatio-temporal alignment of the V2X data.

4   Evaluation Realization

Required components and interfaces within the ITS communication architecture are specified in the ETSI ITS standards, like [5,6,8]. The ezCar2X framework [34,35] provides an implementation of these standards, which is used in the following. The facility entities for the position and time providers are located within the ezCar2X framework, too. Only the sensor data received at the CAN interface is directly handed over to the ADAS.

Usage of the evaluation methodology from Sect. 3 for the evaluation of two dedicated ADASs (ICRW and LCRW) is described in the following.

4.1   Evaluation Environments

The design goal was to generate the smallest possible effort for a change between simulation environment and real test vehicles, as outlined above. Hence, we optimized the software architecture regarding the reusability of most of the components in both environments. To obtain the required flexibility, all interfaces to external entities are realized with abstract classes and different implementations. For example, three implementations are used for GPS/GNSS and V2X input: file (i.e. playback), real hardware, simulation based.

According to [5,8] all time stamps within the protocol stack have to be aligned to their respective GPS coordinates. Thus, the ezCar2X position and time providers were implemented in a coupled way ensuring mutual consistency of both time and position data.

In our trajectory provider we used cubic splines with third order polynomials from [28] for the prediction of all trajectories. The forecasting horizon of the trajectory providers was set to 10 s, but only the next 2 s were considered within the collision risk estimation, as the number of false detections increases with a longer time span. The vehicle tracker uses the MHT approach as described in [41]. It is implemented using the library from [38] for vehicle tracking, using the Dempster-Shafer theory of evidence.

The used collision detector implementation is illustrated in Fig. 3. Trajectory prediction is combined with a piecewise approximation of the trajectories as described in [29]. Therefore, the two trajectories of ego and other vehicle are examined in each time step of the prediction period. A bounding box is generated around each vehicle's reference position using its width and length (taken from the vehicle's CAM) [7]. Additionally, for the V2X data the bounding box has to be expanded by adding the current uncertainty of the GPS position (also taken from the CAM) in a circular shape. This expansion is justified by the nature of GPS positions, which are defined as a position and a circular region around it. Within that circle, 95% of the measured positions are located. Thus, a GPS position should not be taken as an exact value. Instead, the whole confidence circle has to be considered during collision detection. The bounding boxes increase with the degradation of the position accuracy. Finally, the two bounding boxes are used to identify whether they overlap. If this is the case, a collision is detected and a warning message is generated, which is displayed on the HMI.

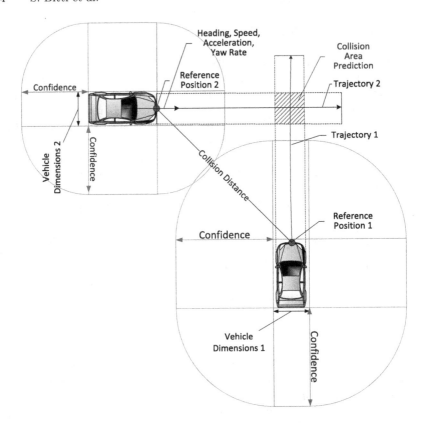

Fig. 3. Collision detection mechanism.

The simulation environment is described next. Then, details about the field test setup are given. Finally, tracing and playback mechanisms are introduced.

Simulation Environment. We use a simulation environment consisting of three dedicated state of the art frameworks. These are SUMO for microscopic traffic flow simulation, the ns-3 network simulator for channel as well as layer 1 and 2 simulation, together with the ezCar2X framework providing the remaining protocol layers in accordance with current ETSI ITS standards [14,33,35]. A detailed description of this combination of dedicated tools and their coupling process can be found in [34].

Vehicles' position information within the simulation environment is perfect, i.e., each vehicle can determine its global position without any error. Positions of other vehicles are only known from received messages. Therefore, Gaussian noise is added to the positions generated in the traffic simulation to study the impact of position uncertainty.

SUMO can be run in parallel or in advance of the remaining simulators. Running it in advance and using generated vehicle traces for the other simulators significantly increases simulation speed. However, this can only be done for use cases with no interaction between ADAS output and vehicle behavior, as vehicle trajectories are fixed in this case. For example, this can be done to determine the time of issued warnings. Therefore, the input data for an ADAS, as described in Sect. 3.1, is then derived from the output file of the SUMO traffic simulator. One of the vehicles within the traffic scenario is defined as the ego vehicle whose positions are then passed to the position interface for the ego position provider.

Two options for receiving position data from other vehicles were realized. To obtain a best case scenario (i.e., no packet loss), positions of all other vehicles are handled as received CAMs at the ITS-G5 interface. This leads to the best possible information quality for the ADAS. A more realistic option is to use ns-3 channel simulation to model real message exchange. Corresponding radar data can be obtained from a radar sensor model within the simulation environment.

Real World Test Setup. The test setup is targeted for ADAS avoiding front or side collisions, like in [3,4]. Thus, the test vehicle is equipped with three radar sensors (depicted by Short Range Radar (SRR) and Long Range Radar (LRR)) acting as object sensors, as shown in Fig. 4.

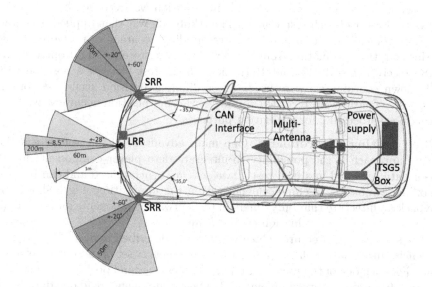

Fig. 4. Test vehicle equipment and radar sensor coverage.

Radar sensors where chosen due to their wide spread use in the automotive domain. They ensure robust object detection, a high resolution and measurement accuracy. The LRR is placed in front of the vehicle to cover the forward-facing street over a distance of 200 m. The two SRRs face to the front-left and front-right with an angle of 35° from the longitudinal axis of the vehicle. The SRRs have shorter coverage of about 50 m, but a wider input opening angle of 60° coverage. All three radars were connected via CAN buses to the notebook running the software framework and ADAS.

The test vehicle was additionally equipped with two multi-frequency antennas for VANET communication within the 5.9 GHz frequency band, an ITS-G5 compliant communication box from Cohda Wireless [21] and a dual-frequency GNSS software receiver that was also connected to the notebook.

In the test scenario a second vehicle was involved that was equally equipped, except for one front LRR. To identify the impact of the position accuracy on the VANET based ICRW scenario the following test scenario was used: the ego vehicle approaches an intersection with a constant velocity where the last possible stop line is marked with a pylon. Afterwards, the ego vehicle is going back to the start position and repeats the maneuver. The second test vehicle approaches the intersection from the left direction in the first run and from the right direction in the second run. This scenario is repeated several times to have comparable sensor data and to compensate possible effects caused by a deviant driver behavior. Thereby, the scenario leads to two collision situations that were detected by the radar based reference implementation.

We look at pure GNSS based positioning as we use no extra data input, i.e., via a mobile network. During the test, information on the satellites' position and pseudo ranges were collected. This information was recorded by the GNSS software receiver of each test vehicle. The rtklib [42] was then applied, to determine the vehicle's position and the corresponding accuracy in post-processing. To this end, the information from the GNSS software receiver and from a second GNSS receiver, installed about 100 m next to the test place at a position with well known coordinates, were used. These different position accuracies for the same test scenario were recorded to have a varying position accuracy available during post-processing for evaluation purposes.

Offline Testing Environment. One main advantage of our software architecture approach is the possibility to integrate data player components for all sensor data recorded in a field test. Data sets are synchronized and can be used for offline evaluation of enhancements within the application. Fig. 5 shows the playback architecture that only differs in the software components for data input to the simulation and evaluation environment.

The software architecture, which was already described in Sect. 3.1, is reused for synchronized offline playback together with another set of data input components. For each one of the radar data, ITS-G5 messages and position data stream a playback component was implemented. These components read one data block out of the recorded stream and replay the data on their corresponding interface on the exact time offset as recorded. All the player components register at a central signal handler during the initialization phase of the software framework. When all components are ready the RUN signal is send which guarantees a synchronized start of the playback.

The playback architecture provides an offline testing environment where small changes of the ADAS application implementations can be tested under consistent conditions for a realistic driving scenario with much lower effort as for a test drive and much faster evaluation speed as the playback system doesn't

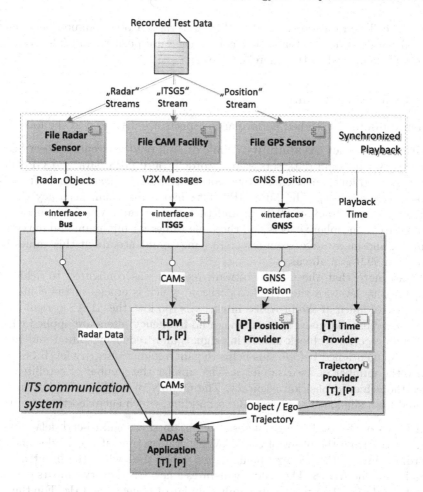

Fig. 5. Synchronized information flow in the playback architecture for offline usage.

have to run in real time. All components after the data player are executed in multiple threads, like done in live testing mode.

Reference System. To compare the evaluated collision detection system with the radar based reference system the following reference values were examined:

1. Difference in the time of collision detection t^{Diff}
2. Distance of the ego and second vehicle at the time of collision detection d
3. Number of false positive and negative detections *#false pos* and *#false neg*

The time of collision detection is used to determine the time lag for the V2X based ADAS in comparison to the radar based reference system. If a collision was detected by the V2X based ADAS, the difference of the two reference positions (in the center of the front bumper) of the ego and the second vehicle is calculated and compared to the distance measured by the reference system at the time when the collision was detected. The third criteria is a very common benchmark for collision detection systems, that counts the false positive and false negative detections compared to the reference system.

4.2 Evaluation Results

Obtained results for studied performance aspects are discussed in the following.

Position Accuracy Availability. While driving the test scenario described in Sect. 4.1, two position data sets were recorded: pure GPS data and GPS data with a reduced ionosphere effect (by incorporation of the measurement from the extra GNSS receiver). The pure GPS data had a minimum accuracy of 7.58 m, maximum accuracy of 20.39 m and the average value was 14.41 m. GNSS with reduced ionosphere effects had the same values for minimum and maximum accuracy, and an average accuracy of 13.7 m. Confidence about this values was more than 99.9% in all cases.

Please note that the GNSS software receiver was configured to achieve a high integrity level for the vehicle's position, which is crucial in the context of collision avoidance. If the provided integrity is too low, the ADAS system might suffer from false alarms. Thus, narrow radio frequency filters are applied within the GNSS receiver to block interfering signals and frequency shifted multi-path effects. As mentioned before, this helps to gain a higher integrity level, but may reduce the number of used satellites. The smaller the number of satellites, the worse the achieved position accuracy. The number of satellites in view further suffered from shadowing by trees and a building, which encircled the test area.

End-to-End Delay. During our tests we obtained the end-to-end delay, which is calculated from the offset of the CAM generation time at the sender and the recording time at the receiver being equal to the time when the input data is available for the ADAS. The sender generates a new CAM every time its position fix gets updated. The positioning unit's update frequency is 1 Hz. For the ego vehicle we found the following delays: minimum 14 ms, maximum 1.001 s and an average delay of 127 ms. The second vehicle obtained a minimum of 14 ms, maximum of 510 ms and an average delay of 70 ms.

End-to-end delay is used in the V2X data pre-processing component, as described in Sect. 3.2, for the spatio-temporal alignment of the second vehicle's position. As this mechanism is meant to predict the vehicle status for the time span of the communication delay, the impact of this end-to-end delay on the ADAS application is reduced to a minimum. In detail, the impact depends on two factors: the accuracy of the prediction mechanism itself and the accuracy of the position data used to calculate the prediction.

Impact of the Position Accuracy. The accuracy of the communicated positions has to be considered within collision detection. Thus, the position data can not be taken as an exact value the applicability of V2X data input strongly depends on the position accuracy received from other vehicles. As described in Sect. 4.1, the bounding boxes of the ego and second vehicle used for collision risk estimation are expanded by adding the current uncertainty of the position value. Thus, collisions were detected earlier as in the reference system, but with less reliability. The difference can be quantified by the position accuracy value itself. To cope with lower position accuracy, advanced approaches for the usage of bounding boxes during collision risk estimation have to be examined.

Analysis of the V2X Based ADAS. The accuracy of the ego and other positions were varied during test runs in the playback environment to analyze the impact of the ego and second vehicle's position accuracy separately. The ego position confidence was set to 0.25 m during one set of test runs, as this is equal to the measurement accuracy of the used radar sensors, and also to the originally derived value in a second test setup. Respectively, the confidence of the second vehicle was set to 0.25 m or to the recorded value in separate test runs.

Our analysis of the results from numerous test runs shows that the best combination of weights for the ego and the second vehicle's position accuracy is to take the confidence for the second vehicle from the CAM and a fixed value of 0.25 m for the ego vehicle. The resulting collision detection within the ADAS is depicted in Fig. 6 for the radar reference system and the V2X based detection.

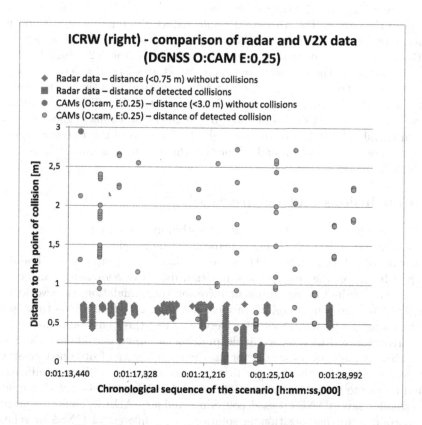

Fig. 6. Comparison of radar and V2X based ADAS

At the horizontal axis, the diagram shows the chronological sequence of the results for the driven test scenario. A detected collision based on radar data is marked with rectangular shapes. The results of the V2X communication based detection are labeled with circular shapes. The collision scenario with the second vehicle approaching from right hand side can be recognized in the middle of the sequence for the reference system. The one collision situation was detected over a period of three prediction steps by the reference system. Future positions of the ego vehicle and the second vehicle are predicted every 100 ms for a time span of 2 s into the future.

The vertical axis of the diagram represents the predicted future distance between the reference positions of the ego and the second vehicle in case of a predicted collision. Since the distances measured by the radar sensors have a much higher accuracy, we took only distances less than 0.75 m into account for the illustration of the results from the radar based collision detection.

As marked with an extra line at 0.25 m in Fig. 6, we added a threshold for the distance of the reference positions of the ego and second vehicle. Thereby, a detected collision by V2X data is only incorporated, if the predicted distance in the future is less than 0.25 m. We recommend to use this threshold as one extra element in a V2X based ADAS to decide whether a warning message is actually passed to the driver. This threshold would reduce the number of false positive collision detections but is does not compensate the low position accuracy.

By means of the defined benchmark parameters, the V2X based collision detection achieved the following results: t^{Diff} is at 1.74 s, $d = 5.65$ cm, *#false pos* = 0 and *#false neg* = 2. The two false negative detections represent the delay of the V2X based application. During a detected collision, the distance of the two reference positions of the ego and the second vehicle goes below the defined threshold not before the third prediction step of the reference system. Since the collision situation ends after this third step, it is only detected once by the V2X based application.

5   Conclusions and Future Work

VANETs are in the wake of mass roll out within upcoming years. To achieve the aim of increased safety of driving, advanced methodologies for development and evaluation of VANET based ADAS are required. The proposed methodology, which is based on the concept of simulation driven development, can be used to realize an evaluation environment incorporating simulation, real world tests and offline testing. Therefore, we propose an approach which allows to keep the effort to a minimum when switching between these three environments.

Moreover, we evaluated available position accuracy of today's GNSS systems with parameters optimized for maximum integrity of obtained positions. Achieved results show that the position accuracy of pure GNSS is not sufficiently high, in order to provide location data for every driving situation with a quality that enables save VANET based collision avoidance ADAS. Thus, we propose to incorporate further positioning solutions, i.e., differential GNSS or relative positioning approaches, to improve the position accuracy.

Future work can obtain additional realistic test scenarios to be used within the evaluation environment. One can also look at integrated test scenarios featuring the availability of simulated traffic information during real test drives.

Acknowledgment. This work was supported by the project "Möglichkeiten und Grenzen des Multi-GNSS RAIM für zukünftige Safety-of-Life Anwendungen" (Multi RAIM II), funded by the German Federal Ministry of Economics and Technology (BMWi) and administered by the Project Management Agency for Aeronautics Research of the German Space Agency (DLR) in Bonn, Germany (grant no. 50NA1313). The authors want to thank Hanno Beckmann, Kathrin Frankl and Bernd Eissfeller from UAF Munich for their support during work on the presented topics.

References

1. Memorandum of Understanding for OEMs within the CAR 2 CAR Communication Consortium on Deployment Strategy for cooperative ITS in Europe, v 4.0102, June 2011
2. Intelligent Transport Systems (ITS); V2X Applications; Part 1: Road Hazard Signalling (RHS) application requirements specification, August 2013
3. Intelligent Transport Systems (ITS); V2X Applications; Part 2: Intersection Collision Risk Warning (ICRW) application requirements specification, August 2013
4. Intelligent Transport Systems (ITS); V2X Applications; Part 3: Longitudinal Collision Risk Warning (LCRW) application requirements specification, November 2013
5. Intelligent Transport Systems (ITS); Users and applications requirements; Part 2: Applications and facilities layer common data dictionary, September 2014
6. Intelligent Transport Systems (ITS); Vehicular Communications; Basic Set of Applications; Local Dynamic Map (LDM), September 2014
7. Intelligent Transport Systems (ITS); Vehicular Communications; Basic Set of Applications; Part 2: Specification of Cooperative Awareness Basic Service, November 2014
8. Intelligent Transport Systems (ITS); Vehicular Communications; GeoNetworking; Part 4: Geographical Addressing and Forwarding for Point-to-Point and Point-to-Multipoint Communications; Sub-part 1: Media-Independent Functionality, July 2014
9. Intelligent Transport Systems; Vehicular Communications; Basic Set of Applications; Use case definitions, August 2015
10. The Collaborative GNSS Encyclopaedia (2016). www.navipedia.net. Accessed Aug 2016
11. Broggi, A., et al.: High performance multi-track recording system for automotive applications. Int. J. Automot. Technol. **13**(1), 123–132 (2012)
12. Strasser, B., et al.: Networking of test and simulation methods for the development of advanced driver assistance systems (ADAS). In: 4. Tagung Sicherheit durch Fahrerassistenz (2010)
13. Beck, K.: Test Driven Development: By Example. Addison-Wesley Longman, Boston (2002)
14. Behrisch, M., Bieker, L., Erdmann, J., Krajzewicz, D.: SUMO - simulation of urban mobility: an overview. In: The Third International Conference on Advances in System Simulation, pp. 63–68, October 2011

15. Berger, C., Rumpe, B.: Engineering autonomous driving software. In: Experience from the DARPA Urban Challenge, pp. 243–272. Springer (2012). https://www. amazon.com/Experience-DARPA-Urban-Challenge-Christopher/dp/0857297716

16. Bittl, S.: Towards solutions for current security related issues in ETSI ITS. In: Mendizabal, J., et al. (eds.) Nets4Cars/Nets4Trains/Nets4Aircraft 2016. LNCS, vol. 9669, pp. 136–148. Springer, Cham (2016). https://doi.org/10.1007/978-3-319-38921-9_15

17. Bittl, S., Gonzalez, A.A., Myrtus, M., Beckmann, H., Sailer, S., Eissfeller, B.: Emerging attacks on VANET security based on GPS time spoofing. In: IEEE Communications and Network Security Conference, pp. 344–352, September 2015

18. Blackman, S.S.: Multiple hypothesis tracking for multiple target tracking. IEEE Aerosp. Electron. Syst. Mag. **19**(1), 5–18 (2004)

19. Chrisofakis, E., Junghanns, A., Kehrer, C., Rink, A.: Simulation-based development of automotive control software with Modelica. In: 8th International Modelica Conference (2011)

20. Christopher, T.: Analysis of dynamic scenes: application to driving assistance. Ph.D. thesis, Institute National Polytechnique de Grenoble-INPG (2009)

21. Cohda Wireless: MK4a V2X Evaluation Kit, June 2013. http://cohdawireless.com/ Portals/0/PDFs/CohdaWirelessMK4a.pdf. Accessed July 2016

22. Hanzlik, A.: Simulation-based application software development in time-triggered communication systems. Int. J. Softw. Eng. Appl. **4**(2), 75–92 (2013)

23. Harding, J., et al.: Vehicle-to-vehicle communications: readiness of V2V technology for application. Technical report DOT HS 812 014, Washington, DC. National Highway Traffic Safety Administration, August 2014

24. Hofmann-Wellenhof, B., Lichtenegger, H., Wasle, E.: GNSS - Global Navigation Satellite Systems: GPS, GLONASS, Galileo, and More. Springer, Heidelberg (2007). https://doi.org/10.1007/978-3-211-73017-1

25. Huckle, T., Schneider, S.: Numerische Methoden: Eine Einführung für Informatiker, Naturwissenschaftler, Ingenieure und Mathematiker, vol. 2. Springer, Heidelberg (2006). https://doi.org/10.1007/3-540-30318-9

26. Joerer, S., Segata, M., Bloessl, B., Cigno, R.L., Sommer, C., Dressler, F.: To crash or not to crash: estimating its likelihood and potentials of beacon-based IVC systems. In: IEEE Vehicular Networking Conference, pp. 25–32 (2012)

27. Kaplan, E., Hegarty, C.: Understanding GPS: Principles and Applications. Artech House, Norwood (2005)

28. Kluge, T.: Library for Cubic Spline interpolation in C++ (2011). http://kluge.in-chemnitz.de/opensource/spline/. Accessed Aug 2016

29. Lytrivis, P., Thomaidis, G., Amditis, A.: Cooperative path prediction in vehicular environments. In: 11th International Conference on Intelligent Transportation Systems, pp. 803–808 (2008)

30. Rondinone, M., et al.: iTETRIS: a modular simulation platform for the large scale evaluation of cooperative ITS applications. Simul. Model. Pract. Theory **34**, 99–125 (2013)

31. Madas, D., et al.: On path planning methods for automotive collision avoidance. In: Intelligent Vehicles Symposium, pp. 931–937 (2013)

32. Misra, P., Enge, P.: Global Positioning System: Signals, Measurements, and Performance. Ganga-Jamuna Press, Lincoln (2001)

33. Riley, G.F., Henderson, T.R.: The NS-3 network simulator. In: Wehrle, K., Günes, M., Gross, J. (eds.) Modeling and Tools for Network Simulation, pp. 15–34. Springer, Heidelberg (2010). https://doi.org/10.1007/978-3-642-12331-3

34. Roscher, K., Bittl, S., Gonzalez, A.A., Myrtus, M., Jiru, J.: ezCar2X: rapid-prototyping of communication technologies and cooperative ITS applications on real targets and inside simulation environments. In: 11th Conference Wireless Communication and Information, pp. 51–62, October 2014

35. Roscher, K., Jiru, J., Gonzalez, A., Heidrich, W.: ezCar2X: a modular software framework for rapid prototyping of C2X applications. In: 9th ITS European Congress, June 2013

36. Schuenemann, B.: V2X simulation runtime infrastructure VSimRTI: an assessment tool to design smart traffic management systems. Comput. Netw. **55**(14), 3189–3198 (2011)

37. Seydel, D., et al.: An evaluation methodology for VANET applications combining simulation and multi-sensor experiments. In: 2nd International Conference on Vehicular Intelligent Transport Systems, pp. 213–224, April 2016

38. Sharp, G., Lin, P.C., Komsuoglu, H.: Ground Truth Measurement System (2003). http://sourceforge.net/projects/gtms. Accessed Aug 2016

39. Sommer, C., German, R., Dressler, F.: Bidirectionally coupled network and road traffic simulation for improved IVC analysis. IEEE Trans. Mob. Comput. **10**(1), 3–15 (2011)

40. Stiller, C., Färber, G., Kammel, S.: Cooperative cognitive automobiles. In: IEEE Intelligent Vehicles Symposium (2007)

41. Streit, R.L., Luginbuhl, T.E.: Probabilistic multi-hypothesis tracking. Technical report, Naval Underwater Systems Center Newport RI (1995)

42. Takasu, T.: RTKLIB: Open Source Program Package for RTK-GPS. FOSS4G (2009)

43. Tan, H.S., Huang, J.: DGPS-based vehicle-to-vehicle cooperative collision warning: engineering feasibility viewpoints. IEEE Trans. Intell. Transp. Syst. **7**(4), 415–428 (2006)

44. Voigtländer, P.: ADTF: framework for driver assistance and safety systems. In: FISTA World Automotive Congress (2008)

45. Winner, H., Hakuli, S., Wolf, G.: Handbuch Fahrerassistenzsysteme: Grundlagen, Komponenten und Systeme für aktive Sicherheit und Komfort. Vieweg+Teubner (2009)

46. Yan, G., Olariu, S., Weigle, M.C.: Providing VANET security through active position detection. Comput. Commun. Mobil. Protoc. ITS/VANET **31**(12), 2883–2897 (2008)

47. Ziegler, J., Bender, P., Dang, T., Stiller, C.: Trajectory planning for bertha - a local, continuous method. In: Intelligent Vehicles Symposium, pp. 450–457 (2014)

Author Index

Printed in the United States
By Bookmasters